T0228023

KAWASAKI
KLR650 • 1987-2007

WHAT'S IN YOUR TOOLBOX?

More information available at haynes.com
Phone: 805-498-6703

J H Haynes & Co. Ltd.
Haynes North America, Inc.

ISBN-10: 1-59969-225-2
ISBN-13: 978-1-59969-225-8
Library of Congress: 2008929211

Author: *Jay Bogart*
Technical Illustrations: *Mitzi McCarthy*
Wiring Diagrams: *Bob Meyer*
Cover: *Mark Clifford Photography at www.markclifford.com.*

M474-3, 10U1, 16-344

1

2

3

4

5

6

7

8

9

10

11

12

13

14

15

16

17

Common spark plug conditions

NORMAL
Symptoms: Brown to grayish-tan color and slight electrode wear. Correct heat range for engine and operating conditions.
Recommendation: When new spark plugs are installed, replace with plugs of the same heat range.

WORN
Symptoms: Rounded electrodes with a small amount of deposits on the firing end. Normal color. Causes hard starting in damp or cold weather and poor fuel economy.
Recommendation: Plugs have been left in the engine too long. Replace with new plugs of the same heat range. Follow the recommended maintenance schedule.

TOO HOT
Symptoms: Blistered, white insulator, eroded electrode and absence of deposits. Results in shortened plug life.
Recommendation: Check for the correct plug heat range, over-advanced ignition timing, lean fuel mixture, intake manifold vacuum leaks, sticking valves and insufficient engine cooling.

CARBON DEPOSITS
Symptoms: Dry sooty deposits indicate a rich mixture or weak ignition. Causes misfiring, hard starting and hesitation.
Recommendation: Make sure the plug has the correct heat range. Check for a clogged air filter or problem in the fuel system or engine management system. Also check for ignition system problems.

PREIGNITION
Symptoms: Melted electrodes. Insulators are white, but may be dirty due to misfiring or flying debris in the combustion chamber. Can lead to engine damage.
Recommendation: Check for the correct plug heat range, over-advanced ignition timing, lean fuel mixture, insufficient engine cooling and lack of lubrication.

ASH DEPOSITS
Symptoms: Light brown deposits encrusted on the side or center electrodes or both. Derived from oil and/or fuel additives. Excessive amounts may mask the spark, causing misfiring and hesitation during acceleration.
Recommendation: If excessive deposits accumulate over a short time or low mileage, install new valve guide seals to prevent seepage of oil into the combustion chambers. Also try changing gasoline brands.

HIGH SPEED GLAZING
Symptoms: Insulator has yellowish, glazed appearance. Indicates that combustion chamber temperatures have risen suddenly during hard acceleration. Normal deposits melt to form a conductive coating. Causes misfiring at high speeds.
Recommendation: Install new plugs. Consider using a colder plug if driving habits warrant.

OIL DEPOSITS
Symptoms: Oily coating caused by poor oil control. Oil is leaking past worn valve guides or piston rings into the combustion chamber. Causes hard starting, misfiring and hesitation.
Recommendation: Correct the mechanical condition with necessary repairs and install new plugs.

DETONATION
Symptoms: Insulators may be cracked or chipped. Improper gap setting techniques can also result in a fractured insulator tip. Can lead to piston damage.
Recommendation: Make sure the fuel anti-knock values meet engine requirements. Use care when setting the gaps on new plugs. Avoid lugging the engine.

GAP BRIDGING
Symptoms: Combustion deposits lodge between the electrodes. Heavy deposits accumulate and bridge the electrode gap. The plug ceases to fire, resulting in a dead cylinder.
Recommendation: Locate the faulty plug and remove the deposits from between the electrodes.

MECHANICAL DAMAGE
Symptoms: May be caused by a foreign object in the combustion chamber or the piston striking an incorrect reach (too long) plug. Causes a dead cylinder and could result in piston damage.
Recommendation: Repair the mechanical damage. Remove the foreign object from the engine and/or install the correct reach plug.

CONTENTS

CHAPTER FIFTEEN
BODY 315

INDEX 319

WIRING DIAGRAMS 325

QUICK REFERENCE DATA

MOTORCYCLE INFORMATION

MODEL:_____YEAR:_____

VIN NUMBER:_____

ENGINE SERIAL NUMBER:_____

CARBURETOR SERIAL NUMBER OR I.D. MARK:_____

FUEL, LUBRICANTS AND FLUIDS

Fuel type	Unleaded gasoline; 87 octane minimum
Fuel tank capacity	23 liters (6.1 U.S. gal.)
Engine oil	SG rated, four-stroke engine oil
	SAE 10W-40, 20W-40, 10W-50 or 20W-50
Engine oil capacity	
Without filter change	2.2 liters (2.3 U.S. qt.)
With filter change	2.5 liters (2.6 U.S. qt.)
Cooling system capacity	1.3 liters (1.4 U.S. qt.)
Coolant type	Ethylene glycol containing anti-corrosion
	inhibitors for aluminum engines
Coolant mixture	50/50 (antifreeze/distilled water)
Drive chain	O-ring type chain lubricant
Fork oil grade	Kayaba G-10, or equivalent 10-weight fork oil
Fork oil capacity (each leg)	
Oil change	355 cc (12.0 U.S. oz.)
Fork rebuild (all parts dry)	416-424 cc (14.0-14.3 U.S. oz.)
Fork tube oil level (from top edge	
of inner tube)	188-192 mm (7.40-7.56 in.)
Brake fluid type	DOT 4
Control cables	Cable lube
Air filter	Foam air filter oil

ROUTINE CHECKS AND ADJUSTMENTS

Brake light switch	On after 15 mm (0.6 in.) of pedal travel
Brake pad lining minimum thickness	1.0 mm (0.040 in.)
Brake pedal height	At or near level with top of peg
Choke lever free play	2-3 mm (0.08-0.12 in.)
Clutch lever free play	2-3 mm (0.08-0.12 in.) at pivot
Drive chain play	50-60 mm (2.0-2.4 in.)
Drive chain length wear limit (20 links/21 pins)	323 mm (12.7 in.)
Fork air pressure	0 (atmospheric pressure)
Radiator cap relief pressure	93-123 kPa (13.5-17.8 psi)
Rim runout (radial and lateral)	2.0 mm (0.08 in.)
Throttle grip free play	2-3 mm (0.08-0.12 in.)
Tire pressure	
Front	150 kPa (22 psi)
Rear	
Loads up to 97.5 kg (215 lb.)	150 kPa (22 psi)
Loads 97.5-182 kg (215-400 lb.)	200 kPa (29 psi)

TUNE-UP SPECIFICATIONS

Battery	12 volt, 14 amp-hour
Compression	530-855 kPa (77-124 psi)
Compression ratio	9.5:1
Idle speed	1200-1400 rpm
Ignition timing	
Idle	10° BTDC at 1300 rpm
Advanced	30° BTDC at 3300 rpm
Pilot mixture screw turns out	1 3/8
Spark plug	
Type	NGK DP8EA-9
	NGK DPR8EA-9
	ND X24EP-U9
	ND X24EPR-U9
Gap	0.8-0.9 mm (0.032-0.036 in.)
Valve clearance	
Intake	0.10-0.20 mm (0.004-0.008 in.)
Exhaust	0.15-0.25 mm (0.006-0.010 in.)

MAINTENANCE TORQUE SPECIFICATIONS

	N•m	in.-lb.	ft.-lb.
Axle nut			
Front	78	–	58
Rear	93	–	69
Fork caps	29	–	21
Handlebar clamp bolts	24	–	18
Oil drain plug	23	–	17
Spark plug	14	–	10
Spokes	2-4	17-35	–
Coolant drain plug	10	88	–

CHAPTER ONE

GENERAL INFORMATION

This detailed and comprehensive manual covers the 1987-2007 Kawasaki KLR650.

The text provides complete information on maintenance, tune-up, repair and overhaul. Hundreds of photos and drawings guide the reader through every job.

All procedures are in step-by-step form and designed for the reader who may be working on the motorcycle for the first time.

MANUAL ORGANIZATION

A shop manual is a tool and as in all Clymer manuals, the chapters are thumb tabbed for easy reference. Main headings are listed in the table of contents and the index. Frequently used specifications and capacities from the tables at the end of each individual chapter are listed in the *Quick Reference Data* section at the front of the manual. Specifications and capacities are provided in U.S. standard and metric units of measure.

During some of the procedures there will be references to headings in other chapters or sections of the manual. When a specific heading is called out in a step it will be *italicized* as it appears in the manual. If a sub-heading is indicated as being "in this section" it is located within the same main heading. For example, the sub-heading *Handling Gasoline Safely* is located within the main heading *SAFETY.*

This chapter provides general information on shop safety, tools and their usage, service fundamentals and shop supplies. **Tables 1-7**, at the end of the chapter, list the following:

Table 1 lists the year and model code numbers for the models covered in this manual.

Table 2 lists general dimensions and weight.

Table 3 lists technical abbreviations.

Table 4 lists general torque specifications.

Table 5 lists conversion formulas.

Table 6 lists metric tap and drill sizes.

Table 7 lists decimal and metric equivalents.

Chapter Two provides methods for quick and accurate diagnosis of problems. Troubleshooting procedures present typical symptoms and logical methods to pinpoint and repair the problem.

Chapter Three explains all routine maintenance and recommended tune-up procedures necessary to keep the motorcycle running well.

Subsequent chapters describe specific systems such as engine, transmission, clutch, drive system, fuel and exhaust systems, suspension and brakes. Each disassembly, repair and assembly procedure is discussed in step-by-step form.

WARNINGS, CAUTIONS AND NOTES

The terms WARNING, CAUTION and NOTE have specific meanings in this manual.

A WARNING emphasizes areas where injury or even death could result from negligence. Mechanical damage may also occur. WARNINGS *are to be taken seriously.*

A CAUTION emphasizes areas where equipment damage could result. Disregarding a CAUTION could cause permanent mechanical damage, though injury is unlikely.

A NOTE provides additional information to make a step or procedure easier or clearer. Disregarding a NOTE could cause inconvenience, but would not cause equipment damage or injury.

SAFETY

Professional mechanics can work for years and never sustain a serious injury or mishap. Follow these guidelines and practice common sense to safely service the motorcycle.

1. Do not operate the motorcycle in an enclosed area. The exhaust gasses contain carbon monoxide, an odorless, colorless, and tasteless poisonous gas. Carbon monoxide levels build quickly in small enclosed areas and can cause unconsciousness and death in a short time. Properly ventilate the work area, or operate the motorcycle outside.

2. *Never* use gasoline or any extremely flammable liquid to clean parts. Refer to *Cleaning Parts* and *Handling Gasoline Safely* in this section.

3. *Never* smoke or use a torch in the vicinity of flammable liquids, such as gasoline or cleaning solvent.

4. If welding or brazing on the motorcycle, move the fuel tank to a safe distance away from the work area.

5. Use the correct type and size of tools to avoid damaging fasteners.

6. Keep tools clean and in good condition. Replace or repair worn or damaged equipment.

7. When loosening a tight fastener, be guided by what would happen if the tool slips.

8. When replacing fasteners, check that the new fasteners are of the same size and strength as the original ones.

9. Keep the work area clean and organized.

10. Wear eye protection *anytime* eye injury is possible. This includes procedures involving drilling, grinding, hammering, compressed air and chemicals.

11. Wear the correct clothing for the job. Tie up or cover long hair so it cannot get caught in moving equipment.

12. Do not carry sharp tools in clothing pockets.

13. Always have an approved fire extinguisher available. Check that it is rated for gasoline (Class B) and electrical (Class C) fires.

14. Do not use compressed air to clean clothes, the motorcycle or the work area. Debris may be blown into the eyes or skin. *Never* direct compressed air at yourself or others. Do not allow children to use or play with any compressed air equipment.

15. When using compressed air to dry rotating parts, hold the part so it cannot rotate. Do not allow the force of the air to spin the part. The air jet is capable of rotating parts at extreme speed. The part may be damaged or disintegrate, causing serious injury.

16. Do not inhale the dust created by brake pad and clutch wear. In most cases these particles contain asbestos. In addition, some types of insulating materials and gaskets may contain asbestos. Inhaling asbestos particles is hazardous to health.

17. When placing the motorcycle on a stand, check that it is secure.

18. When performing maintenance procedures or working on the fuel system, observe the following:

   a. Turn the fuel valve off.

   b. Never work on a hot engine.

   c. Wipe up fuel and solvent spill immediately.

Handling Gasoline Safely

Gasoline is a volatile, flammable liquid and is one of the most dangerous items in the shop. Keep in mind, when working on a motorcycle, gasoline is always present in the fuel tank, fuel line and carburetor. To avoid an accident when working around the fuel system, carefully observe the following precautions:

1. *Never* use gasoline to clean parts. See *Cleaning Parts* in this section.

2. When working on the fuel system, work outside or in a well-ventilated area.

3. Do not add fuel to the fuel tank or service the fuel system while the motorcycle is near open flames,

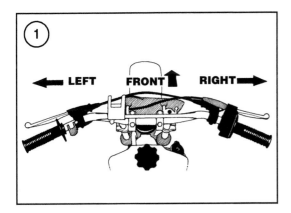

sparks or where someone is smoking. Gasoline vapor is heavier than air. It collects in low areas and more easily ignites than liquid gasoline.

4. Allow the engine to cool completely before working on any fuel system component.

5. When draining the carburetor, catch the fuel in a plastic container and then pour it into an approved gasoline storage device.

6. Do not store gasoline in glass containers. If the glass breaks, a serious explosion or fire may occur.

7. Immediately wipe up spilled gasoline. Store the contaminated shop cloths in a metal container with a lid until they can be properly disposed, or place them outside in a safe place for the fuel to evaporate.

8. Do not pour water onto a gasoline fire. Water spreads the fire and makes it more difficult to put out. Use a class B, BC or ABC fire extinguisher to extinguish the fire.

9. Always turn off the engine before refueling. Avoid spilling fuel onto the engine or exhaust system. Do not overfill the fuel tank. Leave an air space at the top of the tank to allow room for the fuel to expand due to temperature fluctuations.

Cleaning Parts

Cleaning parts is one of the more time-consuming jobs performed in the home shop. There are many types of chemical cleaners and solvents available for shop use. Most are poisonous and extremely flammable. To prevent chemical exposure, vapor buildup, fire and serious injury, observe each product warning label and note the following:

1. Read and observe the entire product label before using any chemical. Always know what type of chemical is being used and whether it is poisonous and/or flammable.

2. Do not use more than one type of cleaning solvent at a time. If mixing chemicals is called for, measure the proper amounts according to the manufacturer.

3. Work in a well-ventilated area.

4. Wear chemical-resistant gloves.

5. Wear safety glasses.

6. Wear a vapor respirator when necessary.

7. Wash hands and arms thoroughly after cleaning parts.

8. Keep chemical products away from children and pets.

9. Thoroughly wipe all oil, grease and cleaner residue from any part that must be heated.

10. Use a nylon brush when cleaning parts. Wire brushes may cause a spark.

11. When using a parts washer, only use the solvent recommended by the manufacturer. Check that the parts washer is equipped with a metal lid that will lower in case of fire.

Warning Labels

Most manufacturers attach information and warning labels to the motorcycle. These labels contain instructions that are important to personal safety when operating, servicing, transporting and storing the motorcycle. Refer to the owner's manual for the description and location of labels. Order replacement labels from the manufacturer if they are missing or damaged.

BASIC SERVICE METHODS

Most of the procedures in this manual are straightforward and can be performed by anyone reasonably competent with tools. However, consider personal capabilities carefully before attempting any operation involving major disassembly of the engine.

1. Front, in this manual, refers to the front of the motorcycle. The front of any component is the end closest to the front of the motorcycle. The left and right sides refer to the position of the parts as viewed by the rider sitting on the seat facing forward (**Figure 1**).

2. Whenever servicing an engine or suspension component, secure the motorcycle in a safe manner.

3. Tag all similar parts for location and mark all mating parts for position. Record the number and thickness of any shims when removing them. Identify parts by placing them in sealed and labeled plastic bags.

4. Tag disconnected wires and connectors with masking tape and a marking pen.

5. Protect finished surfaces from physical damage or corrosion. Keep gasoline and other chemicals off painted surfaces.

6. Use penetrating oil on frozen or tight bolts. Avoid using heat where possible. Heat can warp, melt or affect the temper of parts. Heat also damages the finish of paint and plastics.

7. When a part is a press-fit or requires a special tool for removal, the information or type of tool is identified in the text. Otherwise, if a part is difficult to remove or install, determine the cause before proceeding.

8. Cover all openings to prevent objects or debris from falling into the engine.

9. Read each procedure thoroughly and compare the illustrations to the actual components before starting the procedure. Perform the procedure in sequence.

10. Recommendations are occasionally made to refer service to a dealership or specialist. In these cases, the work can be performed more economically by the specialist than by the home mechanic.

11. The term *replace* means to discard a defective part and replace it with a new part. *Overhaul* means to remove, disassemble, inspect, measure, repair and/or replace parts as required to recondition an assembly.

12. Some operations require using a hydraulic press. If a press is not available, have these operations performed by a shop equipped with the necessary equipment. Do not use makeshift equipment that may damage the motorcycle.

13. Repairs are much faster and easier if the motorcycle is clean before starting work. Degrease the motorcycle with a commercial degreaser; follow the directions on the container for the best results. Clean all parts with cleaning solvent as they are removed.

CAUTION
Do not apply a chemical degreaser to an O-ring drive chain. These chemicals will damage the O-rings. Use kerosene to clean O-ring type chains.

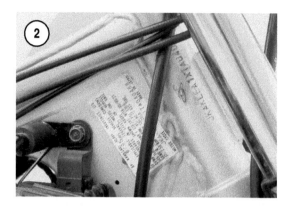

CAUTION
Do not direct high-pressure water at steering bearings, carburetor hoses, wheel bearings, suspension and electrical components, or O-ring drive chains. The water will force the grease out of the bearings and possibly damage the seals.

14. If special tools are required, have them available before starting the procedure. When special tools are required, they will be described in the procedure.

15. Make diagrams of similar-appearing parts. For instance, crankcase bolts are often not the same lengths. Do not rely on memory alone. It is possible that carefully laid out parts will become disturbed, making it difficult to reassemble the components correctly without a diagram.

16. Check that all shims and washers are reinstalled in the same location and position.

17. Whenever rotating parts contact a stationary part, look for a shim or washer.

18. Use new gaskets if there is any doubt about the condition of old ones.

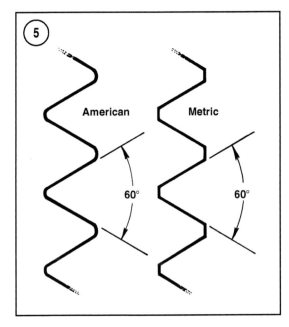

The engine number is stamped on the right crank-case, below the exhaust pipe (**Figure 4**).

Table 1 lists year and model code numbers.

FASTENERS

Proper fastener selection and installation is important to ensure that the motorcycle operates as designed and can be serviced efficiently. Check that replacement fasteners meet all the same requirements as the originals.

Threaded Fasteners

Threaded fasteners secure most of the components on the motorcycle. Most are tightened by turning them clockwise (right-hand threads). If the normal rotation of the component being tightened would loosen the fastener, it may have left-hand threads. If a left-hand threaded fastener is used, it is noted in the text.

Two dimensions are required to match the thread size of the fastener: the number of threads in a given distance and the outside diameter of the threads.

Two systems are currently used to specify threaded fastener dimensions: the U.S. Standard System and the metric system (**Figure 5**). Pay particular attention when working with unidentified fasteners; mismatching thread types can damage threads.

CAUTION
To ensure that fastener threads are not mismatched or become cross-threaded, start all fasteners by hand. If a fastener is hard to start or turn, determine the cause before tightening with a wrench.

The length (L), diameter (D) and distance between thread crests (pitch) (T) (**Figure 6**) classify metric screws and bolts. A typical bolt may be identified by the numbers, 8-1.25 × 130. This indicates the bolt has a diameter of 8 mm. The distance between thread crests is 1.25 mm and the length is 130 mm. Always measure bolt length as shown in **Figure 7** to avoid purchasing replacements of the wrong length.

The numbers located on the top of the fastener (**Figure 6**) indicate the strength of metric screws

19. If self-locking fasteners are used, replace them with new ones. Do not install standard fasteners in place of self-locking ones.

20. Use grease to hold small parts in place if they tend to fall out during assembly. Apply grease to electrical or brake components.

SERIAL NUMBERS

Serial numbers are stamped onto the frame and engine. Record these numbers in the *Quick Reference Data* section at the front of the book. Have these numbers available when ordering parts.

The frame number is indicated on a decal and stamped on the frame at the steering head (**Figure 2**). The number is also on a sticker located on the front frame tube, next to the radiator (**Figure 3**).

and bolts. The higher the number, the stronger the fastener. Unnumbered fasteners are the weakest.

Many screws, bolts and studs are combined with nuts to secure particular components. To indicate the size of a nut, manufacturers specify the internal diameter and the thread pitch.

The measurement across two flats on a nut or bolt indicates the wrench size.

> *WARNING*
> *Do not install fasteners with a strength classification lower than what was originally installed by the manufacturer. Doing so may cause equipment failure and/or damage.*

Torque Specifications

The materials used in the manufacture of the motorcycle may be subjected to uneven stresses if the fasteners of the various subassemblies are not installed and tightened correctly. Fasteners that are improperly installed or work loose can cause extensive damage. It is essential to use an accurate torque wrench, described in this chapter, with the torque specifications in this manual.

Specifications for torque are provided in Newton-meters (N•m), foot-pounds (ft.-lb.) and inch-pounds (in.-lb.). Refer to **Table 5** for torque conversions and **Table 4** for general specifications. To use **Table 4**, first determine the size of the fastener as described in *Threaded Fasteners* in this section. Torque specifications for specific components are at the end of the appropriate chapters. Refer to *Basic Tools* in this chapter for torque wrenches.

Self-Locking Fasteners

Several types of bolts, screws and nuts incorporate a system that creates interference between the two fasteners. Interference is achieved in various ways. The most common type is the nylon insert nut and a dry adhesive coating on the threads of a bolt.

Self-locking fasteners offer greater holding strength than standard fasteners, which improves their resistance to vibration. Most self-locking fasteners cannot be reused. The materials used to form the lock become distorted after the initial installation and removal. It is a good practice to discard and replace self-locking fasteners after their removal.

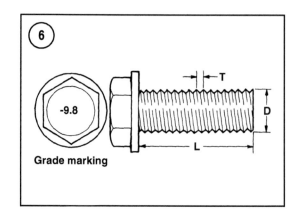

Grade marking

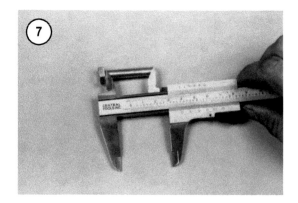

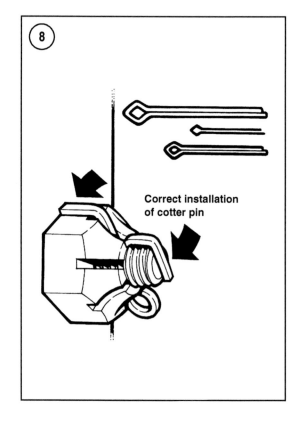

Correct installation of cotter pin

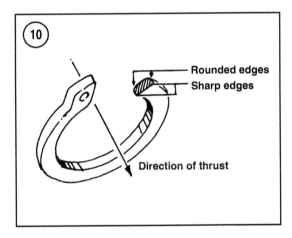

Internal snap ring   Plain circlip

External snap ring   E-clip

Rounded edges
Sharp edges

Direction of thrust

Do not replace self-locking fasteners with standard fasteners.

Washers

There are two basic types of washers: flat washers and lockwashers. Flat washers are simple discs with a hole to fit a screw or bolt. Lockwashers are used to prevent a fastener from working loose. Washers can be used as spacers and seals, or to help distribute fastener load and to prevent the fastener from damaging the component.

As with fasteners, when replacing washers check that the replacement washers are of the same design and quality.

Cotter Pins

A cotter pin is a split metal pin inserted into a hole or slot to prevent a fastener from loosening. In certain applications, such as the rear axle on an ATV or motorcycle, the fastener must be secured in this way. For these applications, a cotter pin and castellated (slotted) nut is used.

To use a cotter pin, first make sure the diameter is correct for the hole in the fastener. After correctly tightening the fastener and aligning the holes, insert the cotter pin through the hole and bend the ends over the fastener (**Figure 8**). Unless instructed to do so, never loosen a torqued fastener to align the holes. If the holes do not align, tighten the fastener just enough to achieve alignment.

Cotter pins are available in various diameters and lengths. Measure length from the bottom of the head to the tip of the shortest pin.

Snap Rings and E-clips

Snap rings (**Figure 9**) are circular-shaped metal retaining clips. They are required to secure parts and gears in place on parts such as shafts, pins or rods. External type snap rings are used to retain items on shafts. Internal type snap rings secure parts within housing bores. In some applications, in addition to securing the component(s), snap rings of varying thickness also determine end play. These are usually called selective snap rings.

Two basic types of snap rings are used: machined and stamped snap rings. Machined snap rings can be installed in either direction, since both faces have sharp edges. Stamped snap rings (**Figure 10**) are manufactured with a sharp edge and a round edge. When installing a stamped snap ring in a thrust application, install the sharp to edge face away from the part producing the thrust.

Observe the following when installing snap rings:
1. Remove and install snap rings with snap ring pliers. See *Snap Ring Pliers* in this chapter.
2. In some applications, it may be necessary to replace snap rings after removing them.
3. Compress or expand snap rings only enough to install them. If overly expanded, they lose their retaining ability.
4. After installing a snap ring, check that it seats completely.

5. Wear eye protection when removing and installing snap rings.

E-clips are used when it is not practical to use a snap ring. Remove E-clips with a flat blade screwdriver by prying between the shaft and E-clip. To install an E-clip, center it over the shaft groove and push or tap it into place.

SHOP SUPPLIES

Lubricants and Fluids

Periodic lubrication helps ensure long service-life for any type of equipment. Using the correct type of lubricant or fluid is as important as performing the service. The following section describes the types of lubricants most often required. Follow the manufacturer's recommendations for lubricant and fluid types.

Engine oil

Engine oil is classified by two standards: the American Petroleum Institute (API) service classification and the Society of Automotive Engineers (SAE) viscosity rating. This information is on the oil container label. Two letters indicate the API service classification. The number or sequence of numbers and letter (10W-40 for example) is the oil's viscosity rating. The API service classification and the SAE viscosity index are not indications of oil quality.

The service classification indicates that the oil meets specific lubrication standards. The first letter in the classification *S* indicates that the oil is for gasoline engines. The second letter indicates the standard the oil satisfies. The classification started with the letter *A* and is currently at the letter *K*. Most motorcycle oils are specifically made to the *G* classification, which is recommended by most motorcycle manufacturers.

Always use an oil with a classification recommended by the manufacturer. Using an oil with a classification different than that recommended can cause engine damage.

Viscosity is an indication of the oil's ability to lubricate and circulate at specific temperatures. Light oils have a lower number while heavy oils have a higher number. Engine oils fall into the 5- to 50-weight range for single-grade oils.

Most manufacturers recommend multigrade oil. These oils perform efficiently across a wide range of operating conditions. Multigrade oils are identified by a *W* after the first number, which indicates the low-temperature viscosity.

Engine oils are most commonly mineral (petroleum) based; however synthetic and semi-synthetic types are frequently used. When selecting engine oil, follow the manufacturer's recommendation for type, classification and viscosity.

Grease

Grease is lubricating oil with thickening agents. The National Lubricating Grease Institute (NLGI) grades grease. Grades range from No. 000 to No. 6, with No. 6 being the thickest. Typical multipurpose grease is NLGI No. 2. For specific applications, manufacturers may recommend water-resistant type grease or one with an additive such as molybdenum disulfide (MoS_2).

Chain lubricant

There are many types of chain lubricants available. Which type of chain lubricant to use depends on the type of chain.

On O-ring (sealed) chains, the lubricant keeps the O-rings pliable and prevents corrosion. The actual chain lubricant is enclosed in the chain by the O-rings. Recommended types include aerosol sprays specifically designed for O-ring chains and conventional engine or gear oils. When using a spray lubricant, check that it is suitable for O-ring chains.

Do not use high-pressure washers, solvents or gasoline to clean an O-ring chain. Only clean with kerosene.

Foam air filter oil

Filter oil is specifically designed to use on foam air filters. The oil is blended with additives making it easy to pour and apply evenly to the filter. Some filter oils include additives that evaporate quickly, making the filter oil very tacky. This allows the oil to remain suspended within the foam pores, trapping dirt and preventing it from being drawn into the engine.

Do not use engine oil as a substitute for foam filter oil. Engine oils do not remain in the filter. Instead, they are drawn into the engine, leaving the filter ineffective.

Brake fluid

Brake fluid is the hydraulic fluid used to transmit hydraulic pressure (force) to the wheel brakes. Brake fluid is classified by the Department of Transportation (DOT). Current designations for brake fluid are DOT 3, DOT 4 and DOT 5. This classification appears on the fluid container.

Each type of brake fluid has its own definite characteristics. Do not intermix different types of brake fluid. DOT 5 fluid is silicone-based. DOT 5 is not compatible with other fluids or in systems for which it was not designed. Mixing DOT 5 fluid with other fluids may cause brake system failure. When adding brake fluid, *only* use the fluid recommended by the manufacturer.

Brake fluid will damage any plastic, painted or plated surface it contacts. Use extreme care when working with brake fluid and remove any spills immediately with soap and water.

Hydraulic brake systems require clean and moisture-free brake fluid. Never reuse brake fluid. Keep containers and reservoirs sealed.

> *WARNING*
> *Never put a mineral-based (petroleum) oil into the brake system. Mineral oil will cause rubber parts in the system to swell and break apart, causing complete brake failure.*

Coolant

Coolant is a mixture of water and antifreeze used to dissipate engine heat. Ethylene glycol is the most common form of antifreeze. Check the manufacturer's recommendations when selecting an antifreeze; most require one specifically designed for use in aluminum engines. These types of antifreezes have additives that inhibit corrosion.

Only mix distilled water with antifreeze. Impurities in tap water may damage internal cooling system passages.

Cleaners, Degreasers and Solvents

Many chemicals are available to remove oil, grease and other residue from the motorcycle.

Before using cleaning solvents, consider how they will be used and disposed of, particularly if they are not water-soluble. Local ordinances may require special procedures for the disposal of many types of cleaning chemicals. Refer to *Safety* and *Cleaning Parts* in this chapter for more information on their use.

Use brake parts cleaner to clean brake system components when contact with petroleum-based products will damage seals. Brake parts cleaner leaves no residue. Use electrical contact cleaner to clean electrical connections and components without leaving any residue. Carburetor cleaner is a strong solvent used to remove fuel deposits and varnish from fuel system components. Use this cleaner carefully, as it may damage finishes.

Generally, degreasers are strong cleaners used to remove heavy accumulations of grease from engine and frame components.

Most solvents are designed to be used in a parts washing cabinet for individual component cleaning. For safety, use only nonflammable or high flash-point solvents.

Gasket Sealant

Sealants are used in combination with a gasket or seal and are occasionally used alone. Follow the manufacturer's recommendation when using sealants. Use extreme care when choosing a sealant. Choose sealants based on their resistance to heat, various fluids and their sealing capabilities.

One of the most common sealants is RTV, or room temperature vulcanizing sealant. This sealant cures at room temperature over a specific time period. This allows the repositioning of components without damaging gaskets.

Moisture in the air causes the RTV sealant to cure. Always install the tube cap as soon as possible after applying RTV sealant. RTV sealant has a limited shelf life and does not cure properly if the shelf life has expired. Keep partial tubes sealed and discard them if they have surpassed the expiration date.

Applying RTV sealant

Clean all old gasket residue from the mating surfaces. Remove all gasket material from blind threaded holes; it can cause inaccurate bolt torque. Spray the mating surfaces with aerosol parts cleaner and then wipe with a lint-free cloth. The area must be clean for the sealant to adhere.

Apply RTV sealant in a continuous bead 2-3 mm (0.08-0.12 in.) thick. Circle all the fastener holes unless otherwise specified. Do not allow any sealant to enter these holes. Assemble and tighten the fasteners to the specified torque within the time frame recommended by the RTV sealant manufacturer.

Gasket Remover

Aerosol gasket remover can help remove stubborn gaskets. This product can speed up the removal process and prevent damage to the mating surface that may be caused by using a scraping tool. Most of these types of products are very caustic. Follow the manufacturer's instructions for use.

Threadlocking Compound

Threadlocking compound is a fluid applied to the threads of fasteners. After tightening the fastener, the fluid sets and becomes a solid filler between the threads. This makes it difficult for the fastener to work loose from vibration, or heat expansion and contraction. Some threadlocking compounds also provide a seal against fluid leakage.

Before applying threadlocking compound, remove any old compound from both thread areas and clean them with aerosol parts cleaner. Use the compound sparingly. Excess fluid can run into adjoining parts.

Threadlocking compounds are available in a wide range of compounds for various strength, temperature and repair applications. Follow the manufacturer's recommendations regarding compound selection.

BASIC TOOLS

Most of the procedures in this manual can be carried out with simple hand tools and test equipment familiar to the home mechanic. Always use the cor-

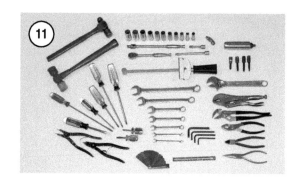

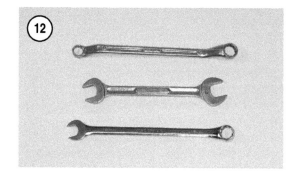

rect tools for the job at hand. Keep tools organized and clean. Store them in a tool chest with related tools organized together.

Quality tools are essential. The best are constructed of high-strength alloy steel. These tools are light, easy to use and resistant to wear. Their working surface is devoid of sharp edges and the tool is carefully polished. They have an easy-to-clean finish and are comfortable to use. Quality tools are a good investment.

Some of the procedures in this manual specify special tools. In most cases, the tool is illustrated in use. Well-equipped mechanics may be able to substitute similar tools or fabricate a suitable replacement. However, in some cases, the specialized equipment or expertise may make it impractical for the home mechanic to attempt the procedure. When necessary, such operations are identified in the text with the recommendation to have a dealership or specialist perform the task. It may be less expensive to have a professional perform these jobs, especially when considering the cost of the equipment.

When purchasing tools to perform the procedures covered in this manual, consider the tools potential frequency of use. If a tool kit is just now being started, consider purchasing a basic tool set (**Figure 11**) from a large tool supplier. These sets are avail-

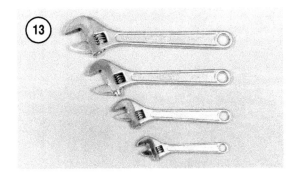

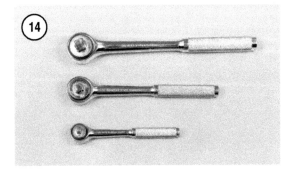

The number stamped on the wrench refers to the distance between the work areas. This size must match the size of the fastener head.

The box-end wrench grips the fastener on all sides. This reduces the chance of the tool slipping. The box-end wrench is designed with either a 6- or 12-point opening. For stubborn or damaged fasteners, the 6-point provides superior holding ability by contacting the fastener across a wider area at all six edges. For general use, the 12-point works well. It allows the wrench to be removed and reinstalled without moving the handle over such a wide arc.

An open-end wrench is fast and works best in areas with limited overhead access. It contacts the fastener at only two points, and is subject to slipping under heavy force, or if the tool or fastener is worn. A box-end wrench is preferred in most instances, especially when breaking loose and applying the final tightness to a fastener.

The combination wrench has a box-end on one end, and an open-end on the other. This combination makes it a very convenient tool.

Adjustable Wrenches

An adjustable wrench or Crescent wrench (**Figure 13**) can fit nearly any nut or bolt head that has clear access around its entire perimeter. Adjustable wrenches are best used as a backup wrench to keep a large nut or bolt from turning while the other end is being loosened or tightened with a box-end or socket wrench.

Adjustable wrenches contact the fastener at only two points, which makes them more subject to slipping off the fastener. The fact that one jaw is adjustable and may loosen only aggravates this shortcoming. Make certain the solid jaw is the one transmitting the force.

Socket Wrenches, Ratchets and Handles

Sockets that attach to a ratchet handle (**Figure 14**) are available with 6-point or 12-point openings (**Figure 15**) and different drive sizes. The drive size indicates the size of the square hole that accepts the ratchet handle. The number stamped on the socket is the size of the work area and must match the fastener head.

As with wrenches, a 6-point socket provides superior holding ability, while a 12-point socket needs

able in many tool combinations and offer substantial savings when compared to individually purchased tools. As work experience grows and tasks become more complicated, specialized tools can be added.

Screwdrivers

Screwdrivers of various lengths and types are mandatory for the simplest tool kit. The two basic types are the slotted tip (flat blade) and the Phillips tip. These are available in sets that often include an assortment of tip sizes and shaft lengths.

As with all tools, use a screwdriver designed for the job. Check that the size of the tip conforms to the size and shape of the fastener. Use them only for driving screws. Never use a screwdriver for prying or chiseling metal. Repair or replace worn or damaged screwdrivers. A worn tip may damage the fastener, making it difficult to remove.

Wrenches

Open-end, box-end and combination wrenches (**Figure 12**) are available in a variety of types and sizes.

to be moved only half as far to reposition it on the fastener.

Sockets are designated for either hand or impact use. Impact sockets are made of thicker material for more durability. Compare the size and wall thickness of a 19 mm hand socket and the 19 mm impact socket (**Figure 16**). Use impact sockets when using an impact driver or air tools. Use hand sockets with hand-driven attachments.

WARNING
Do not use hand sockets with air or impact tools, as they may shatter and cause injury. Always wear eye protection when using impact or air tools.

Various handles are available for sockets. The speed handle is used for fast operation. Flexible ratchet heads in varying lengths allow the socket to be turned with varying force, and at odd angles. Extension bars allow the socket setup to reach difficult areas. The ratchet is the most versatile. It allows the user to install or remove the nut without removing the socket.

Sockets combined with any number of drivers make them undoubtedly the fastest, safest and most convenient tool for fastener removal and installation.

Impact Driver

An impact driver provides extra force for removing fasteners, by converting the impact of a hammer into a turning motion. This makes it possible to remove stubborn fasteners without damaging them. Impact drivers and interchangeable bits (**Figure 17**) are available from most tool suppliers. When using a socket with an impact driver, check that the socket is designed for impact use. Refer to *Socket Wrenches, Ratchets and Handles* in this section.

WARNING
Do not use hand sockets with air or impact tools as they may shatter and cause injury. Always wear eye protection when using impact or air tools.

Allen Wrenches

Allen or set screw wrenches (**Figure 18**) are used on fasteners with hexagonal recesses in the fastener

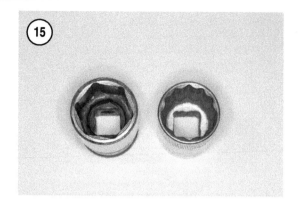

head. These wrenches are available in L-shaped bar, socket and T-handle types. A metric set is required when working on most motorcycles. Allen bolts are sometimes called socket bolts.

Torque Wrenches

A torque wrench is used with a socket, torque adapter or similar extension to tighten a fastener to a measured torque. Torque wrenches come in several drive sizes (1/4, 3/8, 1/2 and 3/4 in.) and have various methods of reading the torque value. The drive size indicates the size of the square drive that ac-

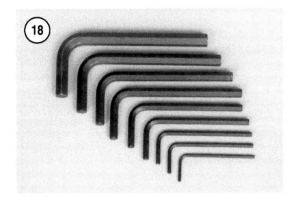

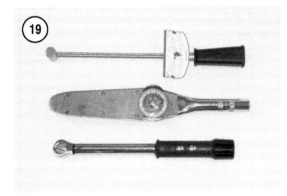

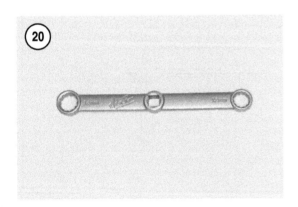

cepts the socket, adapter or extension. Common types of torque wrenches are the deflecting beam, dial indicator and audible click (**Figure 19**).

When choosing a torque wrench, consider the torque range, drive size and accuracy. The torque specifications in this manual provide an indication of the range required.

A torque wrench is a precision tool that must be properly cared for to remain accurate. Store torque wrenches in cases or separate padded drawers within a toolbox. Follow the manufacturer's instructions for their care and calibration.

Torque Adapters

Torque adapters or extensions extend or reduce the reach of a torque wrench. The torque adapter shown in **Figure 20** is used to tighten a fastener that cannot be reached due to the size of the torque wrench head, drive, and socket. If a torque adapter changes the effective lever length (**Figure 21**), the torque reading on the wrench will not equal the actual torque applied to the fastener. It is necessary to recalibrate the torque setting on the wrench to compensate for the change of lever length. When using a torque adapter at a right angle to the drive head, calibration is not required, since the effective length has not changed.

To recalculate a torque reading when using a torque adapter, use the following formula, and refer to **Figure 21**.

$$TW = \frac{TA \times L}{L + A}$$

TW is the torque setting or dial reading on the wrench.

TA is the torque specification and the actual amount of torque that will be applied to the fastener.

A is the amount that the adapter increases (or in some cases reduces) the effective lever length as measured along the centerline of the torque wrench (**Figure 21**).

L is the lever length of the wrench as measured from the center of the drive to the center of the grip.

The effective length of the torque wrench measured along the centerline of the torque wrench is the sum of L and A.

Example:

TA = 20 ft.-lb.

A = 3 in.

L = 14 in.

$$TW = \frac{20 \times 14}{14 + 3} = \frac{280}{17} = 16.5 \text{ ft. lb.}$$

In this example, the torque wrench would be set to the recalculated torque value (TW = 16.5 ft.-lb.). When using a beam-type wrench, tighten the fastener until the pointer aligns with 16.5 ft.-lb. In this example, although the torque wrench is preset to 16.5 ft.-lb., the actual torque is 20 ft.-lb.

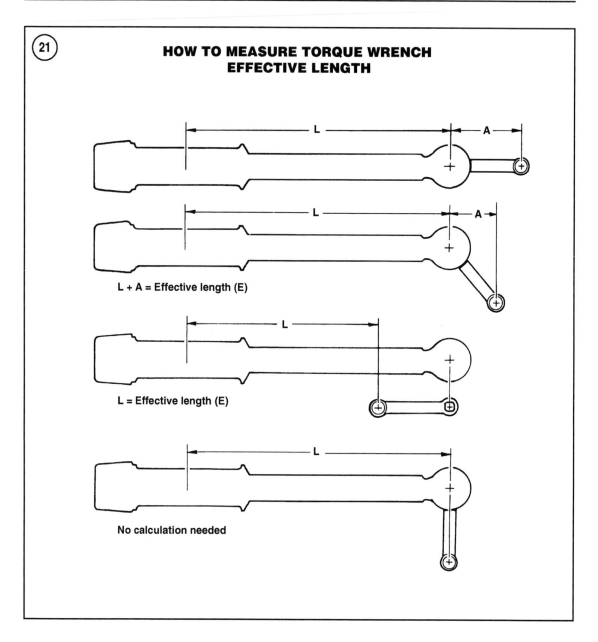

(21)

HOW TO MEASURE TORQUE WRENCH
EFFECTIVE LENGTH

L + A = Effective length (E)

L = Effective length (E)

No calculation needed

Pliers

Pliers come in a wide range of types and sizes. Pliers are useful for holding, cutting, bending, and crimping. Do not use them to turn fasteners. **Figure 22** and **Figure 23** show several types of useful pliers. Each design has a specialized function. Slip-joint pliers are general-purpose pliers used for gripping and bending. Diagonal cutting pliers are needed to cut wire and can be used to remove cotter pins. Needlenose pliers are used to hold or bend small objects. Locking pliers (**Figure 23**), some-

times called Vise Grips, are used to hold objects very tightly. They have many uses ranging from holding two parts together, to gripping the end of a broken stud. Use caution when using locking pliers, as the sharp jaws will damage the objects they hold.

Snap Ring Pliers

Snap ring pliers (**Figure 24**) are specialized pliers with tips that fit into the ends of snap rings to remove and install them.

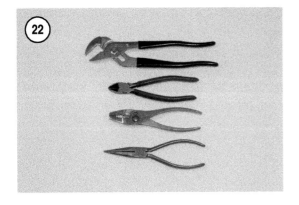

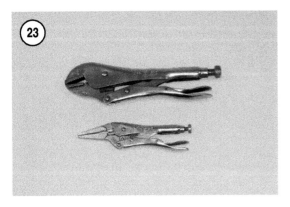

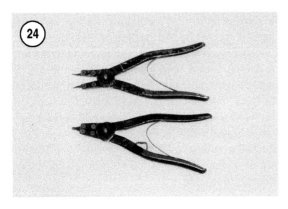

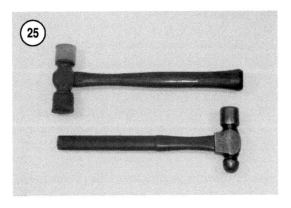

Snap ring pliers are available with a fixed action (either internal or external) or convertible (one tool works on both internal and external snap rings). They may have fixed tips or interchangeable ones of various sizes and angles. For general use, select a convertible type plier with interchangeable tips.

WARNING
Snap rings can spring from the pliers during removal or installation. Also, the snap ring plier tips may break. Always wear eye protection when using snap ring pliers.

Hammers

Various types of hammers (**Figure 25**) are available to fit a number of applications. A ball-peen hammer is used to strike another tool, such as a punch or chisel. Soft-faced hammers are required when a metal object must be struck without damaging it. *Never* use a metal-faced hammer on engine and suspension components, as damage will occur in most cases.

Always wear eye protection when using hammers. Check that the hammer face is in good condition and the handle is not cracked. Select the correct hammer for the job and check that it strikes the object squarely. Do not use the handle or the side of the hammer to strike an object.

PRECISION MEASURING TOOLS

The ability to accurately measure components is essential to many of the procedures in this manual. Equipment is manufactured to close tolerances, and obtaining consistently accurate measurements is essential to determining which components require replacement or further service.

Each type of measuring instrument is designed to measure a dimension with a certain degree of accuracy and within a certain range. Always use a measuring tool that is designed for the task.

As with all tools, measuring tools provide the best results if cared for properly. Improper use can damage the tool and cause inaccurate results. If any measurement is questionable, verify the measurement using another tool. A standard gauge is usually provided with measuring tools to check accuracy and calibrate the tool if necessary.

Precision measurements can vary according to the experience of the person performing the procedure. Accurate results are only possible if the mechanic possesses a feel for using the tool. Heavy-handed use of measuring tools will produce less accurate results than if the tool is grasped gently by the fingertips so the point at which the tool contacts the object is easily felt. This feel for the equipment will produce more accurate measurements and reduce the risk of damaging the tool or component. Refer to the following sections for specific measuring tools.

Feeler Gauge

The feeler or thickness gauge (**Figure 26**) is used for measuring the distance between two surfaces.

A feeler gauge set consists of an assortment of steel strips of graduated thickness. Each blade is marked with its thickness. Blades can be of various lengths and angles for different procedures.

A common use for a feeler gauge is to measure valve clearance. Wire (round) type gauges are used to measure spark plug gap.

Calipers

Calipers (**Figure 27**) are used to determine outside and depth measurements. Although not as precise as a micrometer, they allow reasonable precision, typically to within 0.05 mm (0.001 in.). Most calipers have a range up to 150 mm (6 in.).

Calipers are available in dial, vernier or digital versions. Dial calipers have a dial readout that provides convenient reading. Vernier calipers have marked scales that must be compared to determine the measurement. The digital caliper uses an LCD to show the measurement.

Properly maintain the measuring surfaces of the caliper. There must not be any dirt or burrs between the tool and the object being measured. Never force the caliper closed around an object; close the caliper around the highest point so it can be removed with a slight drag. Some calipers require calibration. Always refer to the manufacturer's instructions when using a new or unfamiliar caliper.

To read a vernier caliper, refer to **Figure 28**. The fixed scale is marked in both inch and millimeter increments. In this example, refer to the metric scale. The ten individual lines on the fixed scale equal 1

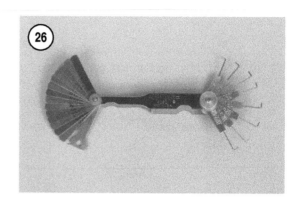

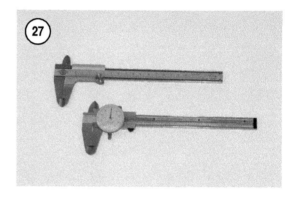

cm. The movable scale is marked in 0.05 mm (hundredth) increments. To obtain a reading, establish the first number by the location of the 0 line on the movable scale in relation to the first line to the left on the fixed scale. In this example, the number is 10 mm. To determine the next number, note which of the lines on the movable scale align with a mark on the fixed scale. A number of lines will seem close, but only one will align exactly. In this case, 0.50 mm is the reading to add to the first number. The result of adding 10 mm and 0.50 mm is a measurement of 10.50 mm.

Micrometers

A micrometer (**Figure 29**) is an instrument designed for linear measurement using the decimal divisions of the inch or meter. While there are many types and styles of micrometers, most of the procedures in this manual call for an outside micrometer. The outside micrometer is used to measure the outside diameter of cylindrical forms and the thickness of materials.

A micrometer's size indicates the minimum and maximum size of a part that it can measure. The

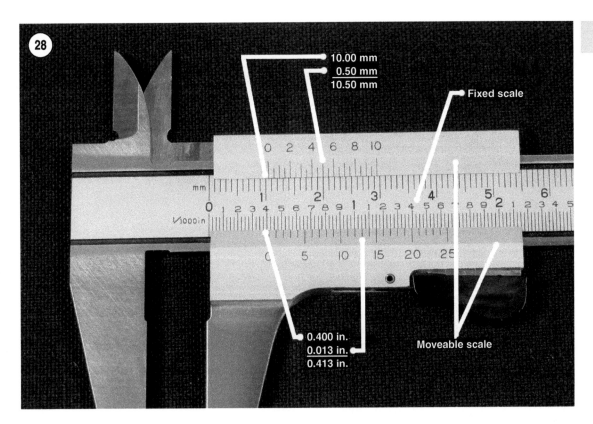

10.00 mm
0.50 mm
10.50 mm

Fixed scale

Moveable scale

0.400 in.
0.013 in.
0.413 in.

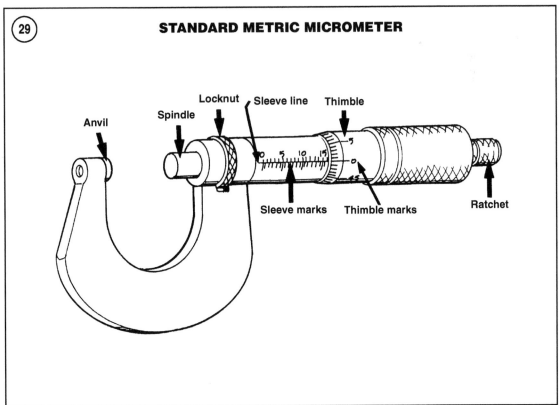

STANDARD METRIC MICROMETER

Anvil

Spindle

Locknut

Sleeve line

Thimble

Sleeve marks

Thimble marks

Ratchet

usual sizes (**Figure 30**) are 0-1 in. (0-25 mm), 1-2 in. (25-50 mm), 2-3 in. (50-75 mm) and 3-4 in. (75-100 mm).

Micrometers that cover a wider range of measurement are available. These use a large frame with interchangeable anvils of various lengths. This type of micrometer offers a cost savings; however, its overall size may make it less convenient.

Reading a Micrometer

When reading a micrometer, numbers are taken from different scales and added together. The following sections describe how to read the measurements of various types of outside micrometers.

For accurate results, properly maintain the measuring surfaces of the micrometer. There cannot be any dirt or burrs between the tool and the measured object. Never force the micrometer closed around an object. Close the micrometer around the highest point so it can be removed with a slight drag. **Figure 29** shows the markings and parts of a standard metric micrometer. Be familiar with these terms before using a micrometer in the follow sections.

Standard inch micrometer

The standard inch micrometer is accurate to one-thousandth of an inch or 0.001. The sleeve is marked in 0.025 in. increments. Every fourth sleeve mark is numbered 1, 2, 3, 4, 5, 6, 7, 8, 9. These numbers indicate 0.100, 0.200, 0.300, and so on.

The tapered end of the thimble has 25 lines marked around it. Each mark equals 0.001 in. One complete turn of the thimble will align its zero mark with the first mark on the sleeve or 0.025 in.

When reading a standard inch micrometer, perform the following steps while referring to **Figure 31**.

1. Read the sleeve and find the largest number visible. Each sleeve number equals 0.100 in.

2. Count the number of lines between the numbered sleeve mark and the edge of the thimble. Each sleeve mark equals 0.025 in.

3. Read the thimble mark that aligns with the sleeve line. Each thimble mark equals 0.001 in.

> *NOTE*
> *If a thimble mark does not align exactly with the sleeve line, estimate the*

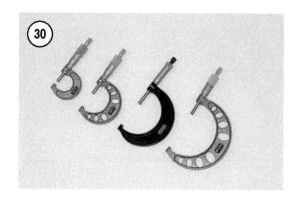

amount between the lines. For accurate readings in ten-thousandths of an inch (0.0001 in.), use a vernier inch micrometer.

4. Add the readings from Steps 1-3.

Metric micrometer

The standard metric micrometer is accurate to one one-hundredth of a millimeter (0.01 mm). The sleeve line is graduated in millimeter and half millimeter increments. The marks on the upper half of the sleeve line equal 1.00 mm. Each fifth mark above the sleeve line is identified with a number. The number sequence depends on the size of the micrometer. A 0-25 mm micrometer, for example, will have sleeve marks numbered 0 through 25 in 5 mm increments. This numbering sequence continues with larger micrometers. On all metric micrometers, each mark on the lower half of the sleeve equals 0.50 mm.

The tapered end of the thimble has 50 lines marked around it. Each mark equals 0.01 mm. One complete turn of the thimble aligns its 0 mark with the first line on the lower half of the sleeve line or 0.50 mm.

When reading a metric micrometer, add the number of millimeters and half-millimeters on the sleeve line to the number of one one-hundredth millimeters on the thimble. Perform the following steps while referring to **Figure 32**.

1. Read the upper half of the sleeve line and count the number of lines visible. Each upper line equals 1 mm.

2. See if the half-millimeter line is visible on the lower sleeve line. If so, add 0.50 mm to the reading in Step 1.

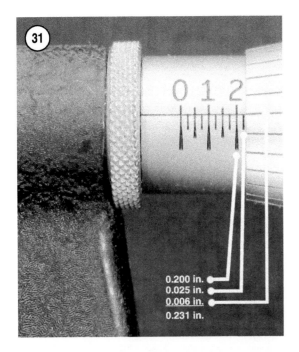

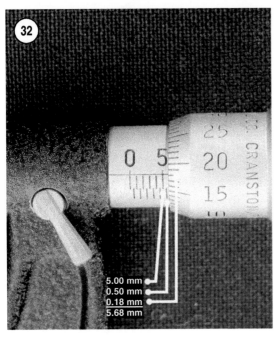

3. Read the thimble mark that aligns with the sleeve line. Each thimble mark equals 0.01 mm.

NOTE
If a thimble mark does not align exactly with the sleeve line, estimate the amount between the lines. For accurate readings in two-thousandths of a

millimeter (0.002 mm), use a metric vernier micrometer.

4. Add the readings from Steps 1-3.

Micrometer Adjustment

Before using a micrometer, check its adjustment as follows:
1. Clean the anvil and spindle faces.
2A. To check a 0-1 in. or 0-25 mm micrometer:
   a. Turn the thimble until the spindle contacts the anvil. If the micrometer has a ratchet stop, use it to ensure that the proper amount of pressure is applied.
   b. If the adjustment is correct, the 0 mark on the thimble will align exactly with the 0 mark on the sleeve line. If the marks do not align, the micrometer is out of adjustment.
   c. Follow the manufacturer's instructions to adjust the micrometer.
2B. To check a micrometer larger than 1 in. or 25 mm use the standard gauge supplied by the manufacturer. A standard gauge is a steel block, disc or rod that is machined to an exact size.
   a. Place the standard gauge between the spindle and anvil, and measure its outside diameter or length. If the micrometer has a ratchet stop, use it to ensure that the proper amount of pressure is applied.
   b. If the adjustment is correct, the 0 mark on the thimble will align exactly with the 0 mark on the sleeve line. If the marks do not align, the micrometer is out of adjustment.
   c. Follow the manufacturer's instructions to adjust the micrometer.

Micrometer Care

Micrometers are precision instruments. They must be used and maintained with great care. Note the following:
1. Store micrometers in protective cases or separate padded drawers in a toolbox.
2. When in storage, make sure the spindle and anvil faces do not contact each other or another object. If they do, temperature changes and corrosion may damage the contact faces.
3. Do not clean a micrometer with compressed air. Dirt forced into the tool will cause wear.

4. Lubricate micrometers with WD-40 to prevent corrosion.

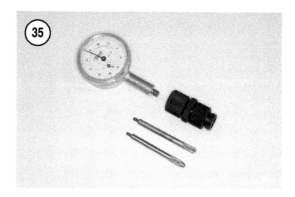

Telescoping and Small-Bore Gauges

Use telescoping gauges (**Figure 33**) and small-hole gauges to measure bores. Neither gauge has a scale for direct readings. An outside micrometer must be used to determine the reading.

To use a telescoping gauge, select the correct size gauge for the bore. Compress the movable post and carefully insert the gauge into the bore. Carefully move the gauge in the bore to check that it is centered. Tighten the knurled end of the gauge to hold the movable post in position. Remove the gauge and measure the length of the posts. Telescoping gauges are typically used to measure cylinder bores.

To use a small-bore gauge, select the correct size gauge for the bore. Carefully insert the gauge into the bore. Tighten the knurled end of the gauge to carefully expand the gauge fingers to the limit within the bore. Do not overtighten the gauge, as there is no built-in release. Excessive tightening can damage the bore surface and damage the tool. Remove the gauge and measure the outside dimension (**Figure 34**). Small-hole gauges are typically used to measure valve guides.

Dial Indicator

A dial indicator (**Figure 35**) is a gauge with a dial face and needle used to measure variations in dimensions and movements. Measuring brake rotor runout is a typical use for a dial indicator.

Dial indicators are available in various ranges and graduations and with three basic types of mounting bases: magnetic, clamp, or screw-in stud. When

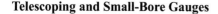

purchasing a dial indicator, select the magnetic stand type, with a continuous dial.

Cylinder Bore Gauge

A cylinder bore gauge is similar to a dial indicator. The gauge set shown in **Figure 36** consists of a dial indicator, handle, and different length adapters (anvils) to fit the gauge to various bore sizes. The bore gauge is used to measure bore size, taper and out-of-round. When using a bore gauge, follow the manufacturer's instructions.

Compression Gauge

A compression gauge (**Figure 37**) measures combustion chamber (cylinder) pressure, usually in psi or kPa. The gauge adapter is either inserted or screwed into the spark plug hole to obtain the reading. Disable the engine so it will not start and hold the throttle in the wide-open position when performing a compression test. An engine that does not have adequate compression cannot be properly tuned.

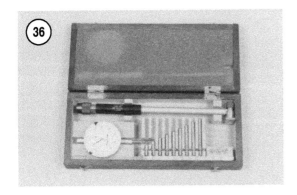

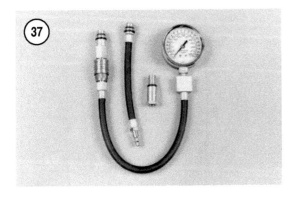

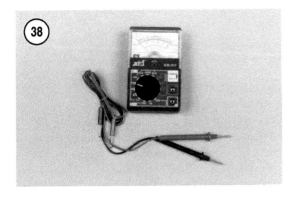

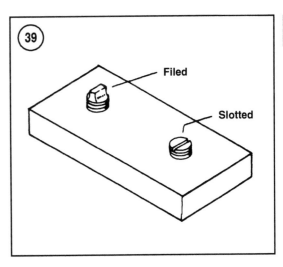

Multimeter

A multimeter (**Figure 38**) is an essential tool for electrical system diagnosis. The voltage function indicates the voltage applied or available to various electrical components. The ohmmeter function tests circuits for continuity, or lack of continuity, and measures the resistance of a circuit.

Some manufacturer's specifications for electrical components are based on results using a specific test meter. Results may vary if using a meter not recommended by the manufacturer is used. Such requirements are noted when applicable.

Removing Frozen Fasteners

If a fastener cannot be removed, several methods may be used to loosen it. First, apply penetrating oil such as Liquid Wrench or WD-40. Apply it liberally and let it penetrate for 10-15 minutes. Strike the fastener several times with a small hammer. Do not hit it so hard as to cause damage. Reapply the penetrating oil if necessary.

For frozen screws, apply penetrating oil as described, then insert a screwdriver in the slot and strike the top of the screwdriver with a hammer. This loosens the rust so the screw can be backed out. If the screw head is too damaged to use this method, grip the head with locking pliers and twist the screw out.

Avoid applying heat unless specifically instructed, as it may melt, warp or remove the temper from parts.

CAUTION
Cases that are made of magnesium alloy can ignite if heated with a flame.

Removing Broken Fasteners

If the head breaks off a screw or bolt, several methods are available for removing the remaining portion. If a large portion of the remainder projects out, try gripping it with locking pliers. If the projecting portion is too small, file it to fit a wrench or cut a slot in it to fit a screwdriver (**Figure 39**).

If the head breaks off flush, use a screw extractor. To do this, center-punch the exact center of the remaining portion of the screw or bolt. Drill a small hole in the screw and tap the extractor into the hole. Back the screw out with a wrench on the extractor (**Figure 40**).

Repairing Damaged Threads

Occasionally, threads are stripped through carelessness or impact damage. Often the threads can be repaired by running a tap (for internal threads on nuts) or die (for external threads on bolts) through the threads (**Figure 41**). To clean or repair spark plug threads, use a spark plug tap.

If an internal thread is damaged, it may be necessary to install a Helicoil or some other type of thread insert. Follow the manufacturer's instructions when installing their insert.

If it is necessary to drill and tap a hole, refer to **Table 6** for metric tap and drill sizes.

Stud Removal/Installation

A stud removal tool is available from most tool suppliers. This tool makes the removal and installation of studs easier. If one is not available, thread two nuts onto the stud and tighten them against each other. Remove the stud by turning the lower nut (**Figure 42**).

1. Measure the height of the stud above the surface.
2. Thread the stud removal tool onto the stud and tighten it, or thread two nuts onto the stud.
3. Remove the stud by turning the stud remover or the lower nut.
4. Remove any threadlocking compound from the threaded hole. Clean the threads with an aerosol parts cleaner.
5. Install the stud removal tool onto the new stud or thread two nuts onto the stud.
6. Apply threadlocking compound to the threads of the stud.
7. Install the stud and tighten with the stud removal tool or the top nut.
8. Install the stud to the height noted in Step 1 or its torque specification.
9. Remove the stud removal tool or the two nuts.

Removing Hoses

When removing stubborn hoses, do not exert excessive force on the hose or fitting. Remove the

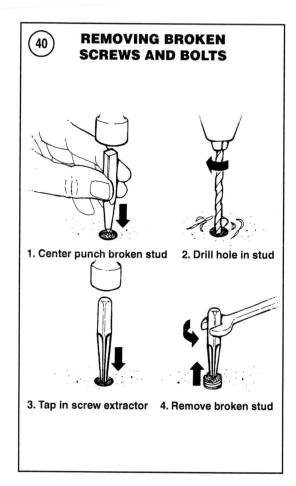

REMOVING BROKEN SCREWS AND BOLTS

1. Center punch broken stud
2. Drill hole in stud
3. Tap in screw extractor
4. Remove broken stud

hose clamp and carefully insert a small screwdriver or pick tool between the fitting and hose. Apply a spray lubricant under the hose and carefully twist the hose off the fitting. Clean the fitting of any corrosion or rubber hose material with a wire brush. Clean the inside of the hose thoroughly. Do not use any lubricant when installing the hose (new or old). The lubricant may allow the hose to come off the fitting, even with the clamp secure.

Bearings

Bearings are used in the engine and transmission assembly to reduce power loss, heat and noise resulting from friction. Because bearings are precision parts, they must be maintained by proper lubrication and maintenance. If a bearing is damaged, replace it immediately. When installing a new bearing, take care to prevent damaging the part. Many bearing replacement procedures are included

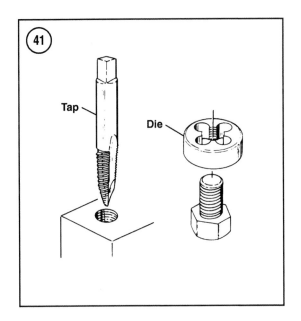

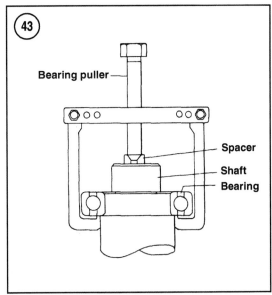

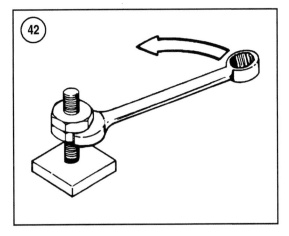

in the individual chapters where applicable; however, use the following sections as a guideline.

NOTE
Unless otherwise specified, install bearings with the manufacturer's mark or number facing outward.

Removal

While bearings are normally removed only when damaged, there may be times when it is necessary to remove a bearing that is in good condition. However, improper bearing removal will damage the bearing and maybe the shaft or case half. Note the following when removing bearings.

1. When using a puller to remove a bearing from a shaft, take care that the shaft is not damaged. Always place a piece of metal between the end of the shaft and the puller screw. In addition, place the puller arms next to the inner bearing race. See **Figure 43**.

2. When using a hammer to remove a bearing from a shaft, do not strike the hammer directly against the shaft. Instead, use a brass or aluminum rod between the hammer and shaft (**Figure 44**) and be sure to support both bearing races with wooden blocks, as shown.

3. The ideal method of bearing removal is with a hydraulic press. Note the following when using a press:

   a. Always support the inner and outer bearing races with a suitable size wooden or aluminum ring (**Figure 45**). If only the outer race is supported, pressure applied against the balls and/or the inner race will damage them.

   b. Always check that the press arm (**Figure 45**) aligns with the center of the shaft. If the arm is not centered, it may damage the bearing and/or shaft.

   c. The moment the shaft is free of the bearing, it will drop to the floor. Secure or hold the shaft to prevent it from falling.

Installation

1. When installing a bearing in a housing, apply
pressure to the *outer* bearing race (**Figure 46**).
When installing a bearing on a shaft, apply pressure
to the *inner* bearing race (**Figure 47**).

2. When installing a bearing as described in Step 1,
some type of driver is required. Never strike the
bearing directly with a hammer or the bearing will
be damaged. When installing a bearing, use a length
of pipe or a driver with a diameter that matches the
bearing race. **Figure 48** shows the correct way to
use a driver and hammer to install a bearing.

3. Step 1 describes how to install a bearing in a case
half or over a shaft. However, when installing a
bearing over a shaft and into a housing at the same
time, a tight fit will be required for both outer and
inner bearing races. In this situation, install a spacer
underneath the driver tool so that pressure is applied
evenly across both races (**Figure 49**). If the outer
race is not supported, the balls will push against the
outer bearing race and damage it.

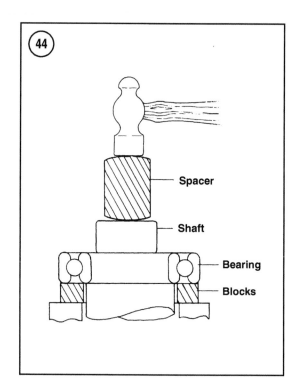

Interference Fit

1. Follow this procedure when installing a bearing
over a shaft. When a tight fit is required, the bearing
inside diameter will be smaller than the shaft. In this
case, driving the bearing onto the shaft using normal methods may cause bearing damage. Instead,
heat the bearing before installation. Note the following:

   a. Secure the shaft so it is ready for bearing in-
      stallation.

   b. Clean all residues from the bearing surface of
      the shaft. Remove burrs with a file or sandpa-
      per.

   c. Fill a suitable container with clean mineral
      oil. Place a thermometer rated above 248° F
      (120° C) in the oil. Support the thermometer
      so that it does not rest on the bottom or side of
      the container.

   d. Remove the bearing from its wrapper and se-
      cure it with a piece of heavy wire bent to hold
      it in the container. Hang the bearing so it does
      not touch the bottom or sides.

   e. Turn the heat on and monitor the thermome-
      ter. When the oil temperature rises to approxi-
      mately 248° F (120° C), remove the bearing
      and quickly install it. If necessary, place a

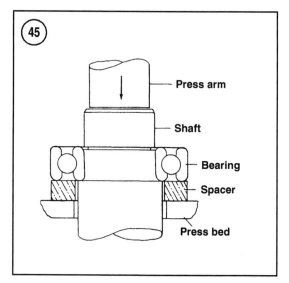

socket on the inner bearing race and tap the
bearing into place. As the bearing chills, it
will tighten on the shaft so installation must
be done quickly. Check that the bearing is installed completely.

2. Follow this step when installing a bearing in a
housing. Bearings are generally installed in a housing with a slight interference fit. Driving the bearing
into the housing using normal methods may damage

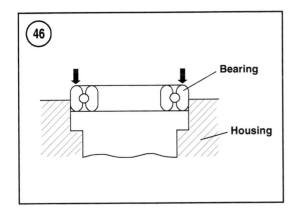

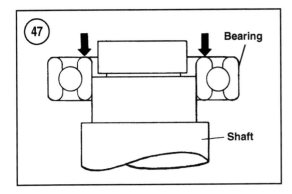

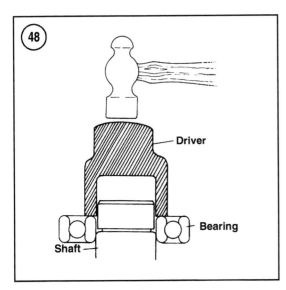

the housing or cause bearing damage. Instead, heat
the housing before the bearing is installed. Note the
following:

CAUTION
Before heating the housing in this pro-
cedure, wash the housing thoroughly

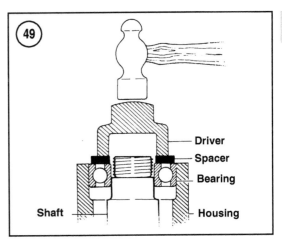

with detergent and water. Rinse and
rewash the cases as required to re-
move all traces of oil and other chem-
ical deposits.

a. Heat the housing to approximately 212° F
   (100° C) in an oven or on a hot plate. An easy
   way to check that it is at the proper temperature is to place tiny drops of water on the
   housing; if they sizzle and evaporate immediately, the temperature is correct. Heat only
   one housing at a time.

CAUTION
Do not heat the housing or bearing
with a propane or acetylene torch.
The direct heat will destroy the case
hardening of the bearing and will
likely warp the housing. Cases that
are made of magnesium alloy can ig-
nite.

b. Remove the housing from the oven or hot
   plate. Wear insulated gloves or use kitchen
   pot holders.

NOTE
Remove and install the bearings with
a suitable-size socket and extension.

c. Hold the housing with the bearing side down
   and tap the bearing out. Repeat for all bearings in the housing.

d. Before heating the bearing housing, place the
   new bearing in a freezer if possible. Chilling a
   bearing slightly reduces its outside diameter
   while the heated bearing housing assembly is

slightly larger due to heat expansion. This will make bearing installation easier.

NOTE
Always install bearings with the manufacturer's mark or number facing outward.

e. While the housing is still hot, install the new bearing(s) into the housing. Install the bearings by hand, if possible. If necessary, lightly tap the bearing(s) into the housing with a socket placed on the outer bearing race (**Figure 46**). Do not install new bearings by driving on the inner bearing race. Install the bearing(s) until it seats completely.

Seal Replacement

Seals (**Figure 50**) are used to contain oil, water, grease or combustion gasses in a housing or shaft. Improperly removing a seal can damage the housing or shaft. Improper installation of the seal can damage the seal. Note the following:
1. Prying is generally the easiest and most effective method of removing a seal from a housing. However, always place a shop cloth under the pry tool to prevent damage to the housing.
2. Pack waterproof grease in the seal lips before installing the seal.
3. Install seals with the manufacturer's numbers or marks facing out.
4. Install seals with a socket placed on the outside of the seal as shown in **Figure 51**. Drive the seal squarely into the housing. Never install a seal by hitting the top of the seal with a hammer.

STORAGE

Several months of non-use can cause a general deterioration of the motorcycle. This is especially true in areas of extreme temperature variations. This deterioration can be minimized with careful preparation for storage. A properly stored motorcycle will be much easier to return to service.

Storage Area Selection

When selecting a storage area, consider the following:

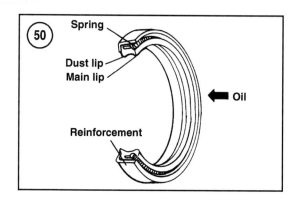

1. The storage area must be dry. A heated area is best, but not necessary. It should be insulated to minimize extreme temperature variation.
2. If the building has large window areas, cover them to keep sunlight off the motorcycle.
3. Avoid storage areas close to saltwater.
4. Consider the area's risk of fire, theft or vandalism. Check with an insurer regarding motorcycle coverage while in storage.

Preparing the Motorcycle for Storage

The amount of preparation a motorcycle should undergo before storage depends on the expected length of non-use, storage area conditions and personal preference. Consider the following list the minimum requirement:
1. Wash the motorcycle. Remove all dirt, mud and road debris.
2. Check the cooling system for proper level and mix ratio.
3. Start the engine and allow it to reach operating temperature. Turn the fuel valve off and run the engine until all the fuel is evacuated from the lines and carburetor, then open the drain on the float chamber to drain any residual fuel.
4. Remove and drain the fuel tank and fuel valve. Pour about 240 cc (8 oz.) of engine oil into the tank, then distribute the oil onto the inner wall of the tank, to prevent corrosion.
5. Drain the engine oil regardless of the riding time since the last service. Fill the engine with the recommended oil.
6. Remove the spark plug and pour a teaspoon of engine oil into the cylinder. Place a shop cloth over the opening and turn the engine over to distribute the oil. Reinstall the spark plug.

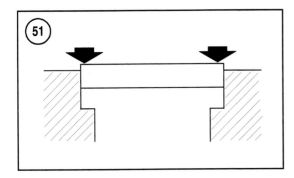

7. Remove the battery. Store the battery in a cool, dry location.

8. Cover the exhaust and intake openings.

9. Reduce the normal tire pressure by 20%.

10. Apply a commercial protectant to the plastic and rubber components, including the tires. Follow the manufacturer's instructions for each type of product being used.

11. Place the motorcycle on a stand so the tires are off the ground. If this is not possible, place a piece of plywood between the tires and the ground. Inflate the tires to the recommended pressure if the motorcycle cannot be elevated.

12. Cover the motorcycle with a drop cloth or similar cover. Do not use plastic covers; these will trap moisture and promote corrosion.

Returning the Motorcycle to Service

The amount of service required to return a motorcycle to operating condition depends on the length of non-use and storage conditions. Follow the above procedure and install/check each area that was prepared at time of storage. Also check that the brakes, clutch, throttle and engine stop switch work properly before operating the motorcycle. Refer to the maintenance and lubrication schedule in Chapter Three and determine which areas require additional service.

Table 1 YEAR AND MODEL CODE NUMBERS

Year and model	Starting frame serial number
1987 KL650-A1	JKAKLEA1 HA000001
1988 KL650-A2	JKAKLEA1 JA008701
1989 KL650-A3	JKAKLEA1 KA013601
1990 KL650-A4	JKAKLEA1 LA016001
1991 KL650-A5	JKAKLEA1 MA018001
1992 KL650-A6	JKAKLEA1 NA021901
1993 KL650-A7	JKAKLEA1 PA026001
1994 KL650-A8	JKAKLEA1 RA030001
1995 KL650-A9	JKAKLEA1 SA032001
1996 KL650-A10	JKAKLEA1 TA040001
1997 KL650-A11	JKAKLEA1 VA042001
1998 KL650-A12	JKAKLEA1 WA045001
1999 KL650-A13	JKAKLEA1 XA050001
2000 KL650-A14	JKAKLEA1 YA057001
2001 KL650-A15	JKAKLEA1 1A070001
2002 KL650-A16	JKAKLEA1 2D075001
2003 KL650-A17	JKAKLEA1 3DA03323
2004 KL650-A18	JKAKLEA1 4DA07245
2005 KL650-A19	JKAKLEA1 5DA13001
2006 KL650-A6F	JKAKLEA1 6DA19201
2007 KL650-A7F	JKAKLEA1 7DA28001

Table 2 GENERAL DIMENSIONS AND WEIGHT

	mm	in.
Ground clearance	240	9.4
Overall length	2205	86.8
Overall width	940	37.0
Overall height	1345	53.0
Wheelbase	1495	58.9
Seat height	890	35.0
Dry weight	153 kg	337 lb.

Table 3 TECHNICAL ABBREVIATIONS

ABDC	After bottom dead center
ATDC	After top dead center
BBDC	Before bottom dead center
BDC	Bottom dead center
BTDC	Before top dead center
C	Celsius (centigrade)
cc	Cubic centimeters
cid	Cubic inch displacement
CDI	Capacitor discharge ignition
cu. in.	Cubic inches
DAI	Direct air induction
ECM	Engine control module
ECT	Engine coolant temperature sensor
EFI	Electronic fuel injection
F	Fahrenheit
ft.	Feet
ft.-lb.	Foot-pounds
gal.	Gallons
H/A	High altitude
hp	Horsepower
IAT	Intake air temperature sensor
ICM	Ignition control module
in.	Inches
in.-lb.	Inch-pounds
I.D.	Inside diameter
kg	Kilograms
kgm	Kilogram meters
km	Kilometer
kPa	Kilopascals
L	Liter
m	Meter
MAG	Magneto
ml	Milliliter
mm	Millimeter
N•m	Newton-meters
O2	Oxygen
O.D.	Outside diameter
oz.	Ounces
PAIR	Pulsed secondary air injection system
PGM-FI	Programmed fuel injection
psi	Pounds per square inch
PTO	Power take off
pt.	Pint
qt.	Quart
rpm	Revolutions per minute
TP	Throttle position sensor

Table 4 GENERAL TORQUE SPECIFICATIONS*

Thread diameter (mm)	N•m	in.-lb.	ft.-lb.
5	5	42	–
6	8	72	–
8	19	–	14
10	34	–	25
12	55	–	40
14	85	–	63
16	135	–	100
18	190	–	140
20	275	–	203

*Use this table as a guide for fasteners that do not have a specification.

Table 5 CONVERSION FORMULAS

Multiply:	By:	To get the equivalent of:
Length		
Inches	25.4	Millimeter
Inches	2.54	Centimeter
Miles	1.609	Kilometer
Feet	0.3048	Meter
Millimeter	0.03937	Inches
Centimeter	0.3937	Inches
Kilometer	0.6214	Mile
Meter	3.281	Feet
Fluid volume		
U.S. quarts	0.9463	Liters
U.S. gallons	3.785	Liters
U.S. ounces	29.573529	Milliliters
Imperial gallons	4.54609	Liters
Imperial quarts	1.1365	Liters
Liters	0.2641721	U.S. gallons
Liters	1.0566882	U.S. quarts
Liters	33.814023	U.S. ounces
Liters	0.22	Imperial gallons
Liters	0.8799	Imperial quarts
Milliliters	0.033814	U.S. ounces
Milliliters	1.0	Cubic centimeters
Milliliters	0.001	Liters
Torque		
Foot-pounds	1.3558	Newton-meters
Foot-pounds	0.138255	Meters-kilograms
Inch-pounds	0.11299	Newton-meters
Newton-meters	0.7375622	Foot-pounds
Newton-meters	8.8507	Inch-pounds
Meters-kilograms	7.2330139	Foot-pounds
Volume		
Cubic inches	16.387064	Cubic centimeters
Cubic centimeters	0.0610237	Cubic inches
Temperature		
Fahrenheit	(°F – 32) × 0.556	Centigrade
Centigrade	(°C × 1.8) + 32	Fahrenheit
Weight		
Ounces	28.3495	Grams
Pounds	0.4535924	Kilograms
Grams	0.035274	Ounces
Kilograms	2.2046224	Pounds
Pressure		
Pounds per square inch	0.070307	Kilograms per square centimeter

Table 6 METRIC TAP AND DRILL SIZES

Metric size	Drill equivalent	Decimal fraction	Nearest fraction
3 × 0.50	No. 39	0.0995	3/32
3 × 0.60	3/32	0.0937	3/32
4 × 0.70	No. 30	0.1285	1/8
4 × 0.75	1/8	0.125	1/8
5 × 0.80	No. 19	0.166	11/64
5 × 0.90	No. 20	0.161	5/32
6 × 1.00	No. 9	0.196	13/64
7 × 1.00	16/64	0.234	15/64
8 × 1.00	J	0.277	9/32
8 × 1.25	17/64	0.265	17/64
9 × 1.00	5/16	0.3125	5/16
9 × 1.25	5/16	0.3125	5/16
10 × 1.25	11/32	0.3437	11/32
10 × 1.50	R	0.339	11/32
11 × 1.50	3/8	0.375	3/8
12 × 1.50	13/32	0.406	13/32
12 × 1.75	13/32	0.406	13/32

Table 7 METRIC, INCH AND FRACTIONAL EQUIVALENTS

mm	in.	Nearest fraction	mm	in.	Nearest fraction
1	0.0394	1/32	26	1.0236	1 1/32
2	0.0787	3/32	27	1.0630	1 1/16
3	0.1181	1/8	28	1.1024	1 3/32
4	0.1575	5/32	29	1.1417	1 5/32
5	0.1969	3/16	30	1.1811	1 3/16
6	0.2362	1/4	31	1.2205	1 7/32
7	0.2756	9/32	32	1.2598	1 1/4
8	0.3150	5/16	33	1.2992	1 5/16
9	0.3543	11/32	34	1.3386	1 11/32
10	0.3937	13/32	35	1.3780	1 3/8
11	0.4331	7/16	36	1.4173	1 13/32
12	0.4724	15/32	37	1.4567	1 15/32
13	0.5118	1/2	38	1.4961	1 1/2
14	0.5512	9/16	39	1.5354	1 17/32
15	0.5906	19/32	40	1.5748	1 9/16
16	0.6299	5/8	41	1.6142	1 5/8
17	0.6693	21/32	42	1.6535	1 21/32
18	0.7087	23/32	43	1.6929	1 11/16
19	0.7480	3/4	44	1.7323	1 23/32
20	0.7874	25/32	45	1.7717	1 25/32
21	0.8268	13/16	46	1.8110	1 13/16
22	0.8661	7/8	47	1.8504	1 27/32
23	0.9055	29/32	48	1.8898	1 7/8
24	0.9449	15/16	49	1.9291	1 15/16
25	0.9843	31/32	50	1.9685	1 31/32

CHAPTER TWO

TROUBLESHOOTING

Diagnosing problems with the motorcycle, either mechanical or electrical, can be relatively easy if the fundamental operating requirements are kept in mind. By doing so, problems can be approached in a logical and methodical manner. The first steps are to:

1. Define the symptoms of the problem.
2. Determine which areas could exhibit those symptoms.
3. Test and analyze the suspect area.
4. Isolate the problem.

Being quick to assume a particular area is at fault can lead to increased problems, lost time and unnecessary parts replacement. The easiest way to keep troubleshooting simple, is to perform the lubrication, maintenance and tune-up procedures described in Chapter Three. The rider will gain a better understanding of the condition and functions of the motorcycle.

Always start with the simple and obvious checks when troubleshooting. This would include: engine stop switch operation, fuel level, fuel valve position and spark plug cap tightness. If the problem cannot be solved, stop and evaluate all conditions prior to the problem.

For removal, installation and test procedures for some components, refer to the specific chapter in the manual. When applicable, tables at the end of each chapter provide specifications and wear limits.

OPERATING REQUIREMENTS

There are three requirements for an engine to run properly. These are: correct air/fuel mixture, compression and properly timed spark. If one of these requirements is not correct, the engine will not run, or will run poorly. A four-stroke engine performs these functions as shown in **Figure 1**.

STARTING THE ENGINE

Before starting the engine, always perform a pre-ride check of the motorcycle, as described in Chapter Three.

Safety Switches

The motorcycle is equipped with safety switches that prevent the engine from starting or running if certain conditions occur. The following describes each switch and when it prevents the engine from running:

1. Neutral switch. With the transmission in any position except neutral, with the clutch engaged (clutch lever out), the engine will not start.
2. Starter lockout switch (located in clutch lever assembly). With the transmission in gear, the engine will not start if the clutch is engaged (clutch lever

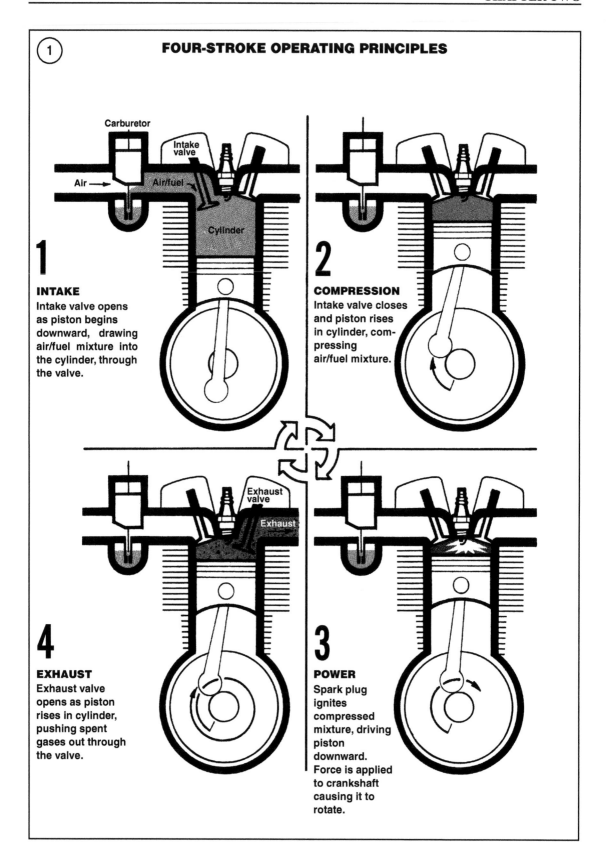

FOUR-STROKE OPERATING PRINCIPLES

1 INTAKE
Intake valve opens as piston begins downward, drawing air/fuel mixture into the cylinder, through the valve.

2 COMPRESSION
Intake valve closes and piston rises in cylinder, compressing air/fuel mixture.

4 EXHAUST
Exhaust valve opens as piston rises in cylinder, pushing spent gases out through the valve.

3 POWER
Spark plug ignites compressed mixture, driving piston downward. Force is applied to crankshaft causing it to rotate.

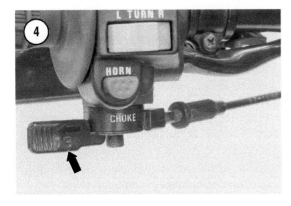

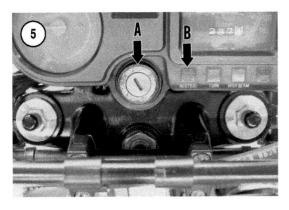

out). If the clutch is disengaged with the transmission in gear, the engine will start.

3. Side stand switch. If the side stand is down with the engine in gear and running, the engine will stop when the clutch lever is released.

4. Engine stop switch (A, **Figure 2**). When moved to the off position (left or right of vertical), the switch will prevent the engine from starting, or, will stop the engine when it is running. The engine will start and run only when the switch is in the run position (vertical).

Engine Is Cold

1. Shift the transmission into neutral.

2. Check that the engine stop switch is in the run position.

3. Turn the fuel valve lever from the off position (**Figure 3**) to the on (vertical) position. An arrow on the fuel valve points to the lever position.

4. Move the choke lever (**Figure 4**) fully to the left to richen the air/fuel mixture.

5. Turn on the ignition switch (A, **Figure 5**).

6. Check that the neutral light (B, **Figure 5**) comes on.

7. Press the starter button (B, **Figure 2**) while keeping the throttle closed.

> *NOTE*
> *The type of choke system used on the KLR is most effective if the throttle remains completely closed during startup.*

8. When the engine starts, gradually move the choke lever to the right as the engine warms up. Allow the engine to warm up at an idle for 1 minute, or until the engine responds smoothly and does not require the choke.

> *CAUTION*
> *Do not race the engine during the warm-up period. Excessive wear and potential engine damage can occur when the engine is not up to operating temperature.*

Engine Is Warm or Hot

1. Shift the transmission into neutral.

2. Check that the engine stop switch is in the run position.

3. Turn the fuel valve lever from the off position (**Figure 3**) to the on (vertical) position. An arrow on the fuel valve points to the lever position.

4. Turn on the ignition switch (A, **Figure 5**).

5. Check that the neutral light (B, **Figure 5**) comes on.

6. Press the starter button (B, **Figure 2**) while keeping the throttle closed.

Engine Is Flooded

If the engine fails to start after several tries (particularly if the choke has been used), it is probably flooded. This occurs when too much fuel is drawn into the engine and the spark plug fails to ignite the air/fuel mixture. The smell of gasoline is often evident when the engine is flooded. If there are no obvious signs of fuel overflow from the carburetor, try starting the engine by fully opening the throttle (no choke) and operating the starter. If the engine starts, keep the engine running at a fast idle until it has burned the excess fuel from the engine.

If the engine does not start, perform the following troubleshooting steps before making other checks:

1. Check that the choke lever is fully to the right.

2. Look for gasoline overflowing from the carburetor or overflow hose.

   a. If gasoline is evident, the float in the carburetor bowl is stuck or adjusted too high.

   b. Remove and repair the float assembly as described in Chapter Eight.

3. Check the air filter for excessive buildup.

4. Remove the spark plug and dry the electrodes. Reinstall the plug and try starting the engine as described above.

5. Perform an engine spark test.

ENGINE SPARK TEST

An engine spark test indicates whether the ignition system is providing power to the spark plug. It is a quick way to determine if a problem is in the electrical system or fuel system.

CAUTION
When performing this test, the spark plug lead must be grounded before cranking the engine. If it is not, it is

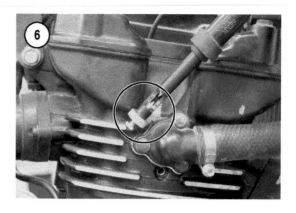

*possible to damage the CDI circuitry. A spark plug can be used for this test, but a spark tester (**Figure 6**) will clearly show if spark is occurring, as well as the strength of the spark. This tester can be purchased at parts supply stores or suppliers of ignition test equipment.*

1. Remove the spark plug. Inspect the spark plug by comparing its condition to the plugs shown in Chapter Three.

2. Connect the spark plug lead to the spark plug or to a spark tester.

3. Ground the plug/tester to bare metal on the engine. Position the plug/tester so the firing end can be viewed (**Figure 7**).

4. Crank the engine and observe the spark. A fat, blue spark should appear at the firing end. The spark should fire consistently as the engine is cranked.

5. If the spark appears weak, or fires inconsistently, check the following areas for the possible cause:

   a. Fouled/improperly gapped spark plug.

   b. Damaged/shorted spark plug lead and cap.

   c. Loose connection in ignition system.

   d. Damaged coil.

   e. Damaged ignition switch.

   f. Dirty/shorted engine stop switch.

   g. Damaged exciter coil or ignition pickup coil (check ignition timing).

   h. Damaged CDI unit.

ENGINE PERFORMANCE

If the engine does not operate at peak performance, the following lists of possible causes may help isolate the problem. The easier checks for each area are listed first. These checks represent what is

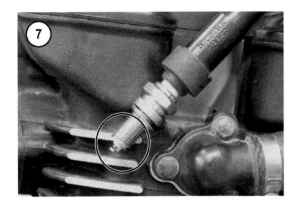

commonly responsible for poor engine performance.

It is recommended to review Chapter Eight (Fuel System) and Chapter Nine (Electrical System) to gain a working knowledge of each of these systems. Throughout the chapters, explanations, photos and diagrams are provided to aid in learning the function of the major components that make up each system.

Engine Will Not Start or Starts and Dies

1. Fuel system:

> *NOTE*
> *For California models, if the fuel tank is overfilled, heat can cause the fuel to expand and overflow into the evaporative emission control unit lines, located near the filler neck. This can cause hard starting and engine hesitation until the fuel is cleared from the system.*

   a. Fuel valve off.
   b. Fuel tank near empty.
   c. Improper operation of choke lever.
   d. Choke plunger stuck open.
   e. Idle speed too low.
   f. Engine flooded.
   g. Clogged fuel tank cap vent.
   h. Contaminated fuel.
   i. Clogged fuel valve, fuel line or carburetor.
   j. Clogged air filter.
   k. Fuel tank diaphragm-valve vacuum hose disconnected or leaking.
   l. Malfunctioning fuel tank diaphragm valve.
   m. Pilot mixture screw misadjusted.

   n. Float valve clogged or sticking.
   o. Air leaks at intake duct.
   p. Wrong pilot jet for altitude.
2. Ignition:
   a. Fouled/improperly gapped spark plug.
   b. Damaged/shorted spark plug lead and cap.
   c. Loose connection in ignition system.
   d. Damaged coil.
   e. Damaged ignition switch.
   f. Dirty/shorted engine stop switch.
   g. Damaged exciter coil or ignition pickup coil (check ignition timing).
   h. Damaged CDI unit.
3. Engine:
   a. Compression release malfunctioning.
   b. No valve clearance.
   c. Leaking cylinder head gasket.
   d. Stuck/seized valve.
   e. Worn piston and/or cylinder.

Poor Idle and Low Speed Performance

1. Fuel system:
   a. Improper operation of choke lever.
   b. Choke plunger stuck open.
   c. Idle speed too low.
   d. Engine flooded.
   e. Clogged fuel tank cap vent.
   f. Contaminated fuel.
   g. Clogged fuel valve, fuel line or carburetor.
   h. Clogged air filter.
   i. Fuel tank diaphragm-valve vacuum hose disconnected or leaking.
   j. Malfunctioning fuel tank diaphragm valve.
   k. Pilot mixture screw misadjusted.
   l. Float valve clogged or sticking.
   m. Air leaks at intake duct.
   n. Loose carburetor diaphragm cover.
   o. Torn or damaged slide diaphragm.
   p. Dragging carburetor slide.
   q. Wrong pilot jet for altitude.
   r. Clogged muffler.
2. Ignition:
   a. Fouled/improperly gapped spark plug.
   b. Damaged/shorted spark plug lead and cap.
   c. Loose connection in ignition system.
   d. Damaged coil.
   e. Damaged ignition switch.
   f. Dirty/shorted engine stop switch.

g. Damaged exciter coil or ignition pickup coil (check ignition timing).

h. Damaged CDI unit.

3. Engine:
   a. Compression release malfunctioning.
   b. Improper valve clearance.
   c. Low compression.
   d. Improper valve/camshaft timing.

Engine Lacks Power and Acceleration

1. Fuel system:
   a. Improper operation of choke lever.
   b. Choke plunger stuck open.
   c. Contaminated fuel.
   d. Clogged fuel valve, fuel line or carburetor jets.
   e. Clogged air filter.
   f. Fuel tank diaphragm-valve vacuum hose disconnected or leaking.
   g. Malfunctioning fuel tank diaphragm valve.
   h. Float valve clogged or sticking.
   i. Air leaks at intake duct.
   j. Loose carburetor diaphragm cover.
   k. Torn or damaged slide diaphragm.
   l. Dragging carburetor slide.
   m. Wrong pilot or main jet for altitude.
   n. Clogged muffler.
2. Ignition:
   a. Fouled/improperly gapped spark plug.
   b. Damaged/shorted spark plug lead and cap.
   c. Loose connection in ignition system.
   d. Damaged coil.
   e. Damaged exciter coil or ignition pickup coil (check ignition timing).
   f. Damaged CDI unit.
3. Engine:
   a. Compression release malfunctioning.
   b. Improper valve clearance.
   c. Low compression.
   d. Improper valve/camshaft timing.
4. Brakes and wheels:
   a. Brake pads dragging on brake disc.
   b. Worn/seized wheel bearings.
   c. Drive chain too tight.
5. Clutch:
   a. Clutch incorrectly adjusted.
   b. Weak clutch springs.
   c. Worn clutch plates and discs.

Poor High Speed Performance

1. Fuel system:
   a. Improper operation of choke lever.
   b. Choke plunger stuck open.
   c. Contaminated fuel.
   d. Clogged fuel valve, fuel line or carburetor jets.
   e. Clogged air filter.
   f. Air leaks at intake duct.
   g. Float valve too high.
   h. Loose carburetor diaphragm cover.
   i. Torn or damaged slide diaphragm.
   j. Dragging carburetor slide.
   k. Worn needle and jet.
   m. Wrong main jet for altitude.
   n. Clogged muffler.
2. Ignition:
   a. Damaged exciter coil or ignition pickup coil (check ignition timing).
   b. Damaged CDI unit.
3. Engine:
   a. Engine oil level too high.
   b. Incorrect valve clearance.
   c. Weak/broken valve spring(s).

Engine Backfires

1. Pilot mixture screw adjusted too lean.
2. Air leaks into exhaust system.
3. Inoperative air cut-off valve (backfiring during deceleration).
4. Damaged exciter coil or ignition pickup coil (check ignition timing).

Engine Overheating

NOTE
Engine overheating can occur when the motorcycle is operated at slow speed at high rpm. This can occur in severe off-road riding conditions. Even though the fan turns on, excessive heat buildup can cause the engine to overheat. When this occurs, stop and allow the engine to cool. If overheating continues after the bike is ridden at moderate speeds and lower rpm, check the motorcycle and determine the cause of overheating.

1. Cooling system:
   a. Coolant level low.
   b. Water in system; no coolant mix.
   c. Air in system.
   d. Radiator clogged.
   e. Radiator cap damaged.
   f. Thermostat damaged.
   g. Fan fuse blown (1990-on).
   h. Fan relay or switch faulty.
   i. Fan shaft seized.
   j. Water pump impeller loose.
   k. Water pump impeller damaged.
   l. Water temperature sending unit faulty.
   m. Water temperature gauge faulty.
2. Engine:
   a. Excessive idling.
   b. Insufficient oil level or viscosity.
   c. Incorrect spark plug heat range.
   d. Clogged crankcase oil strainer.
   e. Excessive carbon buildup on piston/cylinder head.
3. Fuel system (causing lean fuel mixture):
   a. Clogged/pinched fuel tank cap vent hose.
   b. Air leaks at intake duct.
   c. Wrong pilot or main jet for altitude.
   d. Clogged carburetor jets.
   e. Float level too low.
4. Ignition:
   a. Improper spark plug heat range.
   b. Damaged exciter coil or ignition pickup coil (check ignition timing).

ELECTRICAL TESTING

Refer to Chapter Nine for testing the starting system, ignition system, charging system, fan system and switches. Refer to the procedures in *Engine Starting System* in this chapter for troubleshooting typical problems with the starting system. These checks should be made before disassembling and bench testing components. When doing electrical tests, refer to the diagrams in the chapter and at the back of the manual.

Before testing a component, check the electrical connections related to that component. Check for corrosion and bent or loose connectors. Most of the connectors have a lock mechanism molded into the connector body. If these are not fully locked, a connection may not be made. If there are connectors that are not locked, pull the connector apart and clean the fittings before reassembling.

STARTING SYSTEM

Starter Turns Slowly

1. Weak battery.
2. Poorly connected/corroded battery terminals and cables.
3. Loose starter motor cable.
4. Worn or damaged starter.

Starter Turns, But Does Not Crank Engine

1. Worn or damaged starter clutch.
2. Damaged teeth on starter motor shaft or starter gears.

Starter Does Not Operate

When the starter does not operate, refer to *Starting System* in Chapter Nine for the system diagram and component testing procedures. Begin testing at the starter relay.

ENGINE NOISE

Noise is often the first indicator that something is wrong with the engine. In many cases, damage can be avoided or minimized if the rider immediately stops the motorcycle and diagnoses the source of the noise. Anytime engine noise is ignored, even when the bike seems to be running correctly, the rider risks causing more damage and injury.

Pinging During Acceleration

1. Poor quality or contaminated fuel.
2. Lean fuel mixture.
3. Excessive carbon buildup in combustion chamber.
4. Damaged exciter coil or ignition pickup coil (check ignition timing).

Knocks, Ticks or Rattles

1. Engine top end:

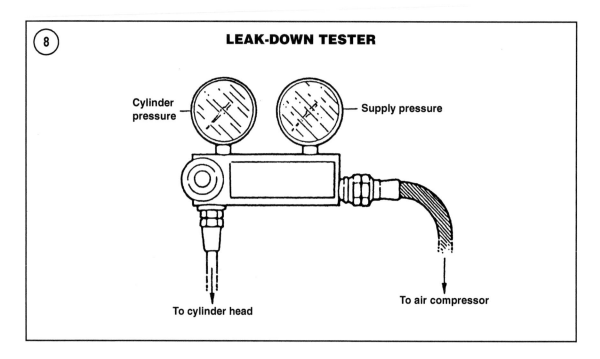

LEAK-DOWN TESTER

Cylinder pressure — Supply pressure

To cylinder head

To air compressor

a. Incorrect valve clearance.
b. Broken or weak valve spring.
c. Damaged compression release.
d. Loose cam chain/damaged tensioner.
e. Worn piston pin or piston pin bore.
f. Worn connecting rod small end.
g. Worn piston, rings and/or cylinder.
2. Engine bottom end:
a. Loose balancer chain.
b. Worn connecting rod bearing.
c. Worn crankshaft bearings.
d. Worn balancer bearings.
e. Worn transmission bearings.
f. Worn or damaged transmission gears.
g. Incorrect installation of balancer.

MOTORCYCLE NOISE

The following possible causes of noise will likely occur only when the bike is in motion.
1. Excessively loose drive chain.
2. Worn chain sliders.
3. Loose exhaust system.
4. Loose/missing body fasteners.
5. Loose skid plate.
6. Loose shock absorber.
7. Loose engine mounting bolts.
8. Brake pads dragging on brake disc.
9. Worn/seized wheel bearings.

ENGINE LEAKDOWN TEST

The condition of the piston rings and valves can accurately be checked with a leakdown tester. With all valves in the closed position, the tester is screwed into the spark plug hole and air pressure is applied to the combustion chamber. The gauge on the tester is then observed to determine the rate of leakage from the combustion chamber. An air compressor is required to use the leakdown tester (**Figure 8**).
1. Start the engine and allow it to warm up.
2. Shut off the engine and remove the carburetor and exhaust pipe.
3. Remove the spark plug.
4. Set the piston to TDC on the compression stroke.
5. Install the leakdown tester following the manufacturer's instructions. The tester must not leak around the spark plug threads.
6. Make the test following the manufacturer's instructions for the tester. When pressure is applied to the cylinder, check that the engine remains at TDC. If necessary, put the transmission in gear.
7. While the cylinder is under pressure, listen for air leakage at the following areas.
a. Exhaust port. If there is leakage, the exhaust valves are leaking.
b. Intake port. If there is leakage, the intake valves are leaking.

c. Crankcase breather tube. If there is leakage, the piston rings are leaking.

8. A cylinder with a leakdown of 5 percent or less is ideal. A cylinder with more than 10 percent leakage should be inspected to determine if leakage is caused by normal wear or damage. Inspection of the parts will then indicate what action should be taken.

CLUTCH

The two main clutch problems are clutch slip (clutch does not fully engage) and clutch drag (clutch does not fully disengage). These problems are often caused by incorrect clutch adjustment or a damaged/unlubricated cable. Perform the following checks before removing the right crankcase cover to troubleshoot the clutch:

1. Check the clutch cable routing from the handlebar to the engine. Check that the cable is free when the handlebar is turned lock to lock, and that the cable ends are installed correctly.
2. With the engine off, pull and release the clutch lever. If the lever is hard to pull, or the action is rough, check for the following:
   a. Damaged/kinked cable.
   b. Incorrect cable routing.
   c. Cable not lubricated.
   d. Worn/unlubricated lever at the handlebar.
   e. Damaged release lever at the engine.
3. If no damage was detected in the previous steps, and the lever moves without excessive roughness or binding, check the clutch adjustment as described in Chapter Three. Note the following:
   a. If the clutch cannot be adjusted to the specifications in Chapter Three, the clutch cable is stretched or damaged.
   b. If the clutch cable is in good condition and adjustment is correct, the clutch plates may be worn or warped.

Clutch Slipping

When the clutch slips, the engine accelerates faster than what the actual forward speed indicates. When continuous slipping occurs between the clutch plates, excessive heat quickly builds up in the assembly. This causes plate wear, warp and spring fatigue. One or more of the following can cause the clutch to slip:

1. Clutch wear or damage:
   a. Incorrect clutch adjustment.
   b. Weak or damaged clutch springs.
   c. Loose clutch springs.
   d. Worn friction plates.
   e. Warped clutch (steel) plates.
   f. Worn/damaged release lever assembly.
   g. Damaged pressure plate.
   h. Clutch housing and hub unevenly worn.
2. Clutch/engine oil:
   a. Excessive oil in crankcase.
   b. Incorrect oil viscosity.
   c. Oil additives.

Clutch Dragging

When the clutch drags, the plates are not completely separating. This causes the motorcycle to creep or lurch forward when the transmission is put into gear. Once underway, shifting is difficult. If this condition is not corrected, it can cause transmission gear and shift fork damage, due to the abnormal grinding and impacts on the parts. One or more of the following can cause the clutch to drag:

1. Clutch wear or damage:
   a. Worn/damaged release lever assembly.
   b. Warped clutch (steel) plates.
   c. Swollen friction plates.
   d. Warped pressure plate.
   e. Incorrect clutch spring tension.
   f. Galled clutch housing bushing.
   g. Uneven wear on clutch housing grooves or clutch hub splines.
   h. Incorrectly assembled clutch.
2. Clutch/engine oil:
   a. Low oil level.
   b. Incorrect viscosity oil.
   c. Oil additives.

Clutch Noise

Clutch noise is usually caused by worn or damaged parts, and is more noticeable at idle or low engine speeds. Clutch noise can be caused by the following conditions:

1. Wear in the clutch lifter bearing and/or lifter.
2. Excessive axial play in the clutch housing.
3. Excessive friction plate to clutch housing clearance.
4. Excessive wear between the clutch housing and primary drive gear.

(9) DISC BRAKE TROUBLESHOOTING

| Disc brake fluid leakage | Check:
• Loose or damaged line fittings
• Worn caliper piston seals
• Scored caliper piston or bore
• Loose banjo bolts
• Damaged oil line washers
• Leaking master cylinder diaphragm
• Leaking master cylinder secondary seal
• Cracked master cylinder housing
• Too high brake fluid level
• Loose or damaged master cylinder |

| Brake overheating | Check:
• Warped brake disc
• Incorrect brake fluid
• Caliper piston and/or brake pads hanging up
• Riding brakes during riding |

| Brake chatter | Check:
• Warped brake disc
• Incorrect caliper alignment
• Loose caliper mounting bolts
• Loose front axle nut and/or clamps
• Worn wheel bearings
• Damaged hub
• Restricted brake hydraulic line
• Contaminated brake pads |

| Brake locking | Check:
• Incorrect brake fluid
• Plugged passages in master cylinder
• Caliper piston and/or brake pads hanging up
• Warped brake disc |

| Insufficient brakes | Check:
• Air in brake lines
• Worn brake pads
• Low brake fluid
• Incorrect brake fluid
• Worn brake disc
• Worn caliper piston seals
• Glazed brake pads
• Leaking primary cup seal in master cylinder
• Contaminated brake pads and/or disc |

| Brake squeal | Check:
• Contaminated brake pads and/or disc
• Dust or dirt collected behind brake pads
• Loose parts |

5. Worn or damaged clutch housing and primary drive gear teeth.

GEAR SHIFT LINKAGE AND TRANSMISSION

Transmission problems are often difficult to distinguish from problems with the clutch and gear shift linkage. Often, the problem is symptomatic of one area, while the actual problem is in another area. For example, if the gears grind during shifting, the problem may be caused by a dragging clutch or a component of the shift linkage, not a damaged transmission. Of course, if the damaged part is not repaired, the transmission eventually becomes damaged, too. Therefore, evaluate all the variables that exist when the problem occurs, and always start with the easiest checks before disassembling the engine.

When the transmission exhibits abnormal noise or operation, drain the engine oil and check it for contamination or metal particles. Examine a small quantity of oil under bright light. If a metallic cast or pieces of metal are seen, excessive wear and/or part failure is occurring.

Difficult Shifting

1. Clutch:
   a. Improper clutch operation.
   b. Incorrect clutch adjustment.
   c. Incorrect oil viscosity.
2. Shift shaft:
   a. Loose/stripped shift lever.
   b. Bent/damaged shift shaft.
   c. Worn pawl plate engagement points.
   d. Damaged pawl plate or pawl spring.
   e. Damaged return spring or loose spring post.
3. Lever:
   a. Damaged lever.
   b. Broken lever spring.
   c. Loose lever bolt.
4. Shift drum and shift forks:
   a. Worn/loose shift cam.
   b. Worn shift drum grooves.
   c. Worn shift forks/guide pins.
   d. Worn shift drum bearings.

Gears Do Not Stay Engaged

1. Lever:
   a. Damaged lever.
   b. Broken lever spring.
2. Shift drum and shift forks:
   a. Worn/loose shift cam.
   b. Worn shift drum grooves.
   c. Worn shift forks/guide pins.
3. Transmission:
   a. Worn gear dogs and mating recesses.
   b. Worn gear grooves for shift forks.
   c. Worn/damaged shaft snap rings or thrust washers.

BRAKES

The disc brakes are critical to riding performance and safety. Inspect the brakes frequently and replace worn or damaged parts immediately. The brake system used on this motorcycle uses DOT 4 brake fluid in both brakes. Always use new fluid, from a sealed and closed container. The troubleshooting checks in **Figure 9** assist in isolating the majority of disc brake problems.

When checking brake pad wear, check that the pads in each caliper squarely contact the disc. Uneven pad wear on one side of the disc can indicate a warped or bent disc, damaged caliper or pad pins.

STEERING AND HANDLING

Correct poor steering and handling immediately after it is detected, since loss of control is possible. Check the following areas:

1. Excessive handlebar vibration:
   a. Incorrect tire pressure.
   b. Unbalanced tire and rim.
   c. Loose/broken spokes.
   d. Damaged rim.
   e. Loose or damaged handlebar clamps.
   f. Loose steering stem nut.
   g. Worn or damaged front wheel bearings.
   h. Bent or loose axle.
   i. Cracked frame or steering head.
2. Handlebar is hard to turn:
   a. Tire pressure too low.
   b. Incorrect cable routing.
   c. Steering stem adjustment too tight.
   d. Bent steering stem.

e. Improperly lubricated or damaged steering bearings.
3. Handlebar pulls to one side:
    a. Bent fork leg.
    b. Fork oil levels uneven.
    c. Bent steering stem.
    d. Bent frame or swing arm.
4. Shock absorption too soft:
    a. Low fork oil level.
    b. Fork oil viscosity too low.
    c. Shock absorber rebound damping too low.

d. Shock absorber spring preload too low.
e. Shock absorber spring weak.
5. Shock absorption too hard:
    a. Air pressure in fork.
    b. High tire pressure.
    c. High fork oil level.
    d. Fork oil viscosity too high.
    e. Bent fork.
    f. Shock absorber rebound damping too high.
    g. Shock absorber spring preload too high.

CHAPTER THREE

LUBRICATION, MAINTENANCE AND TUNE-UP

This chapter provides information and procedures for properly lubricating, fueling and adjusting the motorcycle. Refer to **Table 1** for the recommended service intervals and those components that require inspection, lubrication or adjustment.

Refer to the sections in this chapter for performing many of the maintenance procedures described in **Table 1**. For services that require extensive disassembly of a component, refer to the appropriate chapter(s) in the manual for inspection and repair.

Refer to *Safety* and *Basic Service Methods* in Chapter One.

PRE-RIDE INSPECTION

Routinely perform the following checks before riding the motorcycle. When riding the motorcycle on extended travel and over rough terrain, perform the checks at least once daily. Perform the checks when the engine is *cold*. Refer to the procedures and tables in this chapter for information concerning fuel, lubricants, tire pressure and component adjustments. Start the motorcycle as described in *Starting the Engine* in Chapter Two.

1. Check fuel lines and fittings for leakage.
2. Check fuel level.
3. Check engine oil level.
4. Check coolant level.
5. Check brake operation and lever/pedal free play.
6. Check throttle operation and free play.
7. Check clutch operation and free play.
8. Check for air pressure in the fork legs.
9. Check steering for smooth operation and no cable binding.
10. Check tire condition and air pressure.
11. Check wheel condition and spoke tightness.
12. Check axle nut tightness.
13. Check for loose nuts and bolts.
14. Check exhaust system for tightness.
15. Check drive chain condition and adjustment.
16. Check rear sprocket for tightness.
17. Check air filter for dirt/debris buildup.
18. Check suspension for leakage and proper settings for riding conditions.
19. Check engine stop switch for proper operation.
20. Check lights and signals for proper operation.

ENGINE BREAK-IN

If the engine bearings, crankshaft, piston, piston rings or cylinder have been serviced, perform the following break-in procedure. The break-in period requires 1000 miles (1600 km) of engine operation.

1. Make the following checks and observations when breaking in the engine:

   a. Operate the bike on hard ground. Do not run in sand, mud or up steep hills. This will overload and possibly overheat the engine.

   b. Vary the throttle position. Do not keep the throttle in the same position for extended periods.

   c. Check that a new spark plug has been installed.

   d. Check that the air filter is clean and oiled.

   e. Check that the engine is filled with the proper amount and grade of engine oil.

   f. Check that the cooling system is filled.

2. Start the engine and allow it to warm up for at least 2 minutes. Do not race the engine while it is warming up. During this time, check for proper idle speed and leaks.

3. For the first 500 miles (800 km) do not operate the motorcycle over 4000 rpm. During this time, vary the engine speed and do not lug the engine. At the end of the 500 mile period, perform the maintenance procedures listed in **Table 1**.

4. For the second 500 miles (800 km) do not operate the motorcycle over 6000 rpm. During this time, vary the throttle speed and do not lug the engine.

TUNE-UP AND SERVICE INTERVALS

The maintenance and lubrication intervals in **Table 1** are based on equal use of the bike, both on and off the road. If the motorcycle is regularly operated in extreme weather conditions, or subjected to water or sand, perform the service procedures more frequently.

Record when each service is performed in the maintenance log at the end of this manual.

BATTERY

When new, the motorcycle is equipped with a 12 volt, 14 amp-hour battery. If the motorcycle has not been used for at least two weeks, the battery should be charged to prevent sulfation of the battery plates.

If necessary, refer to *Charging System* in Chapter Nine for additional battery and charging system tests.

Removal/Installation

1. Check that the ignition switch is off.

2. Remove the side covers and seat (Chapter Fifteen).

3. Disconnect the negative cable from the battery (A, **Figure 1**). Do not allow the loose cable to touch the frame or other metal part of the motorcycle.

4. Remove the insulator cover from the positive cable, then remove the cable from the battery (B, **Figure 1**).

5. Remove the vent tube, then the screw securing the battery holder (**Figure 2**).

6. Clean and check the components for damage.

7. Reverse this procedure to install the battery. Note the following:

   a. Check that the battery terminals face the rear of the motorcycle.

   b. To prevent corrosion, apply a thin coating of dielectric grease to the battery terminals and cable ends.

   c. Tighten the cables firmly. Do not apply excessive force.

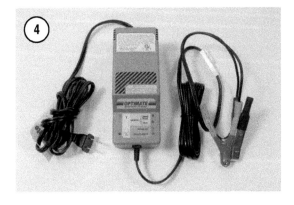

1. Connect a voltmeter to the negative and positive terminals. If a voltmeter is not available, use a syringe hydrometer and measure the specific gravity of the electrolyte (**Figure 3**).

2. Measure the voltage.

    a. A fully charged battery will have a minimum of 12.6 volts (1.265 hydrometer reading).

    b. A battery that is approximately 75 percent charged will have a minimum of 12.4 volts (1.210 hydrometer reading).

    c. A battery that is approximately 50 percent charged will have a minimum of 12.1 volts (1.160 hydrometer reading).

3. If battery charging or replacement is required, refer to the procedures in this section.

Charging

When recharging the battery, a charger (**Figure 4**) with a variable amperage output is recommended. Do not use a charger that is rated higher than 1.4 amps output. Also, do not use an automotive-type charger. The charge rates are too high and will overheat the battery and damage the battery plates.

CAUTION
To prevent possible electrical system damage, always remove the cables from the battery before charging the battery.

1. Remove the battery from the motorcycle as described in this section.

2. Connect the positive and negative leads of the charger to the positive and negative terminals on the battery.

3. Set the charger to 12 volts. If the charger has a variable charge rate, select a low setting. Using a 1 amp, constant-current charger, the suggested charge rates for a conventional battery are:

    a. 75 percent charge: 1 amp for 4 hours.

    b. 50 percent charge: 1 amp for 9 hours

4. Turn on the charger and allow the battery to charge for the specified time.

5. After the battery is charged, turn off the charger and remove the leads from the terminals.

6. If necessary, add distilled water to any cells that are below the upper level mark.

7. Check battery voltage as described in this section. If the battery voltage does not remain stable for

Electrolyte Check

The battery electrolyte level for each cell must be maintained between the upper and lower level marks, indicated on the *front* of the battery. The battery should be removed from the bike so the level of each cell can be viewed. The upper and lower level indicator on the battery *holder* only shows the condition of the cell at the left end of the battery. This should not be relied upon for the condition of all cells.

If the electrolyte level is too low for any cell, remove the cap above the cell, then add *distilled water* to raise the level to the upper mark. Do not over fill the battery. Replace the cap.

Voltage Test

For a conventional (original equipment) battery, check the unloaded voltage using a hydrometer or voltmeter. An unloaded test will indicate the basic state of charge. To determine whether the battery is adequate to operate the motorcycle, perform a battery load test, as described in Chapter Nine.

at least one hour, or continues to be undercharged, replace the battery.

FUEL AND LUBRICANTS

Fuel Requirements

The engine is designed to operate on unleaded pump-grade gasoline, with an octane rating of 87 (R + M/2) or 91 RON (Research Octane Number). The use of gasoline containing alcohol is not recommended for continual use.

> *NOTE*
> *For California models, do not fill the tank up to the neck of the fuel tank. Heat can cause the fuel to expand and overflow into the evaporative emission control unit lines, located near the filler neck. This can cause hard starting and engine hesitation until the fuel is cleared from the system.*

Engine Oil

The engine requires a multigrade oil with an SAE viscosity rating of 10W-40, 20W-40, 10W-50 or 20W-50. The oil should have a service classification of SG. The oil viscosity rating and classification are indicated on the oil container.

Since multigrade oil is blended with chemical polymers that change the characteristics of the oil as it is heated or cooled, the SAE numerical ratings indicate the equivalent *weight* of the oil at 0° F (-18° C) and 210° F (99° C). The numbers do not indicate an actual thickness, but rather the oils ability to circulate and lubricate at these extremes.

For this engine, the engine oil is also used to lubricate the transmission. Over time, the transmission will *shear* the polymers in the oil, reducing its lubrication abilities. Therefore, always change the oil and filter at the intervals indicated in **Table 1**. If possible, use the same brand of oil at each oil change.

> *CAUTION*
> *Do not use oils or oil additives that contain graphite or molybdenum. These additives can cause clutch slippage and erratic operation.*

Fork Oil

The fork uses 10-weight fork oil in the fork legs. Refer to **Table 2** for the recommended fork oil.

Grease

Use a good quality, lithium-base grease to lubricate components requiring grease. Some components require the extreme-pressure qualities of molybdenum disulfide grease, or the protective qualities of waterproof grease. When these greases are required, it is indicated in the procedures throughout the manual. Grease components frequently, as this purges water and grit from the component and extends its life.

Chain Lubricant

Use a good-quality chain lubricant that is compatible with the type of chain installed on the bike. An O-ring chain was standard equipment for the years covered by this manual. If an O-ring chain is installed, use chain lubricant that is specifically for

O-ring chains. Since the links of an O-ring chain are permanently lubricated and sealed, O-ring chain lubricant is formulated to prevent exterior corrosion of the chain and to condition the O-rings. It is not tacky and resists the adhesion of dirt. Avoid lubricants that are tacky and for conventional chains. These lubricants attract dirt and subject the O-rings to unnecessary abrasion.

Control Cable Lubricant

Use lithium grease to lubricate the control cable pivots. Lubricate the cable with light oil or a commercial cable lubricant.

Air Filter Oil

Use a commercial air filter oil, specifically for foam filters. This type of oil stays adhered to the foam and traps dust effectively.

PERIODIC LUBRICATION

Engine Oil Level Check

The oil level is checked at the sight glass (**Figure 5**) and replenished at the cap (**Figure 6**), located on the right side of the engine.

1. When checking the oil level, note the following:
    a. To achieve an accurate measurement, check the oil after the engine has been allowed to stand. Preferably, this would be prior to starting a cold engine.
    b. If the engine has been running, allow the engine to stand for several minutes so the oil can drain to the crankcase.

c. If the oil was changed prior to this check, start the engine and allow the oil to circulate throughout the engine. Failure to run the engine prior to checking the oil level causes an inaccurate reading.

2. Park the bike on level ground.
3. Hold the bike upright (off the sidestand) so the engine is level.
4. Inspect the oil level at the sight glass.
    a. The oil level should be between the upper and lower level marks that are to each side of the sight glass (**Figure 7**). Read the oil level at the center of the sight glass, not at the edges.
    b. Preferably, keep the oil level near the upper mark.
5. If the oil level is too low, remove the cap (**Figure 6**)on the right crankcase cover and add the appropriate grade of oil to bring the level to the upper mark. Add oil in small quantities and check the level often. Do not overfill the crankcase.
6. Screw the cap into place. If oil leakage is evident around the cap, replace the O-ring on the cap.

Engine Oil and Filter Change

Change the oil and filter at the intervals recommended in **Table 1**. If the bike is used in extreme conditions (hot, cold, wet or dusty) change the oil more often and use the appropriate grade of oil as recommended in *Fuel and Lubricants*.

Always change the oil when the engine is warm. Contaminants will remain suspended in the oil and it will drain more completely and quickly.

WARNING
Prolonged contact with used engine oil may cause skin cancer. Minimize contact with the engine oil.

1. Support the motorcycle so it is level and secure.
2. Wipe the area around the oil fill cap, then remove the cap from the engine (A, **Figure 8**).
3. Place a drain pan below the engine drain plug, then remove the drain plug (**Figure 9**). Allow the oil to drain from the engine.
4. Remove the bolts from the oil filter cover (B, **Figure 8**), then remove the cover.

NOTE
An O-ring seals the cover to the engine. When removing the cover, grasp

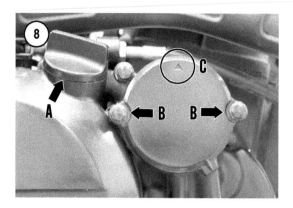

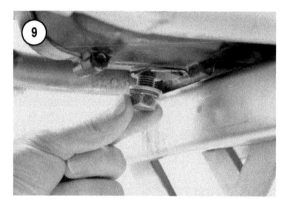

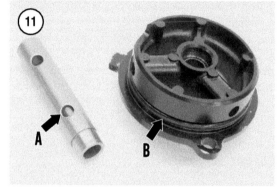

the tab on the cover and work the cover straight out. Avoid binding, or using tools to pry the cover.

5. Remove the filter and mounting pin (**Figure 10**).

6. Clean and inspect the filter housing and parts.

   a. The oil filter bypass valve is inside the mounting pin (A, **Figure 11**). When cleaning the pin, check that the spring-loaded valve is clean and free to open. Press the valve open from the narrow (stepped) end of the mounting pin.

   b. Install a new seal washer on the drain plug. Replace the drain plug if the threads are damaged or excessively worn.

   c. Install a new, lubricated O-ring onto the filter cover (B, **Figure 11**).

   d. If leakage is evident around the oil fill plug, replace the O-ring on the plug.

7. Wipe dirt and oil from around the drain plug hole, then install and torque the drain plug to 23 N•m (17 ft.-lb.).

8. Install the mounting pin and new oil filter as follows:

CAUTION
The mounting pin must be installed in the correct direction.

   a. Lubricate the grommets at both ends of the filter, then insert the mounting pin into the filter. The filter grommets must rest on the wide part of the mounting pin and not on the narrow end (**Figure 12**).

   b. Insert the filter and mounting pin into the filter housing. The narrow end of the pin must be inserted first (**Figure 13**). Fully seat the

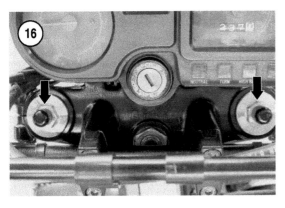

pin into the collar at the back of the filter housing.

   c. Align and install the oil filter cover squarely against the engine cover. Check that the arrow on the cover (C, **Figure 8**) points up and that the O-ring is not pinched.

   d. Finger-tighten the mounting bolts, then tighten the bolts in several passes to avoid binding the cover.

9. Fill the crankcase with the required quantity and grade of engine oil.

   a. Refer to **Table 2** for engine oil capacity.

   b. Note that oil capacity is greater when the filter is changed.

10. Screw the fill cap into place.

11. Check the engine oil level as described in *Engine Oil Level Check* in this section.

12. Check all fittings for leakage.

13. Dispose the used engine oil in an environmentally safe manner.

Fork Oil Replacement

The fork legs *do not* need to be removed from the motorcycle to change the oil. If fork seal leakage or other damage is evident, refer to Chapter Twelve for seal replacement and filling procedures, with the fork legs off the bike.

1. For each fork leg, remove the cap and depress the air valve (**Figure 14**) to release any air pressure.

2. Remove the drain screw and O-ring (**Figure 15**) from both fork legs and allow the oil to drain. Pump the fork to assist in draining. When completed, install a new O-ring and tighten the drain screw.

3. Support the motorcycle so it is stable and the front wheel is off the ground.

> *CAUTION*
> *Do not attempt to disassemble the fork without the front wheel off the ground. If not properly supported, when both fork caps are removed, the fork legs are free to collapse. This will cause the bike to fall forward.*

4. Remove the fork caps (**Figure 16**).

5. Remove the spring spacer (**Figure 17**) from each leg.

6. Remove the spring seat and spring (**Figure 18**) from each leg. If the springs are progressively wound, make note of which end of the springs is up.

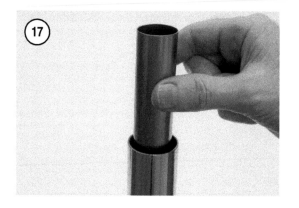

7. Clean and inspect the parts.

8. Refer to **Table 2** for the recommended oil and quantity.

9. Measure and pour the oil into the fork legs (**Figure 19**).

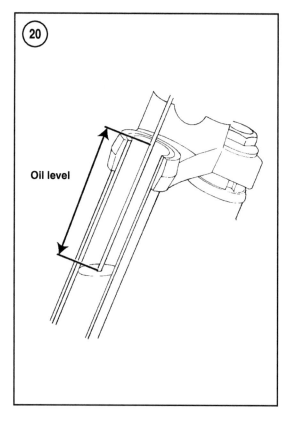

10. Measure and set the oil level as follows:

    a. Compress the fork legs to the bottom of their travel. If necessary, have assistance in handling and stabilizing the motorcycle. An alternative is to completely pull the wheel up, then place a support under the wheel.

    b. Use a fork oil level tool to set the oil to the specification in **Table 2**. The specification represents the distance from the top edge of the fork leg to the surface of the oil (**Figure 20**).

    c. If necessary, add or remove oil to bring the levels to the required distance. Both legs must have identical levels.

NOTE
The oil level establishes the air pocket that is above the oil. Since the air pocket contributes to fork damping, slight variation of the oil level is permissible to meet the riding conditions. Excess oil in the fork leg (small air pocket), will make damping harder. This condition can also lead to fork seal leakage. Too little oil in the fork leg (large air pocket), will make damping softer. This condition can cause the fork to bottom and possibly damage fork components.

11. Extend the fork legs, then install the spring, spring seat and spacer.

12. Install a new, lubricated O-ring on each fork cap.

Oil level

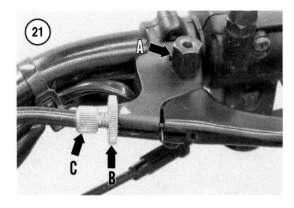

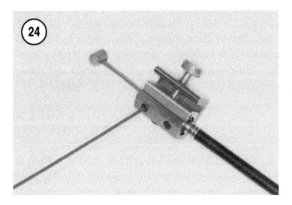

13. Tighten the fork caps to 29 N•m (21 ft.-lb.).

Cable Lubrication

If there is binding or drag in the throttle or clutch, this can indicate a lack of cable lubrication or worn parts. Since the throttle uses two cables, also check the adjustment. Use lithium grease to lubricate the control cable pivots. Lubricate the cables with light oil or cable lubricant. If the clutch or throttle cables continue to operate poorly after lubrication, disconnect the cable(s) at both ends and check for binding or drag. Replace the cable(s) if necessary. If the cable(s) is in good condition, check for binding or drag in the carburetor/clutch.

Clutch cable and starter lockout switch

1. Remove the clutch lever hand guard. Carefully flex the guard to remove it from the clutch lever assembly.
2. Remove the pivot bolt (A, **Figure 21**) and lower nut from the lever.
3. Loosen the clutch cable locknut (B, **Figure 21**), then turn the adjuster (C) in until the cable and lever can be removed from the slots in the adjuster assembly (**Figure 22**).
4. Remove the cable end at the release lever (**Figure 23**).
5. Clean the levers and cable ends. Also clean the pivot and clutch lever housing.
6. Attach a cable lubricator, then lubricate the cable with an aerosol cable lubricant (**Figure 24**). Keep the cable in a vertical position so the lubricant can pass to the opposite end. Move the cable in the housing to help distribute the oil. Stop lubrication when oil is seen at the opposite end of the cable.
7. Lubricate the lever pivots and cable ends with lithium grease, then attach the cable to the release lever and clutch lever.

NOTE
*When lubricating the clutch cable and lever assembly, occasionally inspect the condition of the starter lockout switch, located on the bottom of the clutch lever housing. Open the switch (**Figure 25**) and clean dirt and lubricant that has accumulated in the switch housing. Clean the switch contacts with electrical contact cleaner,*

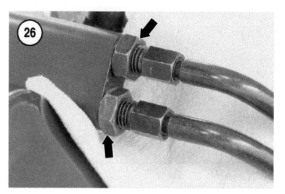

then wipe a small amount of dielectric grease on the parts to prevent corrosion. When installed, the switch housing is open to the bottom side of the clutch lever. Any dirt, grease or moisture that passes under the clutch lever can migrate into the switch. Whenever the switch has been disassembled, check for proper operation after assembling the switch.

8. Install the lever assembly. Install the hand guard after the cable is adjusted.

9. Adjust the clutch cable as described in this chapter.

Throttle cables

The throttle uses two cables. One cable pulls the throttle open during acceleration, while the other pulls, the throttle closed during deceleration. In operation, the cables always move in opposite directions to one another. Use the following procedure to disassemble, clean and lubricate the cables and throttle assembly.

1. At the handlebar, remove the rubber boots from the throttle cable adjusters, then loosen the locknuts (**Figure 26**). The top cable is the accelerator cable.

2. At the carburetor, loosen the locknuts and remove the cables from the holders and throttle valve (**Figure 27**). The left cable is the accelerator cable.

3. Remove the screws from the housing and separate the parts (**Figure 28**).

4. Remove the cable ends from the throttle drum, then remove the cable guide from the upper housing (**Figure 29**).

5. Remove the cable adjusters, then remove the cable ends from the housing. Remove the grip assembly from the handlebar.

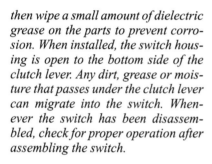

6. Clean the throttle assembly and cable ends.

7. Attach a cable lubricator, then lubricate each cable with an aerosol cable lubricant (**Figure 24**). Keep the cable in a vertical position so the lubricant can pass to the opposite end. Move the cable in the housing to help distribute the oil. Stop lubrication when oil is seen at the opposite end of the cable.

8. Lubricate the throttle drum, cable ends and cable guide with lithium grease.

9. Install the grip on the handlebar.

10. Pass the cable ends through the upper housing, then pull all slack out of the cables. Check that the cables are in their correct hole. Temporarily tape the cable housings to the handlebar fittings. This prevents the housings from falling out of the fittings and eases in handling the cables and throttle assembly. Unless assistance is available, do not apply more tape than can be removed with one hand.

11. Route the accelerator cable over the drum and the decelerator cable under the drum (**Figure 28**). Seat the cable ends in the drum.

12. Route the cables through the cable guide, then hold the assembly together while the tape from the handlebar is removed. **Figure 30** shows how the ca-

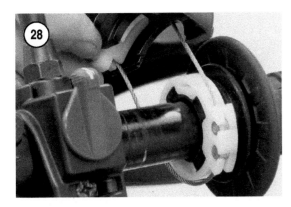

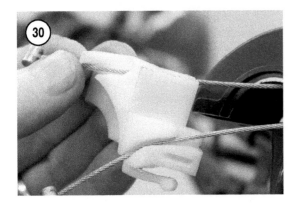

bles must route through the cable guide. The top cable is the accelerator cable.

13. Fit the cables and guide into the upper housing, then seat the assembly onto the lower housing. Hold the parts together until the screws can be installed.

14. At the carburetor, hold the cable ends and check that they move in the correct direction when the throttle is operated.

15. Connect the cables at the carburetor, then adjust the cables as described in this chapter.

Drive Chain Cleaning and Lubrication

The bike is equipped with an O-ring chain that requires routine cleaning and lubrication. If the chain has been replaced with a standard chain, it too requires regular cleaning, lubrication and adjustment for long life. Although O-ring chains are internally lubricated and sealed, the O-rings need to be kept clean and lubricated, to prevent them from drying out and disintegrating.

Never clean chains with high-pressure water sprays or strong solvents. This is particularly true for O-ring chains. If water is forced past the O-rings, water will be trapped inside the links. Strong solvents can soften the O-rings so they tear or damage easily.

Although chains are often lubricated while they are installed on the bike, the chain should periodically be removed from the bike and thoroughly cleaned. The following procedure describes the preferred method for cleaning and lubricating the chain.

1. Refer to *Drive Chain Removal and Installation* in Chapter Eleven to remove the chain.

2. Immerse the chain in kerosene and work the links so dirt is loosened.

3. Lightly scrub the chain with a soft-bristle brush.

> *CAUTION*
> *Brushes with coarse or wire bristles can damage O-rings.*

4. Rinse the chain with clean kerosene and wipe dry.

> *NOTE*
> *While the chain is removed, check that it is still within the wear limit as described in **Drive Chain and Sprockets Inspection** in this chapter.*

5. Lubricate the chain with chain lubricant. Lubricate an O-ring chain with lubricant specifically for O-ring chains.

> *CAUTION*
> *Since the links of an O-ring chain are permanently lubricated and sealed, O-ring chain lubricant is formulated to prevent exterior corrosion of the chain and to condition the O-rings. It is not tacky and resists the adhesion of dirt. Avoid lubricants that are tacky*

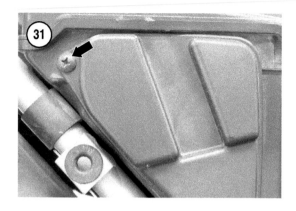

and for conventional chains. These lubricants attract dirt and subject the O-rings to unnecessary abrasion.

6. Install the chain (Chapter Eleven).

7. Adjust the chain as described in this chapter.

Air Filter Cleaning and Lubrication

The engine is equipped with a reusable, foam air filter. Do not operate the bike without the air filter, or with a damaged air filter. Performance will not be enhanced and rapid engine wear will occur.

1. Remove the right side cover (Chapter Fifteen).

2. Remove the screw and cover (**Figure 31**) from the air filter housing.

3. Remove the retaining bolt and washer (**Figure 32**) from the air filter assembly, then remove the assembly from the engine.

4. Remove the frame from the air filter (**Figure 33**).

5. Wash all parts in solvent (kerosene), a commercial filter wash, or hot soapy water. *Squeeze* the cleaner from the filter. Do not wring the filter, as tearing may occur. Shake the filter of any particles that may remain on the filter.

6. Allow the filter to completely dry.

7. Apply filter oil to the filter, squeezing the filter so the oil is distributed evenly. Squeeze out the excess oil. When performing this step, handle the filter with disposable gloves, or put the filter in a plastic bag to squeeze and distribute the oil. Follow the manufacturer's instructions when oiling the filter.

NOTE
Oil specifically formulated for foam filters should be used. This type of oil

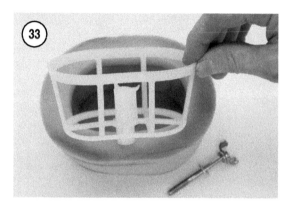

stays adhered to the foam and traps dust effectively.

8. Install the frame into the filter, seating it under the foam lip (**Figure 34**).

9. Apply lithium grease around the perimeter of the filter lip. This helps seal the filter against the housing.

10. Clean the housing and filter sealing surface. Check that the drain in the bottom of the housing is open.

11. Clean the housing cover and inspect the gasket (**Figure 35**). The gasket seals the cover, and minimizes vibration of the cover.

12. Install the filter, checking that it seats against the filter housing. Install the washer and retaining bolt.

13. Align and fit the filter assembly into the air filter housing. Check that the filter is completely seated against the housing, then tighten the bolt.

14. Install the cover, fitting the tab on the right side into the housing.

15. Screw the cover into place.

16. Install the right side cover.

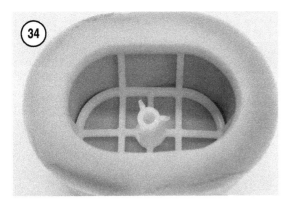

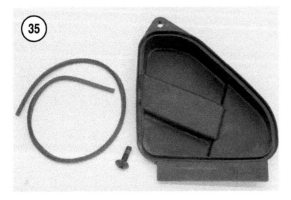

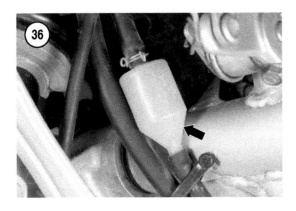

MAINTENANCE AND INSPECTION

Evaporative Emissions Control System Inspection (California Models Only)

No adjustments are required for the evaporative emissions system. Visually inspect the hoses and connections as recommended in **Table 1**. Refer to Chapter Eight for a schematic view of the system and its components.

Fastener Inspection

Inspect all fasteners on the motorcycle for tightness and condition.
1. Retorque nuts, bolts and screws as specified in the tables at the end of each chapter. Refer to **Table 4** in Chapter One when a torque value is not specified.
2. Check that all cotter pins are secure and undamaged.
3. Check that tie straps, used to secure cables and electrical wiring, are not broken or missing.

Muffler Cleaning

The muffler is approved by the U.S. Forest Service for spark arresting. The muffler also limits exhaust noise to 80 decibels. This information is stamped into the muffler near the rear heat shield. In order for the muffler to perform correctly and not affect engine performance, the internal baffle should be purged of carbon buildup. This is particularly important if the engine has been running too rich.

> *WARNING*
> *Do not spray solvents or other combustible liquids into the muffler to aid in removing buildup. This can result in an explosion and/or fire.*

Air Filter Housing Drain

Whenever the air filter has been lubricated, regularly check the air filter housing drain (**Figure 36**) for excess oil that has drained from the filter. The drain is located behind the rear brake pedal. If oil is in the reservoir, remove the plug from the drain hose and allow the oil to drain. The drain hose is not attached to the drain hole that is visible in the air filter housing. The drain hose attaches to a fitting that is in the housing duct, near the air filter.

1. Park the motorcycle in an open area, away from combustible materials.
2. Remove the two plugs on the back side of the muffler (**Figure 37**) and the plug near the right passenger peg (**Figure 38**).
3. Start the engine.
4. Wearing gloves, use a rubber mallet to tap on the surface of the muffler as the engine speed is raised and lowered. Also, momentarily place a folded shop

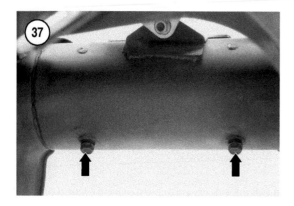

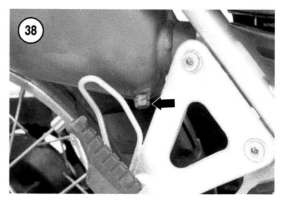

cloth over the end of the muffler to force exhaust pressure out of the plug openings.

5. When no more carbon particles are purged, stop the engine and install the plugs.

Balancer Chain Tensioner Adjustment

As the balancer chain and guide wear, the tensioner assembly takes excess slack out of the chain. Since the tensioner assembly is spring-loaded, it automatically tightens the chain when the tensioner bolt is loosened. When the bolt is tightened, it locks the tensioner into place until the bolt is again loosened, usually at the next scheduled maintenance.

1. If desired, remove the skid plate for easier access to the tensioner bolt. The bolt is located below the alternator cover (left side of engine), near the frame (**Figure 39**).

2. Remove the rubber plug.

3. Loosen the tensioner bolt by turning it no more than two turns counterclockwise.

4. Retighten the tensioner bolt.

5. Install the rubber plug.

6. Install the skid plate, if removed.

CAUTION
When starting the engine after performing the tensioner adjustment, listen for any abnormal engine noise coming from the balancer chain assembly. Do not race the engine. If noise is heard, immediately shut off the engine and inspect the condition of the tensioner assembly. It is possible for shaft lever or the tensioner spring to break, prior to the adjustment. The tensioner bolt will hold the

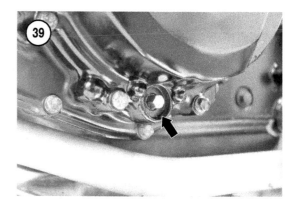

tensioner assembly in place, although the part is broken. At the scheduled maintenance interval, when the tensioner bolt is loosened, the untensioned assembly is free to create chain slack or fall apart. This slack could cause the chain to derail, causing engine damage.

Carburetor Float Chamber Drain

The carburetor float chamber is equipped with a drain so moisture and sediment can be flushed from the chamber. Moisture and sediment can clog carburetor jets and cause poor engine performance. If the motorcycle is put into storage, or not started for an extended period, drain the float chamber.

WARNING
Do not drain the float chamber while the engine is hot or running.

1. Support the bike so it is vertical and level.

2. Connect a length of 25 mm (1/4 in.) ID hose onto the drain (A, **Figure 40**). Route the end of the hose into a suitable container.

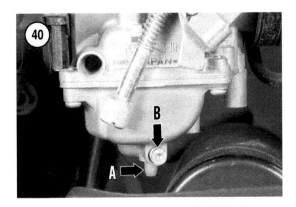

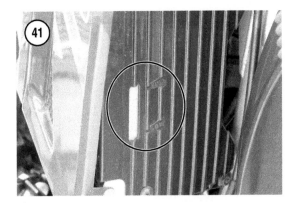

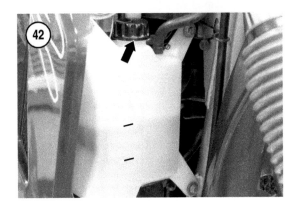

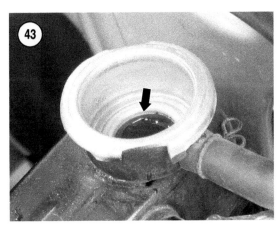

3. Turn the drain screw (B, **Figure 40**) out two turns and allow all fuel in the chamber to flow into the container.

4. Close the drain screw.

5. Disconnect the drain hose.

Coolant Level Inspection

WARNING
Inspect the cooling system when the engine and coolant are cold. Severe

injury could occur if the system is checked while it is hot. If the radiator cap must be removed while the coolant is still warm, cover the cap with a towel and open it slowly. Do not remove the cap until all pressure is relieved. Also, when the engine is hot, the cooling fan can operate even though the key is in the off position. Do not touch or use tools near the fan until the engine has cooled.

Refer to **Table 2** for the recommended coolant and mixing ratio.

1. Support the bike so it is vertical and level.

2. At the reserve tank cover, check the coolant level at the gauge. The level should be between the low and full marks on the cover (**Figure 41**). If the level is too low, do the following:

    a. Remove the two screws securing the reserve tank cover.

    b. Remove the cap from the reserve tank (**Figure 42**) and fill it to the full mark, indicated by the emboss on the reserve tank. Do not allow the tank level to fall below the low mark, also indicated by an emboss on the tank.

    c. Check for leakage at the hoses and reserve tank.

    d. Install the reserve tank cover.

    e. If the reserve tank is empty, check the level in the radiator. The coolant level should be to the bottom of the filler neck on the radiator (**Figure 43**).

    f. If the coolant level is below the filler neck, add coolant mixture to raise the level.

    g. Install the radiator cap.

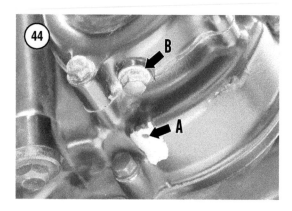

h. Start the engine and inspect for leaks. If there is leakage at the drain hole at the bottom of the water pump (A, **Figure 44**), this indicates that leakage is occurring at the mechanical seal in the water pump. Repair the water pump as described in Chapter Ten. If engine oil is leaking from the hole, the oil seal in the right crankcase cover is leaking. Replace the seal as described in Chapter Six.

CAUTION
If the coolant level continues to drop, pressure test the cooling system. Severe engine damage can occur if the engine is allowed to overheat.

Cooling System Inspection

Annually check the condition of the cooling system, or whenever it is suspected that overheating is occurring. The radiator cap and cooling system are checked individually with a cooling system tester. This tester applies the required pressure to the cooling system and cap. A pressure gauge attached to the tester is observed and leakage can be detected. A Kawasaki dealership can perform this inspection, or a tester can be purchased at an automotive parts supplier. Test the radiator cap and cooling system as follows:

WARNING
Test the cooling system when the engine and coolant are cold. Severe injury could occur if the system is checked while it is hot.

1. Support the bike so it is vertical and level.
2. Remove the radiator cap and check the following:

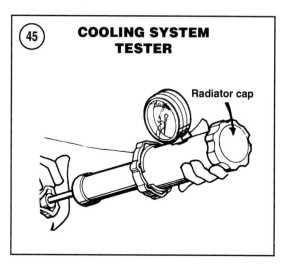

a. Rubber seals. Check for cracks, compression and pliability. Replace the cap if damage is evident.

b. Relief valve. Check for damage. Replace the cap if damage is evident.

3. Determine the cap relief pressure. Wet the seal on the radiator cap, then attach the cap to the tester (**Figure 45**). Apply pressure to the cap. Relief pressure for the cap is 93-123 kPa (13.5-17.8 psi). Observe the pressure gauge and do the following:

a. If the gauge holds pressure up to the relief pressure range, the cap is good.

b. If the gauge does not hold pressure, or the relief pressure is too high or low, replace the cap.

4. Check that the radiator is filled to the bottom of the filler neck (**Figure 43**). Attach the tester to the radiator, then pump the tester to 137 kPa (20 psi).

CAUTION
Do not exceed 137 kPa (20 psi). Excessive pressure can damage the cooling system components.

Observe the pressure gauge and do the following:

a. If the gauge holds the required pressure, the cooling system is in good condition.

b. If the gauge does not hold the required pressure, check for leakage at the radiator and all fittings. If the pressure lowers and then stabilizes, check for swollen radiator hoses. Replace or repair the cooling system components so it maintains the test pressure.

Coolant Draining and Replacement

WARNING
Replace the coolant in the cooling system when the engine and coolant are cold. Severe injury can occur if the system is drained while it is hot.

CAUTION
Do not allow coolant to contact painted surfaces. If contact does occur, immediately wash the surface with water.

1. Remove the skid plate (Chapter Fifteen).
2. Support the bike so it is vertical and level.
3. Remove the fuel tank and radiator covers (Chapter Fifteen).
4. Place a drain pan under the right side of the engine, below the water pump. Remove the drain plug from the bottom of the water pump (B, **Figure 44**).
5. As coolant begins to drain from the engine, slowly loosen and remove the radiator cap so the flow from the engine increases. Be ready to reposition the drain pan, if necessary.
6. Remove the reserve tank cover, then remove the bolts from the corners of the reserve tank (**Figure 46**). When the tank can be inverted, drain the tank and remove the upper and lower hoses.
7. Flush the cooling system and reserve tank with clean water. Check that all water drains from the system.
8. Inspect the condition of:
   a. Radiator hoses. Check for leaks, cracks and loose clamps.
   b. Radiator core. Check for leaks, debris and tightness of mounting bolts.
   c. Radiator fan. Check for damaged wiring and tight connections.

9. Install a new seal washer on the coolant drain plug, then install and tighten the drain plug to 10 N•m (88 in.-lb.).
10. Connect the hoses to the reserve tank and install the tank.
11. Refill the radiator with a coolant mixture as specified in **Table 2**.
   a. Tip the bike from side to side to allow the coolant to flow through the engine, and to purge air from the water jackets.
   b. Refill the radiator as the coolant level goes down.
   c. When the coolant level no longer goes down, fill the radiator to the bottom of the filler neck (**Figure 43**). Install the radiator cap.
   d. Fill the reserve tank to the full mark.
12. Install the fuel tank.
13. Start the engine and allow the coolant to circulate for about 30 seconds. Shut off the engine and do the following:

WARNING
Cover the cap with shop cloths and open it slowly. Do not remove the cap until all pressure is relieved.

   a. Remove the radiator cap and check the level. If necessary, add coolant to bring the level to the bottom of the filler neck. Install the radiator cap.
   b. Check the level in the reserve tank. If necessary, remove the cap from the reserve tank and fill it to the full mark.
14. Rinse and dry the frame and engine where coolant was splashed.
15. Start the engine and allow it to reach operating temperature.
16. Check for leakage at the drain plug, hoses and reserve tank.
17. Install the reserve tank cover and removed body work.
18. Dispose the old coolant in an environmentally-safe manner.

Drive Chain and Sprockets Inspection

A worn drive chain and sprockets is both unreliable and potentially dangerous. Inspect the chain and rear sprocket for wear and replace if necessary. If there is wear, replace both sprockets and the

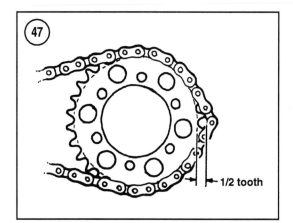

1/2 tooth

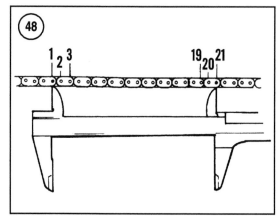

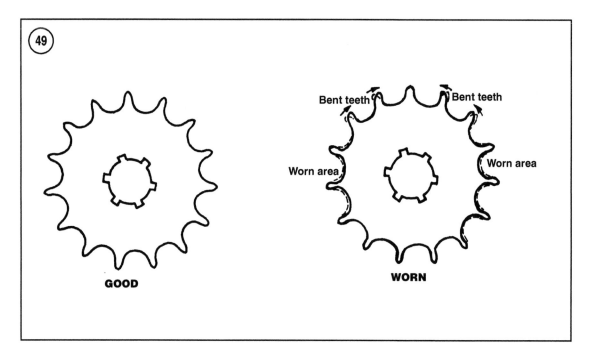

GOOD   WORN

chain. Mixing old and new parts will prematurely wear the new parts.

Determine if the chain should be measured for wear by pulling one chain link away from the sprocket.

Generally, if more than half the height of the sprocket tooth is visible (**Figure 47**), accurately measure the chain for wear. Refer to the following procedure to measure chain wear and inspect the rear sprocket.

1A. If the chain is not removed from the sprockets, loosen the axle nut and turn the chain adjusters equally to take all play out of the chain along its top run.

1B. If the chain is removed from the sprockets, lay the chain on a flat surface and pull the ends of the chain to remove the slack.

2. Measure the length of any 20-link (21 pin) span (**Figure 48**). Measure center-to-center from the pins.

    a. The service limit for the chain is 323 mm (12.7 in.). If the measured distance meets or exceeds the service limit, replace the chain.

    b. If the chain is within the service limit, inspect the inside surfaces of the link plates. The plates should be shiny at both ends of the chain roller. If one side of the chain is worn, the chain has been running out of alignment.

This also causes premature wear of the rollers and pins. Replace the chain if there is abnormal wear.

3. Inspect the teeth on the front and rear sprockets. Compare the sprockets to **Figure 49**. A new sprocket will have symmetrical and uniform teeth. A used sprocket will wear on the back side of each tooth (**Figure 50**). The sprocket shown has some usable life, but should be replaced if other damage

is evident, or if a new chain is being installed. If either sprocket is worn out, replace both sprockets.

Drive Chain Adjustment

The drive chain must have adequate play so it can adjust to the actions of the swing arm when the bike is in use. Too little play can cause the chain to become excessively tight and cause unnecessary wear to the driveline components. Too much play can cause excessive looseness and possibly cause the chain to derail.

1. Support the motorcycle so the rear wheel is off the ground.

2. Rotate the rear wheel and determine when the chain is tightest along its bottom length (least amount of play).

3. Make a small mark on the swing arm, indicating the midpoint between the sprockets.

4. Measure the play in the bottom length of chain (**Figure 51**) as follows:

   a. Place a tape measure so it is stable and vertical, below the swing arm midpoint.

   b. Press the chain down and note where a chain link-pin aligns with the tape measure. Note the measurement.

   c. Push the chain up and note where the same link pin aligns with the tape measure. Note the measurement.

   d. The difference between the two measurements is the chain play.

   e. Refer to **Table 4** for the required amount of play.

5. If necessary, adjust the chain play as follows:

   a. Remove the cotter pin from the axle nut, then loosen the nut (A, **Figure 52**).

   b. Loosen the chain adjuster locknuts (B, **Figure 52**) on both sides of the wheel.

   c. Equally turn the chain adjuster nuts (C, **Figure 52**) until the chain play is correct. Use the adjustment marks (D, **Figure 52**) on the swing arm to equally adjust the chain. If increasing chain play, push the wheel forward to take the play out of the adjusters.

NOTE
If free play cannot be adjusted within the limits of the adjusters, the chain is excessively worn and should be replaced.

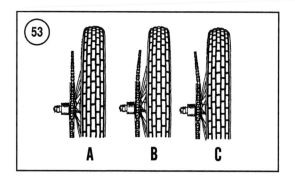

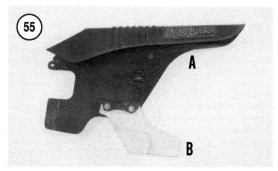

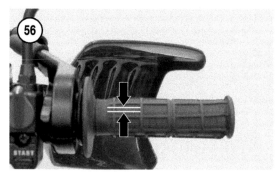

d. When free play is correct, check that the wheel is aligned (A, **Figure 53**). If the chain curves in or out (B and C, **Figure 53**) readjust the chain so the wheel is aligned with the rest of the motorcycle.

e. Tighten the axle nut to lock the setting. Torque the nut to 93 N•m (69 ft.-lb.).

f. Recheck the chain play. Adjust, if necessary.

g. Tighten the adjuster locknuts.

h. Install a new cotter pin.

Drive Chain Slider, Guard and Guide Inspection

Inspect the following parts for wear or damage. The parts support or protect the drive chain.

1. Chain slider (**Figure 54**).
   a. Inspect the upper and lower surface of the slider for wear or damage. The chain will contact the swing arm, possibly causing damage, if the slider is worn or missing.
   b. Check the mounting screws for tightness.

2. Chain guard (A, **Figure 55**).
   a. Inspect the chain guard for proper clearance and damage.
   b. Check the mounting bolts for tightness.

3. Chain guide (B, **Figure 55**).

a. Inspect the sliding surface of the guide for wear or damage. A torn or broken guide can snag the chain, causing noise and possible damage.

b. Check the mounting bolts for tightness.

Throttle Cable Adjustment

The throttle uses two cables. One cable pulls the throttle open during acceleration, while the other pulls the throttle closed during deceleration. In operation, the cables always move in opposite directions to one another. Before adjusting the throttle cables, check that they are in good condition. To achieve accurate cable adjustment, the cables must not bind or drag. Engine idle speed should also be

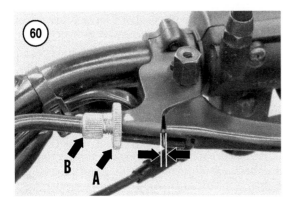

correct before adjusting the cables. If necessary, refer to Chapter Eight for throttle cable replacement procedures.

1. Measure the amount of free play at the throttle grip (**Figure 56**). Correct free play is 2-3 mm (0.08-0.12 in.). If free play is incorrect, adjust the cable as described in the following steps.

2. At the handlebar, pull back the rubber boots from the throttle cable adjusters.

3. Loosen the locknuts (A, **Figure 57**), then turn the cable adjusters (B) completely in.

4. Turn the bottom adjuster (decelerator cable) out only far enough to eliminate play when the throttle is closed. At the carburetor, the decelerator cable is at the right (A, **Figure 58**). Feel the cable for tightness. Do not make the cable excessively tight.

5. Turn the top adjuster (accelerator cable) out until free play is 2-3 mm (0.08-0.16 in.) at the throttle grip. At the carburetor, the accelerator cable is at the left (B **Figure 58**). Feel the cable for play when the carburetor is closed.

6. Operate the throttle and watch the action of the cables and throttle valve at the carburetor. The throttle valve should fully open and close.

7. If correct play cannot be achieved at the handlebar adjusters, adjust the cables at the carburetor as follows:

    a. Loosen the locknuts (**Figure 59**) and reposition the cables so the threads are approximately centered in the holders.

    b. Tighten the locknuts.

    c. Repeat the cable adjustment procedure.

8. When adjustment is correct, tighten the locknuts at the handlebar.

9. Install the rubber boots over the handlebar adjusters.

10. At engine startup, turn the handlebar from side to side as the engine idles. If engine speed varies, check for proper cable adjustment and cable routing.

Clutch Cable Adjustment

The clutch cable must be properly adjusted to ensure smooth shifting, full clutch engagement and minimal wear on the clutch plates.

1. Remove the left guard. Carefully flex the guard to remove it from the clutch lever assembly.

2. Measure the amount of free play at the lever pivot (**Figure 60**). The gap width should be 2-3 mm (0.08-0.12 in.). If free play is incorrect, adjust the cable as described in the following steps.

3. Loosen the locknut (A, **Figure 60**) and turn the cable adjuster (B) to increase/decrease play in the cable and lever.

4. If correct play cannot be achieved at the handlebar adjuster, adjust the cable at the release lever, located on the right crankcase cover. Adjust as follows:

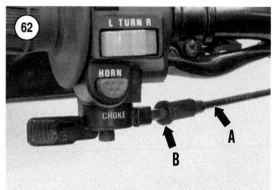

a. Loosen the locknuts (**Figure 61**) and reposition the cable so the threads are approximately centered in the holder.

b. Tighten the locknuts.

c. Repeat the cable adjustment procedure.

5. When adjustment is correct, tighten the locknut at the handlebar.

6. Install the left guard.

7. At engine startup, check that the transmission properly engages and disengages from the clutch.

Choke Cable Adjustment

The choke cable must be properly adjusted to ensure that it fully opens and closes.

To replace the choke cable, refer to *Throttle Cable Replacement* in Chapter 8.

1. Move the choke lever fully to the left (start position), then move it fully to the right (run position) and check that the cable moves freely. If the cable binds, inspect the cable and where it is attached to the carburetor.

2. With the choke lever *fully to the right*, grasp the cable housing and pull it away from where it enters the adjuster (A, **Figure 62**). The amount of cable housing movement is the cable play. The movement should be 2-3 mm (0.08-0.12 in.). If free play is incorrect, adjust the cable as follows:

a. Loosen the locknut (B, **Figure 62**) and turn the cable adjuster to increase/decrease play in the housing.

b. When adjustment is correct, tighten the locknut.

Front Brake Lever Adjustment

There is no routine adjustment required for the front brake lever. If the master cylinder is in good

condition and properly bled, the lever is automatically adjusted. If the brake drags, or brake lever play is unacceptable, inspect for worn or damaged parts in the master cylinder and brake caliper.

Rear Brake Pedal Adjustment

Brake pedal position and free play will be correct if the master cylinder is in good condition, properly installed and adjusted. Also, the pedal must be installed in the correct position on the pedal shaft. Inspect and adjust the pedal and linkage as follows:

1. Check that the pedal is at, or near level with the footpeg (**Figure 63**). If the pedal is not near level, inspect the pedal assembly for:

a. Alignment of the marks on the pedal and pivot shaft (**Figure 64**). If the marks are not aligned, remove, inspect and install the pedal assembly (Chapter Fourteen).

b. A broken or missing pedal shaft spring.

c. A bent brake pedal.

2. Check that the pushrod threads protrude 3-3.5 mm (0.12-0.14 in.) below the nut in the clevis (**Figure 65**). If necessary, adjust the clevis as follows:

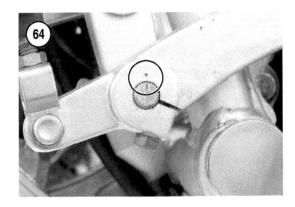

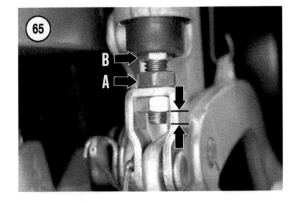

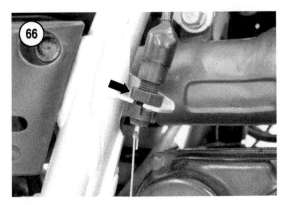

a. Loosen the locknut (A, **Figure 65**) and turn the adjuster (B) to increase/decrease the length of the pushrod.

b. Tighten the locknut.

3. Check the brake pedal adjustment from the riding position. If brake action continues to be incorrect, bleed the brake.

4. With the rear wheel off the ground, turn the wheel and operate the pedal to ensure the brake can be fully engaged and disengaged. If the brake drags, or if brake pedal height and free play cannot be achieved using the specified adjustments, inspect for worn or damaged parts in the master cylinder and caliper.

5. Check brake light operation. If necessary, adjust the switch as described in this section.

Rear Brake Light Switch Adjustment

The rear brake light should turn on when the brake pedal is depressed 15 mm (0.6 in.). When checking operation of the light, the ignition switch must be in the on position. Adjust the brake light switch position by turning the nut on the switch (**Figure 66**) as follows:

1. If the light comes on too late (too much pedal travel), turn the switch adjustment nut and *raise* the switch position.

2. If the light comes on too early (too little pedal travel), turn the switch adjustment nut and *lower* the switch position.

> *NOTE*
> *If the light does not come on after adjustment, check the bulb condition. If necessary, disconnect the switch wires and use an ohmmeter to check for continuity. The switch is in the on position when the switch plunger is extended.*

Brake Fluid Level Inspection

1. Support the motorcycle so the brake fluid reservoir being checked (front or back) is level.

2. Inspect the front reservoir as follows:

a. The fluid level should be between the lower level mark and the top of the sight glass (**Figure 67**).

b. If the fluid level is below the low mark, remove the cover and diaphragm, then add

DOT 4 brake fluid. Replace the diaphragm and cap.

c. Check for master cylinder leaks and worn brake pads.

3. Inspect the rear reservoir (**Figure 68**) as follows:

a. The fluid level should be between the upper and lower level marks, embossed on the reservoir.

b. If the fluid level is below the low mark, remove the guard assembly to access the reservoir (**Figure 69**).

c. Remove the cap, diaphragm plate and diaphragm, then add DOT 4 brake fluid.

d. Replace the diaphragm, diaphragm plate and cap, then screw the guard assembly into place.

e. Check for master cylinder leaks and worn brake pads.

> *NOTE*
> *If the brake fluid is not clear to slightly yellow, the fluid is contaminated and should be replaced. Drain and bleed the brake system (Chapter Fourteen).*

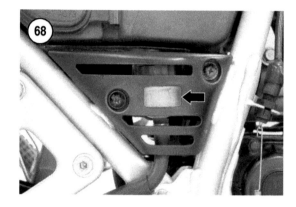

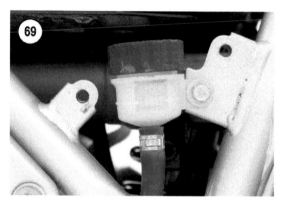

Brake Pad and Disc Inspection

Check the brake discs and pads regularly to ensure they are in good condition. During severe riding conditions, the scoring of a disc can occur rapidly if the brake pads are damaged or have debris lodged in the pad material. If damage is evident for any of the following inspections, refer to Chapter Fourteen for brake pad replacement, disc specifications and service limits.

1. Support the motorcycle so the wheels are off the ground.

2. Visually inspect the front and rear discs for the following:

a. Scoring. The disc should be smooth in the friction area.

b. Runout. Spin the wheel and visually check for lateral movement of the disc. Runout should not be evident.

c. Disc thickness. If the disc shows wear in the friction area, measure the thickness of both discs.

3. Inspect the brake pads. The pads are visible by looking into the caliper, on both sides of the disc. If the front or rear pad material (**Figure 70**) is less than

1 mm (0.04 in.) thick, replace the pad set for that wheel. Some pad sets have a wear indicator, which can be a groove or step at the edge of the pad.

Steering Head Bearing Inspection

The steering head is fitted with tapered roller bearings and should be inspected whenever the steering feels loose or uncontrollable. The bearings must be greased and torqued in order to prevent wear and to maintain proper handling characteristics.

1. Support the bike so the front wheel is off the ground.

2. Inspect the steering head as follows:

a. Turn the handlebar in both directions and feel for roughness or binding.

b. Grasp the fork legs near the axle and check for front-to-back play.

3. If there is roughness, binding or play in the steering head, refer to *Steering Play Check and Adjustment* in Chapter Twelve to adjust the steering head.

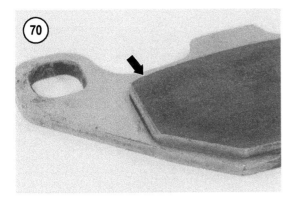

Swing Arm Bearing Inspection

A general lateral inspection of the swing arm bearings can be made in the following procedure. To perform a more thorough inspection, refer to *Swing Arm Bearing Inspection* in Chapter Thirteen. The procedure inspects for lateral and vertical wear in the swing arm bearings and pivots.

1. Support the motorcycle so the rear wheel is off the ground.

2. Have an assistant steady the motorcycle, then grasp the ends of the swing arm and leverage it from side to side. There should be no detectable play. If play is evident, refer to Chapter Thirteen for servicing the swing arm.

Rear Suspension Adjustment

The rear shock absorber is adjustable to meet the requirements of the riding conditions. Refer to *Rear Suspension Adjustment* in Chapter Thirteen to adjust the shock absorber.

Fork Air Release Valves

The fork is designed to operate with a cushion of air at the top of the fork leg. When riding, the air warms and expands, increasing the pressure in the chamber. This increase in pressure will stiffen the action of the fork. Kawasaki recommends *not* adding pressure to the fork before riding. Adding pressure will increase the possibility of oil leakage at the fork seals. The pressure in the fork legs should be released on a regular basis. When regularly riding in extremely rough terrain, release the pressure at least once a day. Release the pressure as follows:

1. Remove the cap from each air release valve (**Figure 71**).

2. Depress the valve core and release the pressure.

3. Tighten the caps onto the valves.

Tire Pressure

The tires must be inflated to meet the demands of the riding conditions. The standard air pressure recommendation is listed in **Table 4**. Slight over- or under-inflation is permissible if the riding conditions justify the change. However, *do not exceed the inflation range embossed on the tire sidewall*. Since inner tubes are used in the tires, too low of air pressure for the riding conditions can cause the tire to slip on the rim. This can bend the valve stem, as shown in **Figure 72**. Running with the valve stem bent can sever the valve stem and deflate the tire. Correct the condition by adjusting the tire and inner tube positions as described in this section.

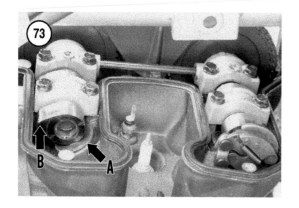

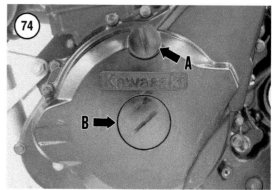

Tube Alignment

When the tube valve stem is bent, as shown in **Figure 72**, the tube must be realigned to prevent valve stem damage. A bent valve stem can sever and deflate the tire. Align the tube as follows:

1. Wash the tire and rim.
2. Remove the valve stem core and deflate the tire.
3. With an assistant steadying the motorcycle, break the tire-to-rim seal completely around both sides of the tire.
4. Support the bike so the wheel is off the ground.
5. Check that the valve stem is loose.
6. Lubricate both tire beads by spraying with soapy water.
7. Have the assistant apply the brake for the wheel being aligned.
8. Grasp the tire, then turn it and the tube until the valve stem is straight.
9. When the tube is correctly positioned, install the valve stem and inflate the tire. If necessary, reapply the soap solution to the beads to help seat the tire on the rim. Check that the beads uniformly seat around the rim.

> *WARNING*
> *Do not over inflate the tire to seat the beads. If the beads do not seat, deflate the tire and relubricate the beads.*

Spoke Tension

Spoke tension should be checked regularly and whenever the wheel has been respoked. During the break-in period, check spoke tension often. Refer to *Rim and Spoke Service* in Chapter Eleven for inspecting and properly tightening the spokes.

ENGINE TUNE-UP

Refer to *Carburetor Systems* in Chapter Eight for information concerning the function of the carburetor jets and jet needle.

Valve Clearance

The engine is designed with two intake valves and two exhaust valves. Valves must be adjusted correctly so they will completely open and close during the combustion cycle. Valves that are out of adjustment can cause poor performance and engine damage. Valve clearance is adjusted by placing the correct size of shim (available from Kawasaki dealerships) on top of each valve lifter, which rests under the camshaft (A, **Figure 73**). Whenever the valve clearance is incorrect, the camshafts must be removed and correctly sized shims inserted to bring the clearance within specification.

Typically, when valve clearance is near the smallest acceptable clearance, the clearance is increased, even though the valve is technically within specification. Valves more often lose clearance than gain

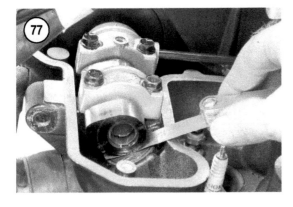

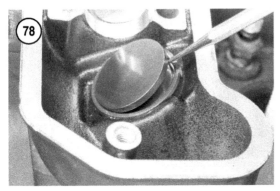

1. Support the bike so it is stable and secure.
2. Remove the cylinder head cover as described in Chapter Four.
3. Remove the timing plug (A, **Figure 74**) and the rotor nut plug (B).
4. Set the engine at TDC as follows:
   a. Fit a socket onto the rotor nut (**Figure 75**) and turn the crankshaft *counterclockwise* until the **T** mark on the rotor is aligned with the index mark in the timing hole (**Figure 76**).
   b. Verify the engine is at TDC by checking the camshaft lobes. If properly set, all cam lobes will be facing *away* from the center of the engine (B, **Figure 73**). If the lobes point toward the center of the engine, rotate the crankshaft one full turn and realign the TDC (bolt) **T** mark.
5. For each valve, use a flat feeler gauge to determine the clearance between the valve shim and cam lobe (**Figure 77**). Clearance is correct if slight resistance is felt when the gauge is inserted and withdrawn. Record the measurement and valve location. This information will be necessary to determine the correct size shim to install, if out of specification.
6. If adjustment is required, refer to Chapter Four for camshaft removal.
7. Remove the shim from the valve lifter(s) (**Figure 78**) that needs adjustment as follows:
   a. Rotate the valve lifter so the notch in the edge of the lifter is accessible.
   b. Wedge a small-tipped tool between the shim and lifter, then tilt the tool back to break the oil adhesion between the parts.
   c. Remove the shim and note its location in the cylinder head.
8. Determine the size of the shim to install on the valve lifter as follows:

NOTE
*Kawasaki shims are available in increments of 0.05 mm (0.002 in.). Shim thickness ranges from 2.0 mm (a No. 200 shim) to 3.2 mm (a No. 320 shim). The number on the shim surface is the original thickness of the shim. **For example**, if the removed shim is marked 255 (**Figure 79**), the shim is 2.55 mm thick. The next larger shim size is 2.60 mm thick and the next smaller size is 2.50 mm thick. The replacement shim should make clearance fall within the valve clearance range. If possible,*

clearance between inspections, and therefore are adjusted toward the larger clearance specification. A small, marginally acceptable clearance may be out of specification by the next inspection interval. However, do not increase clearance beyond the largest specification.

Check the valve clearance when the engine temperature is below 35° C (95° F). Read the entire procedure and understand the skill and equipment required. Refer to **Table 3** for valve clearance specifications and check/adjust the valve clearances as follows:

when removing and installing shims, measure the shims with a caliper to ensure their actual thickness.

a. Refer to **Table 3** for the intake and exhaust valve clearances.

b. Find the difference between the *specified* clearance and the *existing* clearance. This difference is the amount that must be added (loose valve) or subtracted (tight valve) from the value of the shim removed in Step 7.

c. For example: If the existing clearance for an exhaust valve is 0.14 mm, and the specified clearance is 0.15-0.25 mm, the difference is 0.01-0.11 mm. In this example, the replacement shim should be this much *smaller* than the removed shim. If the removed shim is 2.55 mm thick (a No. 255 shim), the replacement shim should be 2.45 mm thick (a No. 245 shim). This would increase clearance by 0.10 mm and make clearance 0.24 mm, which is within the specification.

9. Lubricate the replacement shim with engine oil, then install the shim onto the valve lifter. Place the shim number down so it does not get worn away by the camshaft lobe.

10. Repeat the procedure for the remaining valves that are out of specification.

11. Install the camshaft(s) as described in Chapter Four.

12. Check valve clearance. If clearance is not correct, remove the camshaft(s) and adjust the valves that are out of specification.

13. Install the cylinder head cover as described in Chapter Four.

14. Install the timing plug (A, **Figure 74**) and the rotor nut plug (B).

Carburetor Idle Speed and Mixture Adjustment

The carburetor must be adjusted so the idle speed keeps the engine running, but is also low enough to provide compression braking. Additionally, the pilot mixture screw must be adjusted so throttle response is good from an idle to about 1/4 throttle.

Use the following procedure to adjust the idle speed and pilot mixture screw for the *standard* jets. The jet needle is not adjustable for the CVK40 carburetor.

1. Check the throttle cables for proper adjustment.

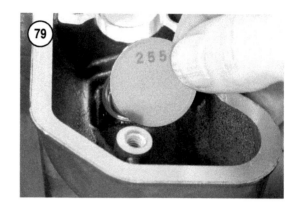

2. Check the air filter for cleanliness.

3. Set the pilot mixture screw (A, **Figure 80**) as follows:

> *NOTE*
> *Unless already removed, the pilot mixture screw will be blocked by a plug. To remove the plug, it is necessary to remove the carburetor from the engine and drill a hole in the plug so it can be pried out. Follow the steps in the carburetor disassembly procedure in Chapter Eight. Also, if the mixture screw is accessible, there is minimal clearance between the pilot screw bore and starter. Unless special tools are available, a small, angled screwdriver with a straight blade will have to be fabricated. Kawasaki does not offer a tool for making this adjustment.*

a. *Lightly* seat the screw, then turn it out 1 3/8 turns. This is a starting point for adjustment.

b. Start the engine and allow it to warm up.

c. Set the engine idle speed to 1200-1400 rpm. Set the idle speed by turning the throttle stop screw (B, **Figure 80**). Raise and lower the engine speed a few times to ensure that it returns to the set idle speed.

d. Mark the position of the pilot air screw or tool. From its initial setting, turn the pilot air screw in and out in small increments to find the points where the engine speed begins to decrease. Set the pilot screw between the two points.

e. Reset the idle speed to bring it within its required setting.

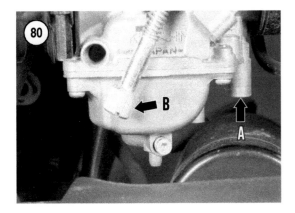

4. Test ride the bike and check throttle response. If throttle response is poor from an idle, adjust the pilot mixture screw out (richer) or in (leaner) by 1/8 turn increments until the engine accelerates smoothly.

5. If necessary, adjust the throttle cables for proper play.

Engine Timing Check

The ignition timing is electronically controlled by the CDI unit. No adjustment is possible to the timing. The timing is checked to verify the CDI unit is functioning properly. If ignition timing needs to be verified, perform the check as described in *Ignition Timing* in Chapter Nine.

Compression Check

A cylinder compression check can help verify the condition of the piston, rings and cylinder head assembly without disassembling the engine. By keeping a record of the compression reading at each tune-up, readings can be compared to determine if normal wear is occurring.

This engine typically has 530-855 kPa (77-124 psi) of compression when broken in properly. Since the compression release opens the right exhaust valve slightly during engine cranking, compression can vary. It is recommended that the owner perform regular compression checks and record the readings. If a current reading is extremely different from a previous reading, troubleshooting can begin to correct the problem. The condition of the compression release should be inspected first. Operating the engine when compression readings are abnormal can lead to severe engine damage.

1. Warm the engine to operating temperature.

2. Remove the spark plug. Insert the spark plug into the cap, then ground the plug to the cylinder.

> *CAUTION*
> *The spark plug must be grounded in order to prevent possible damage to the CDI unit.*

3. Thread a compression gauge into the spark plug hole. The gauge must be fitted airtight in the hole for an accurate reading.

    a. The compression gauge requires a 12 mm spark plug fitting to thread into the cylinder head. This size fitting is not commonly included in compression gauge sets.

    b. Optionally, use a compression gauge with a tapered rubber fitting that can be seated and held in the hole.

4. Hold or secure the throttle fully open.

5. Operate the starter and turn the engine over until the highest gauge reading is achieved.

6. Record the reading. Compare the reading with previous readings, if available. Under normal operating conditions, compression will slowly lower from the original specification, due to wear of the piston rings and/or valve seats.

    a. If the reading is higher than normal, this can be caused by a broken or jammed compression release spring. If the spring cannot retract the compression release weights, no compression will be released when the engine is cranking. Commonly, carbon buildup in the combustion chamber is another cause of high compression. This can cause high combustion chamber temperatures and potential engine damage.

    b. If the reading is lower than normal, this can be caused by worn piston rings, worn valves, a damaged piston, leaking head gasket, or a combination of these parts. The compression release is less likely to be the problem in this case, since the release weights are probably not spinning fast enough to stop the release of compression. This would occur when the engine fired and camshaft speed increased.

    c. To help pinpoint the source of leakage, pour 15 cc (1/2 oz.) of four-stroke engine oil through the spark plug hole and into the cylinder. Turn the engine over to distribute and clear excess oil. Recheck compression. If compression increases, the piston rings are worn or damaged. If compression is the same, the piston, head gasket, valves or compression release are worn or damaged.

SPARK PLUGS

Spark Plug Removal

Careful removal of the spark plug is important in preventing grit from entering the combustion chamber. It is also important to know how to remove a plug that is seized, or is resistant to removal. Forcing a seized plug can damage the threads in the cylinder head.

1. Remove the fuel tank (Chapter Fifteen).
2. Grasp the spark plug cap (**Figure 81**) and twist it loose from the spark plug. There may be slight suction and resistance while removing the cap.
3. Clean dirt from the plug well (**Figure 82**), preferably with compressed air.
4. Fit a spark plug wrench onto the spark plug, then remove it by turning the wrench counterclockwise. If the plug is seized or drags excessively during removal, stop and try the following techniques:

   a. Apply a penetrating lubricant such as Liquid Wrench or WD-40 and allow it to stand for 15 minutes.

   b. If the plug is completely seized, apply moderate pressure in both directions with the wrench. Only attempt to break the seal so lubricant can penetrate under the spark plug and into the threads. If this does not work, and the motorcycle is still operable, install the spark plug cap and fuel tank, then start the engine. Allow it to completely warm up. The heat of the engine may be enough to expand the parts and allow the plug to be removed.

   c. When a spark plug has been loosened, but drags excessively during its removal, apply penetrating lubricant around the spark plug threads. Turn the plug *in* (clockwise) to help distribute the lubricant onto the threads. Slowly remove the plug, working it in and out of the cylinder head as lubricant is added.

   d. Inspect the threads in the cylinder head for damage. Clean and true the threads with a spark plug thread-chaser.

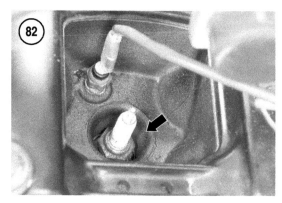

5. Inspect the removed plug to determine if the engine is operating properly.
6. A spark plug that is in good condition and will be reused after inspection should be cleaned with electrical contact cleaner and a shop cloth. Do not use abrasives or wire brushes to clean the plugs.

Spark Plug Gap and Installation

Proper adjustment of the electrode gap is important for reliable and consistent spark. Also, the proper preparation of the spark plug threads will ensure that the plug can be removed easily in the future, without damage to the cylinder head threads.

1. Refer to **Table 3** for the required spark plug gap.
2. Insert a wire feeler gauge (the size of the required gap) between the center electrode and the ground electrode.
3. Pull the gauge through the gap. If there is slight drag, the setting is correct. If the gap is too large or small, adjust the gap by bending the ground electrode (**Figure 83**) to achieve the required gap. Use an adjusting tool (**Figure 84**) to bend the electrode. Do not pry the electrode with a screwdriver or other

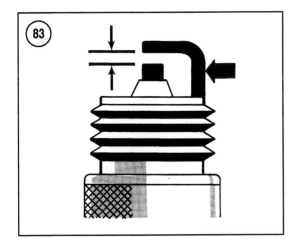

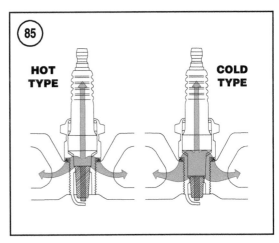

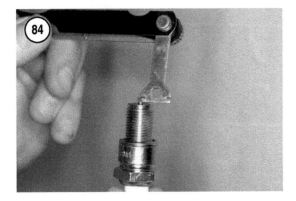

tool. Damage to the center electrode and insulator is possible.

4. Inspect the spark plug to ensure it is fitted with a crush washer.

5. Wipe a small amount of antiseize compound onto the spark plug threads. Do not allow the compound to get on the electrodes.

6. Finger-tighten the spark plug into the cylinder head. This will ensure the plug is not cross-threading.

7. Torque the spark plug to 14 N•m (10 ft.-lb.). If a torque wrench is not available, turn a new spark plug 1/4-1/2 turn from the seated position; a used spark plug 1/8-1/4 turn from the seated position.

8. To help prevent water from migrating under the cap, wipe a small amount of dielectric grease around the interior of the cap. Press and twist the cap onto the spark plug.

Spark Plug Selection

Refer to **Table 3** for the recommended standard or resistor-type spark plug and gap.

> *CAUTION*
> *The following paragraphs provide general information and operation fundamentals that apply to all spark plugs. However, before changing to a plug other than what is recommended by Kawasaki, check with the spark plug manufacturer for specific part numbers and equivalents that apply to this motorcycle. Poor performance or engine damage can occur by installing a spark plug that is not compatible with this engine.*

Heat range

Spark plugs are available in several heat ranges to accommodate the load and performance demands put on the engine. The standard spark plug recommended by manufacturers is usually a medium heat-range plug that operates well over a wide range of engine speeds. As long as engine speeds vary, these plugs will stay relatively clean and perform well.

If the engine is run in hot climates, at high speed or under heavy loads for prolonged periods, a spark plug with a colder heat range is recommended. A colder plug quickly transfers heat away from its firing tip and to the cylinder head (**Figure 85**). This is accomplished by a short path up the ceramic insulator and into the body of the spark plug. By transfer-

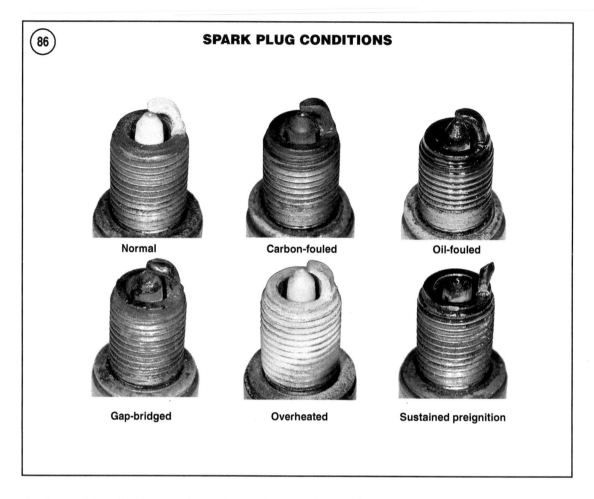

SPARK PLUG CONDITIONS

86

Normal

Carbon-fouled

Oil-fouled

Gap-bridged

Overheated

Sustained preignition

ring heat quickly, the plug remains cool enough to avoid overheating and preignition problems. If the engine is run slowly for prolonged periods, this type of plug may foul and cause poor performance.

If the engine is run in cold climates or at slow speed for prolonged periods, a spark plug with a hotter heat range is recommended. A hotter plug slowly transfers heat away from its firing tip and to the cylinder head. This is accomplished by a long path up the ceramic insulator and into the body of the spark plug (**Figure 85**). By transferring heat slowly, the plug remains hot enough to avoid fouling and buildup. If the engine is run in hot climates or fast for prolonged periods, this type of plug may overheat, cause preignition problems and possibly melt the electrode. Damage to the piston and cylinder assembly is possible.

If choosing to change a spark plug to a different heat range, go one step hotter or colder from the recommended plug. Do not try to correct poor carburetor or ignition problems by using different spark plugs. This can only compound the existing problems and possibly lead to severe engine damage.

Reach

Reach is the length of the threaded portion of the plug. Always use a spark plug that is the correct reach. Too short of a reach can lead to deposits or burning of the exposed threads in the cylinder head. Misfiring can also occur since the tip of the plug is shrouded and not exposed to the fuel mixture. If the reach is too long, the exposed plug threads can burn, causing preignition. It is also possible that the piston may contact the plug on the upstroke, causing severe engine damage.

Spark Plug Reading

The spark plug is an excellent indicator of how the engine is operating. By correctly evaluating the condition of the plug, you can diagnose and pinpoint problems, or potential problems. When removing the spark plug, compare the firing tip with the ones shown in **Figure 86**. The following paragraphs provide a description, as well as common causes for each of the conditions.

CAUTION
In all cases, when the spark plug does not read normal, find the cause of the problem before continuing engine operation. Severe engine damage is possible when abnormal plug readings are ignored.

Normal

The plug has light tan or gray deposits on the tip. No erosion of the electrodes or abnormal gap is evident. This indicates an engine that has properly adjusted carburetion, ignition timing, and proper fuel. This heat range of the plug is appropriate for the conditions in which the engine has been operated. The plug can be cleaned and reused.

Carbon-fouled

The plug is black with a dry, sooty deposit on the entire plug surface. This dry sooty deposit is conductive and can create electrical paths that bypass the electrode gap. This often causes misfiring of the plug.
1. Fuel mixture too rich.
2. Spark plug range too cold.
3. Faulty ignition component.
4. Prolonged idling.
5. Clogged air filter.
6. Poor compression.

Oil-fouled

The plug is wet with black, oily deposits on the electrodes and insulator. The electrodes do not show wear.
1. Incorrect carburetor jetting.
2. Prolonged idling or low idle speed.
3. Spark plug range too cold.
4. Worn valve guides.
5. Worn piston rings.
6. Ignition component failure.

Gap-bridged

The plug is clogged with deposits between the electrodes. The electrodes do not show wear.
1. Incorrect oil being used.
2. Incorrect fuel or fuel contamination.
3. Carbon deposits in combustion chamber.
4. High-speed operation after excessive idling.

Overheated

The plug is dry and the insulator has a white or light gray cast. The insulator may also appear blistered. The electrodes may have a bluish-burnt appearance.
1. Fuel mixture too lean.
2. Spark plug range too hot.
3. Air leak into intake system.
4. No crush washer on plug.
5. Plug improperly tightened.
6. Faulty ignition component.

Preignition

The plug electrodes are severely eroded or melted. This condition can lead to severe engine damage.
1. Faulty ignition component.
2. Spark plug range too hot.
3. Air leak into intake system.
4. Carbon deposits in combustion chamber.

Worn out

The plug electrodes are rounded from normal combustion. There is no indication of abnormal combustion or engine conditions. Replace the plug.

Spark Plug Cap

The spark plug cap should fit tight to the spark plug and be in good condition. A cap that does not seal and insulate the spark plug terminal can lead to flashover (shorting down the side of the plug), particularly when the bike is operated in wet conditions. To help prevent water from migrating under the cap, wipe a small amount of dielectric grease around the interior of the cap before installing it onto the plug.

Table 1 MAINTENANCE AND LUBRICATION SCHEDULE

	Initial 500 miles or 800 km	Every 3000 miles or 5000 km	Every 6000 miles or 10,000 km	Other
Adjust and lubricate drive chain				Every 200 miles/300 km
Inspect/fill/charge battery				Monthly
Inspect brake fluid level	x	x		Or monthly
Inspect brake pedal adjustment	x	x		Or monthly
Inspect brake light adjustment	x	x		Or monthly
Change engine oil	x	x		Or annually
Clean air filter	x	x		Or more often, if required
Replace air filter				Every 5 cleanings
Inspect air filter housing drain	x	x		
Adjust balancer chain tensioner	x	x		
Inspect/adjust steering bearings	x	x		
Check engine idle speed	x	x		
Inspect lights and switch operation	x	x		
Check throttle grip play	x	x		
Inspect evaporative emission system*	x	x		
Inspect/replace coolant hoses	x	x		
Inspect/adjust clutch	x	x		
Inspect/replace brake pads	x	x		
Inspect/tighten spokes	x	x		
Replace engine oil filter	x		x	
Check valve clearance	x		x	
Inspect nuts, bolts and fasteners	x		x	
Inspect/replace spark plug		x		
Inspect drive chain and sprockets		x		
Drain carburetor float chamber		x		
Clean muffler		x		
Perform general lubrication		x		
Inspect/lubricate swing arm linkage			x	
Lubricate steering bearings				Every 12,000 miles/20,000 km
Replace fork oil				Every 18,000 miles/30,000 km
Replace coolant				Every 2 years
Replace brake fluid				Every 2 years
Replace fuel hose				Every 4 years
Replace brake hoses				Every 4 years

*California models only. Inspect/replace hoses, as necessary.

Table 2 FUEL, LUBRICANTS AND FLUIDS

Fuel type	Unleaded gasoline; 87 octane minimum
Fuel tank capacity	23 liters (6.1 U.S. gal.)
Engine oil	SG rated, four-stroke engine oil
	SAE 10W-40, 20W-40, 10W-50 or 20W-50
Engine oil capacity	
Without filter change	2.2 liters (2.3 U.S. qt.)
With filter change	2.5 liters (2.6 U.S. qt.)
Cooling system capacity	1.3 liters (1.4 U.S. qt.)
Coolant type	Ethylene glycol containing anti-corrosion
	inhibitors for aluminum engines
Coolant mixture	50/50 (antifreeze/distilled water)
Drive chain	O-ring type chain lubricant
Fork oil grade	Kayaba G-10, or equivalent 10-weight fork oil
Fork oil capacity (each leg)	
Oil change	355 cc (12.0 U.S. oz.)
Fork rebuild (all parts dry)	416-424 cc (14.0-14.3 U.S. oz.)
Fork tube oil level (from top edge	
of inner tube)	188-192 mm (7.40-7.56 in.)
Brake fluid type	DOT 4
Control cables	Cable lube
Air filter	Foam air filter oil

Table 3 TUNE-UP SPECIFICATIONS

Battery	12 volt, 14 amp-hour
Compression	530-855 kPa (77-124 psi)
Compression ratio	9.5:1
Idle speed	1200-1400 rpm
Ignition timing	
Idle	10° BTDC at 1300 rpm
Advanced	30° BTDC at 3300 rpm
Pilot mixture screw turns out	1 3/8
Spark plug	
Type	NGK DP8EA-9
	NGK DPR8EA-9
	ND X24EP-U9
	ND X24EPR-U9
Gap	0.8-0.9 mm (0.032-0.036 in.)
Valve clearance	
Intake	0.10-0.20 mm (0.004-0.008 in.)
Exhaust	0.15-0.25 mm (0.006-0.010 in.)

Table 4 ROUTINE CHECKS AND ADJUSTMENTS

Brake light switch	On after 15 mm (0.6 in.) of pedal travel
Brake pad lining minimum thickness	1.0 mm (0.040 in.)
Brake pedal height	At or near level with top of peg
Choke lever free play	2-3 mm (0.08-0.12 in.)
Clutch lever free play	2-3 mm (0.08-0.12 in.) at pivot
Drive chain play	50-60 mm (2.0-2.4 in.)
Drive chain length wear limit (20 links/21 pins)	323 mm (12.7 in.)
Fork air pressure	0 (atmospheric pressure)
Radiator cap relief pressure	93-123 kPa (13.5-17.8 psi)
Rim runout (radial and lateral)	2.0 mm (0.08 in.)
Throttle grip free play	2-3 mm (0.08-0.12 in.)
Tire pressure	
Front	150 kPa (22 psi)
Rear	
Loads up to 97.5 kg (215 lb.)	150 kPa (22 psi)
Loads 97.5-182 kg (215-400 lb.)	200 kPa (29 psi)

Table 5 MAINTENANCE TORQUE SPECIFICATIONS

	N•m	in.-lb.	ft.-lb.
Axle nut			
Front	78	–	58
Rear	93	–	69
Fork caps	29	–	21
Handlebar clamp bolts	24	–	18
Oil drain plug	23	–	17
Spark plug	14	–	10
Spokes	2-4	17-35	–
Coolant drain plug	10	88	–

ENGINE TOP END

4

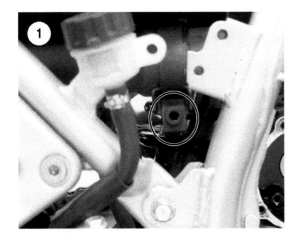

This chapter provides information for the removal, inspection and replacement of the major assemblies that make up the engine top end. This includes the exhaust system, cylinder head, camshafts, valves, cylinder and piston. All the parts can be removed with the engine mounted in the frame. Refer to the tables at the end of this chapter for specifications.

Read this chapter before attempting any repair to the engine top end. Become familiar with the procedures, photos and illustrations to understand the skill and equipment required. Refer to Chapter One for tool usage and techniques. When special tools are required or recommended, the part number is provided.

SHOP CLEANLINESS

Always clean the engine before starting repairs. If the engine will remain in the frame, clean the surrounding framework, cables and harnesses. Avoid letting dirt enter the engine.

Keep the work environment as clean as possible. Store parts and assemblies in well-marked plastic bags and containers. Keep reconditioned parts wrapped until they are installed.

EXHAUST SYSTEM

Removal and Installation

Do not attempt to remove the exhaust pipe or muffler when the engine is hot.

1. Support the motorcycle so it is stable and secure.
2. Remove the side covers and seat (Chapter Fifteen).
3. Remove the rear brake reservoir cover, then unbolt the reservoir to access the clamp behind the reservoir (**Figure 1**). Loosen the clamp bolt.

4. Remove the muffler as follows:
   a. Remove the bolts securing the muffler to the frame (**Figure 2**).
   b. Pull the muffler from the exhaust pipe and out of the frame. Do not hammer on the muffler if it is stuck to the exhaust pipe. Twist the muffler off the pipe. If necessary, apply penetrating oil around the connection.
5. Remove the exhaust pipe as follows:
   a. Remove the exhaust pipe hanger bolt (**Figure 3**), located near the rear brake light switch.
   b. Remove the two nuts that secure the exhaust pipe (**Figure 4**) to the cylinder head. These nuts are often corroded. To prevent damaging the studs, apply penetrating oil as needed during removal.
   c. Pull the exhaust pipe off the cylinder head.
   d. Remove the gasket from the exhaust port.
6. Reverse these steps to install the system. Note the following:
   a. Clean the exhaust port and install a new exhaust pipe gasket, seating it in the port before installing the exhaust pipe (**Figure 5**). The gasket can be held in place with a small amount of antiseize compound. Install the gasket with the round side facing out.
   b. Apply antiseize compound to all bolt threads.
   c. Install and align the entire exhaust system with all bolts finger-tight. Tighten the exhaust pipe squarely into the exhaust port, then tighten the remaining clamp and muffler bolts.

CYLINDER HEAD COVER

Removal and Installation

Refer to **Figure 6**.
1. Support the bike so it is stable and secure.
2. Remove the fuel tank and radiator covers (Chapter Fifteen).
3. Remove the fan assembly as follows:
   a. Disconnect the fan switch wire (**Figure 7**).
   b. Remove the wires from the retainers, then remove the mounting bolts (**Figure 8**).
   c. Secure the fan out of the way.
4. Remove the coil and the leads from the coolant temperature switch and spark plug (**Figure 9**).
5. Remove the bolts securing the upper engine mounting (**Figure 10**).

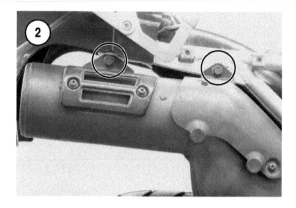
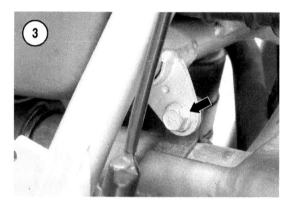
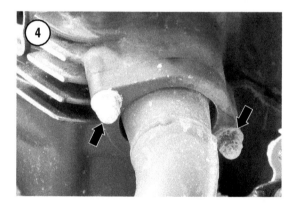
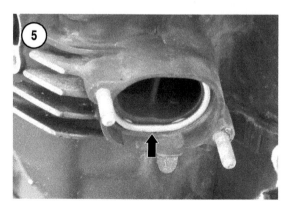

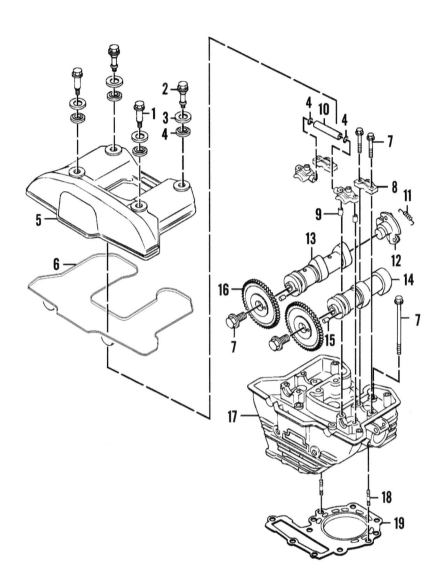

CYLINDER HEAD ASSEMBLY

1. Left cover bolt
2. Right cover bolt
3. Washer
4. O-ring
5. Cylinder head cover
6. Cylinder head cover gasket
7. Bolt
8. Camshaft cap
9. Dowel
10. Oil pipe
11. Spring
12. Compression release
13. Exhaust camshaft
14. Intake camshaft
15. Pin
16. Sprocket
17. Cylinder head
18. Stud
19. Cylinder head gasket

6. Temporarily tape the wires and cables to the frame, providing maximum clearance.

7. Remove the rubber mounts that support the fuel tank, and loosen the four cover bolts (**Figure 11**).

8. Remove the cylinder head cover has follows:

   a. Raise the cylinder head cover and separate the gasket from the cover. This step will provide additional clearance when removing the cover from the engine.

   b. Tilt the left side of the cover up, then remove the cover out the left side of the engine, passing the cover over the cam chain and sprockets. On 1996-on models, clearance is minimal since the cover must also pass over the upper chain guide (A, **Figure 12**). Earlier models have a smaller guide in the cover.

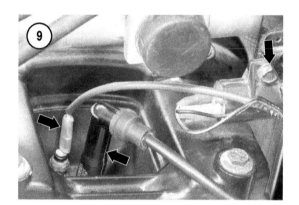

9. If necessary, refer to Chapter Three for performing the valve clearance check and adjustment. If valve adjustment is required, refer to this chapter for camshaft removal.

10. Reverse this procedure to install the cover. Note the following:

   a. Clean the gasket and the gasket surfaces on the cylinder head and cover. Apply RTV silicone sealant to the gasket surface on the left side of the cylinder head (B, **Figure 12**). This will help seal the plugs that are part of the gasket. Install the gasket onto the cylinder head, then seat the cover onto the gasket.

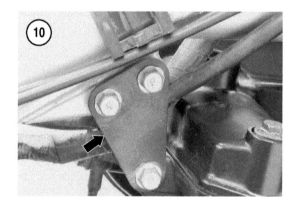

   b. Check that the O-rings on the cover bolts are in good condition. Replace if necessary.

   c. Install the cover bolts and torque the bolts to 8 N•m (71 in.-lb.). Do not overtighten the bolts. The left bolts are threaded into the camshaft caps.

   d. Torque the upper engine mounting bolts to 25 N•m (18 ft.-lb.).

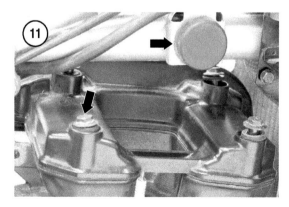

e. Clean all electrical connections and apply dielectric grease when reconnecting.

CAMSHAFTS AND CAM CHAIN TENSIONER

Camshaft and Cam Chain Tensioner Removal

The camshafts and cam chain tensioner can be removed with the engine mounted in the frame. If performing the valve clearance check and adjustment (Chapter Three), the camshafts must be removed only when valve clearance is incorrect.

Refer to **Figure 6**.

1. Remove the cylinder head cover as described in this chapter.

2. Remove the timing plug (A, **Figure 13**) and the rotor nut plug (B).

3. Set the engine at TDC (top dead center) as follows:

    a. Fit a socket onto the rotor nut and turn the crankshaft *counterclockwise* until the **T** mark on the rotor is aligned with the index mark in the timing hole (**Figure 14**).

    b. Verify the engine is at TDC by checking the camshaft lobes. If properly set, all camshaft lobes will point *away* from the center of the engine (**Figure 15**). If the lobes point toward the center of the engine, rotate the crankshaft one full turn and realign the TDC **T** mark.

> *NOTE*
> *Before further disassembly, measure and record the valve clearance (refer to Chapter Three for valve clearance check and adjustment procedures) for all valves. Incorrect clearances can be adjusted during the reassembly process, preventing removal of the camshafts a second time.*

4. Remove the cam chain tensioner (**Figure 16**) from the cam chain tunnel as follows:

    a. Loosen, but do not remove the center bolt (A).

    b. Remove the top and bottom bolts (B).

    c. Remove the tensioner and gasket.

> *CAUTION*
> *Anytime the tensioner mounting bolts are loosened, the tensioner must be*

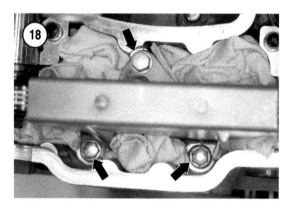

completely removed and reset. Do not partially remove, then retighten the bolts. The ratchet-type plunger will have extended and locked itself. Retightening the bolts causes the cam chain and tensioner to be too tight, possibly causing engine damage if the engine is operated. Also, do not turn the crankshaft when the tensioner is removed from the engine. Camshaft timing could be altered because of the excess slack in the cam chain. This also could cause engine damage if the engine is operated.

5. Remove the camshaft caps (**Figure 17**) as follows:

   a. Stuff clean shop cloths around the cam chain and in the tunnel to prevent parts or debris from entering the engine. If parts do fall into the tunnel, they do not enter the crankcase. The tunnel leads to the left end of the crankshaft, in the left crankcase cover.

   b. Loosen the eight camshaft cap bolts. Loosen the bolts equally in several passes.

   c. Remove the bolt and cap sets from the right intake and exhaust valves. Raise the caps straight up and slowly. Account for the two dowels under each cap. The dowels may be loose and fall from the cap. If the dowels are tight in either the cap or cylinder head, they may be left in place. Keep all sets of parts identified.

CAUTION
If the caps are tight, light tapping with a small, soft mallet can be used to loosen the caps. Do not pry on the caps or damage the machined sur-

faces. Check that all caps are marked with their cylinder head position.

   d. Remove the bolt and cap sets from the left intake and exhaust valves. Since the caps are joined by the oil supply pipe, slowly raise both caps straight up. Account for the two dowels under each cap. The dowels may be loose and fall from the cap. If the dowels are tight in either the cap or cylinder head, they may be left in place. Inspect the oil pipe and O-rings during inspection of the camshaft caps. Keep all sets of parts identified.

6. On 1996-on models, remove the upper cam chain guide (**Figure 18**) as follows:

   a. Check that the cam chain tunnel is stuffed with clean shop cloths.

   b. Loosen the three bolts securing the guide.

   c. Remove the single bolt to the inside of the guide.

   d. Raise the guide and remaining two bolts from the engine (**Figure 19**).

7. Attach a length of wire to the cam chain. Secure the wire (**Figure 20**) so the chain cannot fall into the engine when the camshafts are removed.

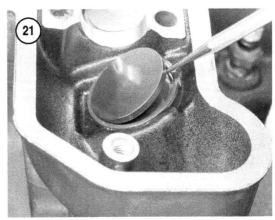

NOTE
If the crankshaft must be turned while the cam chain is loose, keep tension on the chain while turning the crankshaft. This prevents possible chain binding at the crankshaft sprocket.

8. Lift both camshafts out of the cylinder head. It is not necessary to remove the shims or valve lifters (**Figure 21**).

9. Check that the cylinder head openings are properly covered to prevent parts or debris from entering the engine.

10. Inspect the camshafts, caps and cam chain tensioner as described in this section.

    a. Reset the cam chain tensioner as described in the inspection procedure. The tensioner must be reset before installation.

    b. If necessary, refer to Chapter Five for cam chain inspection. Although the chain cannot be removed from the engine, a partial inspection can be performed.

Camshaft and Cam Chain Tensioner Installation

1. Prior to installing the camshafts and cam chain tensioner, note the following:

    a. Adjust any valves that are out of specification as described in Chapter Three. If the valves were reconditioned, install the original shims at this time. Valve clearance will need to be rechecked after the camshafts are installed. It is possible that the camshafts will have to be removed a second time to correctly set the valves.

    b. Check that the cam chain tensioner is disassembled and reset, as described in the inspection procedure.

    c. Lubricate parts with engine oil during assembly.

    d. Refer to **Figure 6** as needed.

2. Inspect the cylinder head and ensure that all surfaces are clean. Remove any shop cloths from the cam chain tunnel.

3. Check that the engine is at TDC and is held in this position. If it is not at TDC, turn the crankshaft *counterclockwise* until the **T** mark on the rotor is aligned with the index mark in the timing hole (**Figure 14**). The engine must remain at TDC when installing and timing the camshafts.

4. Install and time the camshafts. Refer to **Figure 22** as needed. The camshafts are properly installed when the arrow on *both* sprockets point to the front of the engine, and the arrows and marks are parallel to the top edge of the cylinder head. All cam chain slack on the tension side of the chain and across the top of the sprockets must also be eliminated.

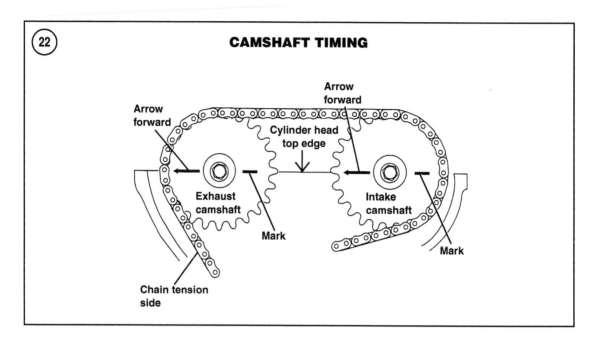

(22) **CAMSHAFT TIMING**

Arrow forward

Arrow forward

Cylinder head top edge

Exhaust camshaft

Intake camshaft

Mark

Mark

Chain tension side

a. Identify the alignment marks on the exhaust camshaft sprocket (**Figure 23**).

b. Pull up on the tension side (front side) of the cam chain and install the exhaust camshaft (with compression release). Check that the shoulder at the left end of the camshaft, seats into the head, then seat the taut chain onto the sprocket. Check that the cam lobes point *away* from the center of the engine.

c. Check that the engine is still at TDC when the sprocket marks are aligned. If necessary, reposition the assembly.

d. Identify the alignment marks on the intake camshaft sprocket (**Figure 23**).

e. Keeping the chain taut, but seated on the exhaust cam sprocket, install the intake camshaft. Check that the shoulder at the left end of the camshaft, seats into the head, then seat the taut chain onto the sprocket. Check that the cam lobes point *away* from the center of the engine.

5. With the engine at TDC, inspect the installation (**Figure 24**).

a. The cam chain should be taut at the front and across the cam sprockets. The excess chain slack at the rear, between the intake camshaft and crankshaft sprocket, will be taken up by the cam chain tensioner.

b. The arrows on both sprockets must point forward.

c. The arrows and alignment marks on both sprockets must be parallel to the top edge of the cylinder head.

6. On 1996-on models, install the upper cam chain guide as follows:

NOTE
The guide must be installed before the camshaft caps. The oil pipe between the left caps will block installation of the guide.

a. Install the two long bolts into the outside holes in the guide (**Figure 19**).

b. Lower the guide into the cam chain tunnel just far enough to start the bolts into their holes. Tighten the bolts with a socket and extension. Check that the socket and extension fit together tightly and cannot separate when lowered into the tunnel. Torque the bolts to 10 N•m (88 in.-lb.).

c. Stuff the cam chain tunnel with clean shop cloths to prevent parts from falling into the engine.

d. Apply threadlocking compound to the remaining guide bolt, then torque the bolt to 8 N•m (71 in.-lb.) (**Figure 18**).

7. Install the camshaft caps and dowels (**Figure 17**) as follows:

a. Install the dowel and cap sets for the left intake and exhaust valves. Since the caps are

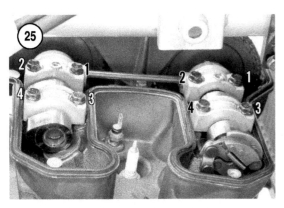

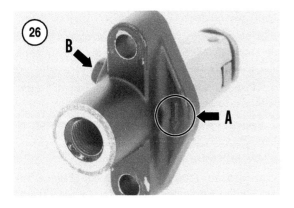

joined by the oil pipe, lower both caps into position equally. Install and finger-tighten the bolts.

b. Install the dowel and cap sets for the right intake and exhaust valves. The arrows on the caps must point forward. Install and finger-tighten the bolts.

c. Seat the camshafts by first lightly tightening bolts 1 and 2 on the left caps (**Figure 25**).

d. For each camshaft, torque the cap bolts in the numerical sequence shown in **Figure 25**. Torque the cap bolts to 12 N•m (106 in.-lb.). Make several passes, increasing the torque evenly on the caps.

CAUTION
Failure to evenly torque the camshaft caps could cause damage to the cylinder head, camshafts or caps.

e. Remove all shop cloths from the cylinder head.

8. Install the cam chain tensioner as follows:

a. Orient the tensioner housing so the arrow on the housing points *down* (A, **Figure 26**) when installed. Later models also have a block cast into the housing (B).

b. Install a new gasket, the housing and the top and bottom bolts (B, **Figure 16**).

c. Insert the spring and washer, then install and tighten the center bolt (A, **Figure 16**).

CAUTION
If for any reason the tensioner mounting bolts are loosened, the tensioner must be completely removed and reset. Do not partially remove, then retighten the bolts. The ratchet-type plunger will have extended and locked itself. Retightening the bolts will cause the cam chain and tensioner to be too tight, possibly causing engine damage if the engine is operated.

9. Turn the crankshaft slowly *counterclockwise* several times, then place it at TDC.

CAUTION
If abnormal resistance is felt when turning the crankshaft, stop and recheck the alignments. Improper alignment can cause engine damage.

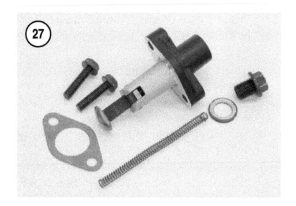

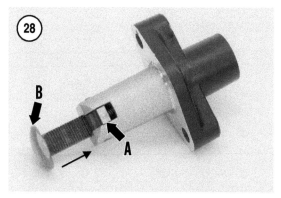

a. Check that the camshaft sprocket arrows point forward and are parallel with the top edge of the cylinder head. If the camshafts are not properly aligned, disassemble the head and realign the camshaft(s).

b. Check valve clearances (Chapter Three). If necessary, remove the camshafts and install the proper size valve lifter shim(s).

10. Install the timing plug and the rotor nut plug.

11. Install the cylinder head cover as described in this chapter.

Cam Chain Tensioner Inspection

The tensioner assembly is a spring-loaded, ratcheting-type tensioner. As the cam chain wears and develops slack, the spring-loaded plunger extends and locks itself against the back of the lower chain guide. The guide then pivots forward and retightens the chain.

Since the tensioner is self-adjusting, there is no routine maintenance required. However, anytime the tensioner is loosened or removed the tensioner must be reset. Do not partially remove, then retighten the bolts. The plunger will have extended and locked itself. Retightening the bolts will cause the cam chain and tensioner to be too tight, possibly causing engine damage if the engine is operated.

1. Remove the center bolt and disassemble the tensioner (**Figure 27**).

   a. Inspect the parts for obvious damage.

   b. Install a new gasket during installation.

2. Reset the plunger as follows:

   a. Press and hold the ratchet release (A, **Figure 28**).

   b. Press and seat the plunger (B, **Figure 28**) into the tensioner housing.

c. Do not install the spring, washer and bolt until the tensioner housing has been mounted and the camshafts and chain have been installed and timed.

Camshaft and Compression Release Inspection

The compression release is located at the right end of the exhaust camshaft. The release slightly opens the right exhaust valve during engine cranking. The reduction in compression makes starting easier. When the engine starts, the release is centrifugally disengaged.

In the following procedure, replace parts that are obviously damaged or not within the specification as listed in **Table 2**.

1. Clean the camshafts in solvent and dry thoroughly.

2. Inspect the camshafts for obvious scoring or damage.

3. For each camshaft, do the following:

   a. Measure the cam lobe heights (**Figure 29**) with a micrometer.

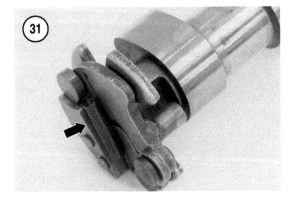

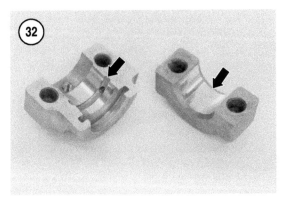

b. Measure the camshaft journal outside diameter (where the cap fits over the cam) (**Figure 30**).

c. Inspect the camshaft sprocket teeth for wear or other damage. The profile of each tooth should be symmetrical. If the sprocket is worn, replace both cam sprockets and the cam chain as a set. When this type of damage occurs, also inspect the crankshaft sprocket, cam chain tensioner and chain guides.

4. Inspect the automatic compression release (**Figure 31**), located on the exhaust camshaft.

a. Inspect the spring for obvious damage.

b. Pivot the weights outward and check for smooth operation. When the weights are pivoted and released, the spring should fully retract the weights. Replace the spring if it is fatigued.

Camshaft Cap Inspection

1. Clean the camshaft caps in solvent and dry with compressed air. For the left intake and exhaust valve caps, pull the oil pipe out of the caps. Prevent damage to the bearing surface of the caps when handling and inspecting.

2. For each camshaft cap (**Figure 32**) do the following:

a. Check all oil passages for cleanliness.

b. Inspect the bearing surface in the cap and the mating bearing in the cylinder head for obvious scoring or damage. If either part is damaged, replace the caps and cylinder head as a set. The parts are only available as a single part number. The caps are machined with the cylinder head, so their dimensions and alignments are unique to that cylinder head.

3. For the left intake and exhaust valve caps, inspect the oil pipe and O-rings (**Figure 33**). The oil pipe is held and sealed by the O-ring in each cap. Since the actual condition of the O-rings cannot be determined without removing them from the caps, periodically replace the O-rings to ensure a good seal around the oil pipe. Leakage past the O-rings will reduce the amount of oil reaching the exhaust camshaft. A convenient time to replace the O-rings would be during a valve adjustment procedure, when the camshafts must be removed.

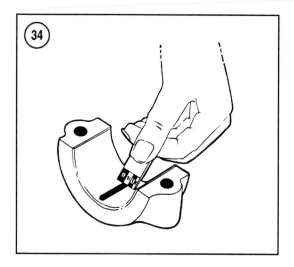

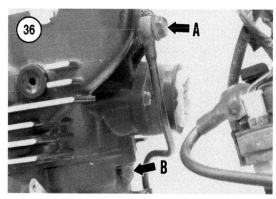

a. Lubricate the ends of the oil pipe and twist it into the caps. If the pipe is firm in both caps, the O-rings can be reused.

b. If the pipe is loose in the caps, or if the O-rings are obviously damaged, replace the O-rings. Use a small, curved pick to remove the O-rings from the caps. Lubricate the new O-rings before installation. Always lubricate and twist the oil pipe into the caps. Pushing the pipe straight in may cause the O-rings to distort and roll out of the grooves.

4. To determine the bearing oil clearance between the camshaft and cap bearing surface, use a Plastigauge kit. To use the kit, the parts must be temporarily installed in the head, placing a small strand of the Plastigauge under each cap. After the parts have been torqued, they are disassembled and the width of the compressed Plastigauge is measured with a scale. **Figure 34** shows a typical view of the material being measured with the Plastigauge scale. The measured width corresponds to a specific clearance between the parts. Perform the clearance check as follows:

a. Stuff clean shop cloths in the cam chain tunnel to prevent parts from entering the engine.

b. Install the camshafts into the cylinder head. The compression release is on the exhaust camshaft. Check that the shoulder at the left end of each camshaft seats into the head. Position the camshafts so the cam lobes are horizontal and not contacting the valve lifters. It is not necessary to install the chain onto the camshafts.

c. Place a strip of Plastigauge onto each camshaft journal, parallel to the camshaft. Install the cap and dowels at that location. On the left journals, align the Plastigauge so it contacts a flat surface on the inside of the cap. Finger-tighten the cap bolts.

d. Seat the camshafts by first lightly tightening bolts 1 and 2 on the left caps (**Figure 25**).

e. For each camshaft, torque the cap bolts in the numerical sequence shown in **Figure 25**. Torque the cap bolts to 12 N•m (106 in.-lb.). Make several passes, increasing the torque evenly on the caps.

CAUTION
Failure to evenly torque the camshaft caps could result in damage to the cylinder head, camshafts or caps.

NOTE
Do not allow the camshaft to turn while the Plastigauge is fitted between the parts.

f. After achieving proper torque, loosen the cap bolts. Loosen the bolts equally in several

passes. Carefully remove the caps, dowels and Plastigauge strips. Keep all strips with their respective cap.

g. Using the scale included with the Plastigauge kit, measure the width of the Plastigauge to determine if clearance is within the specification. If oil clearance is not within specification, measure the camshaft journal to determine which part(s) are worn. Replace the camshaft, cylinder head, or if necessary, both parts.

CYLINDER HEAD

Removal

The cylinder head can be removed with the engine mounted in the frame. Refer to **Figure 6** as needed.

1. Drain the engine coolant (Chapter Three).

2. Remove the coolant hose, thermostat and housing (**Figure 35**) (Chapter Ten).

3. Remove the carburetor (Chapter Eight).

4. Remove the exhaust pipe as described in this chapter.

5. Remove the cylinder head cover, camshafts and cam chain tensioner as described in this chapter.

6. Remove the shims from the valve lifters (**Figure 21**) as follows:

   a. Rotate the valve lifter so the notch in the edge of the lifter is accessible.

   b. Wedge a small-tipped tool between the shim and lifter, then tilt the tool back to break the oil adhesion between the parts.

   c. Remove the shim and mark its location in the cylinder head. The shims must be identified, since they can vary in thickness. Make a drawing of the cylinder head so all parts from each location can be placed in the appropriate position on the guide.

7. Remove the banjo bolt and seal washers from the oil pipe (A, **Figure 36**).

8. Remove the nut from the rear of the cylinder head (B, **Figure 36**).

9. Remove the nut and bolt from the front of the cylinder head (**Figure 37**).

10. On 1987-1995 models, remove the two cylinder head bolts located in the cam chain tunnel. On later models, these are the outer bolts that secure the upper chain guide, which was removed in the camshaft removal procedure.

   a. Stuff the cam chain tunnel with clean shop cloths to prevent parts from entering the engine.

   b. Remove the bolts with a socket and extension. Check that the socket and extension fit together tightly and cannot separate when lowered into the tunnel.

11. Remove the four cylinder head bolts (10 mm), located in the cylinder head (**Figure 38** and **Figure 39**). Loosen the bolts in a crossing pattern and in several passes.

> *CAUTION*
> *Do not remove the four cylinder head bolts before removing the nuts and bolt in the previous steps. Doing so will overstress the smaller fasteners.*

12. Loosen the cylinder head by lightly tapping around its base with a soft mallet. Lift the head out the left side of the engine while routing the cam chain out of the head. Secure the chain so it does not fall into the engine.

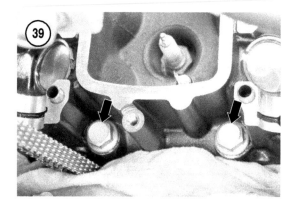

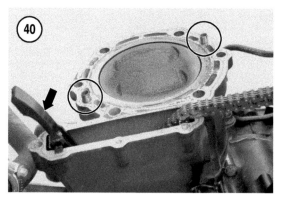

13. Stuff clean shop cloths into the cam chain tunnel, then remove the head gasket, two dowels, front chain guide and rubber damper that are fitted in the cylinder (**Figure 40**).

14. At the workbench, remove the spark plug, water temperature sending unit, carburetor intake duct and rear cam chain guide.

15. Remove the valve lifters (**Figure 41**). Note their location in the head and store them with the shims. If necessary, remove and inspect the valve assembly as described in this chapter.

16. Wash all parts in solvent and dry with compressed air. Note the following:

   a. Remove all gasket residue from the cylinder head and cylinder. Do not scratch or gouge the surfaces.

   b. Remove all carbon deposits from the combustion chamber. Use solvent and a soft brush or hardwood scraper. Do not use sharp-edged tools that could scratch the valves or combustion chamber. If the piston crown is cleaned, keep solvent and carbon deposits out of the gap between the piston and cylinder.

CAUTION
If the valves are removed from the head, the valve seats are exposed and can be damaged from careless cleaning. A scratched or gouged valve seat does not seal properly.

   c. If the cylinder head will be bead-blasted, wash the entire assembly in hot, soapy water after it has been reconditioned. This will remove blasting grit that is lodged in crevices and threads. Clean and chase all threads to ensure no grit remains. Blasting grit that remains in the head will be picked up by the

engine oil and circulated to other parts of the engine. This will damage the bearings, piston and rings.

   d. Check all oil passages for debris or blockage.

   e. Check all parts for obvious wear or damage.

17. Inspect the cylinder head as described in this section.

Inspection

Anytime the cylinder head is removed, the valves should be tested for leakage with a solvent test. This test is quick and easy to perform and helps identify problems in the valve train. Refer to *Valves* in this chapter for the solvent test.

1. Inspect the spark plug hole threads. If the threads are dirty or mildly damaged, use a spark plug thread tap to clean and straighten the threads. Keep the tap lubricated while cleaning the threads.

NOTE
If the threads are galled, stripped or cross-threaded, the cylinder head should be fitted with a steel thread in-

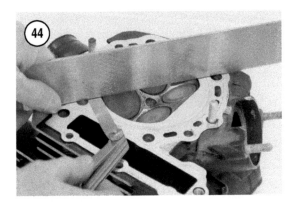

sert, such as a HeliCoil. Thread dam-age can be minimized by applying antiseize compound to the spark plug threads before installation. Do not overtighten the spark plug.

2. If necessary, clean the entire cylinder head after thread repair.

3. Inspect the inside of the cylinder head (**Figure 42**).

  a. Inspect for cracks or damage in the combustion chamber, water jackets and exhaust port.

  b. Inspect the studs for damage or looseness.

4. Inspect the outside of the cylinder head (**Figure 43**).

  a. Inspect for cracks or damage around the holes for the spark plug and water temperature sending unit.

  b. Inspect the camshaft bearing surfaces and mating caps for obvious scoring or damage. If either part is damaged, replace the caps and cylinder head as a set. The parts are only available as a single part number. The caps are machined with the cylinder head, so their dimensions and alignments are unique to that cylinder head.

  c. If cracks are found anywhere in the cylinder head, take the head to a dealership or machine shop to see if the head can be repaired. If not, replace the head and camshaft cap set.

5. Inspect the cylinder head for warp as follows:

  a. Lay a machinist's straightedge across the cylinder head as shown in **Figure 44**.

  b. Try to insert a flat feeler gauge between the straightedge and the machined surface of the head. If clearance exists, record the maximum measurement.

  c. Repeat substeps a and b several times, laying the straightedge both across and diagonally on the head.

  d. Compare the measurements to the warp service limit listed in **Table 2**. If the clearance is not within the service limit, take the cylinder head to a dealership or machine shop for further inspection and possible resurfacing.

6. Inspect the front and rear cam chain guides (**Figure 45**) for wear and damage.

7. Inspect the cylinder fasteners (**Figure 46**) for damaged threads and heads. Replace fasteners that are rusted.

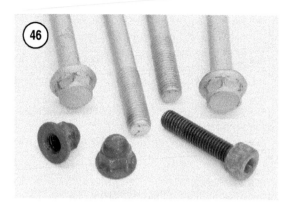

8. Assemble and install the cylinder head as described in this section.

Installation

Refer to **Figure 6**.

1. Note the following:
   a. Check that all gasket residue is removed from all mating surfaces. All cylinder head surfaces must be clean and dry.
   b. The valve lifters and shims can be installed after the cylinder head is installed.

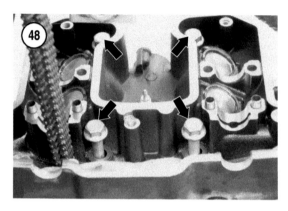

2. Install the rear cam chain guide into the cam chain tunnel.
3. Install the front cam chain guide and rubber damper. Seat the guide in the notch.
4. Install the dowels and a new cylinder head gasket onto the cylinder (**Figure 47**).
5. Lower the cylinder head onto the engine, routing the cam chain through the head.
   a. Keep adequate tension on the cam chain so it does bind at the crankshaft sprocket. Secure the cam chain when the cylinder head is seated.
   b. Be careful to not dislodge the dowels as the head is positioned.
   c. Cover engine openings as needed.
6. Install and torque the four cylinder head bolts (10 mm) (**Figure 48**) as follows:
   a. Apply molydisulfide grease to the bolt threads and to the seating surface on the heads.
   b. Refer to the torque sequence in **Figure 49**.
   c. Torque the bolts in two passes. In the first pass, torque the bolts to 20 N•m (15 ft.-lb.). In the second pass, torque the bolts to 65 N•m (48 ft.-lb.).

7. Install and finger-tighten the nut at the rear of the cylinder head (B, **Figure 36**) and the nut and bolt at the front of the cylinder head (**Figure 37**).
   a. Refer to the torque sequence in **Figure 49**. Note that the nuts are torqued twice before torquing the bolt.
   b. Torque the nuts to 25 N•m (18 ft.-lb.). A 12 mm crowfoot wrench may be preferred. If necessary, fit a torque adapter wrench onto the torque wrench to improve access to the nuts. Refer to Chapter One for adjusting torque readings when adapters are used.
   c. Torque the bolt (8 mm) to 18 N•m (13 ft.-lb.).

NOTE
The nuts and bolt in this step should be retorqued after the engine has been started and allowed to completely cool.

8. On 1987-1995 models, install the two cylinder head bolts (6 mm) located in the cam chain tunnel. On 1996-on models, these are the outer bolts that secure the upper chain guide, which are installed during the camshaft installation procedure.

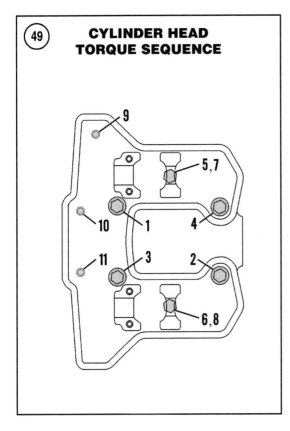

CYLINDER HEAD
TORQUE SEQUENCE

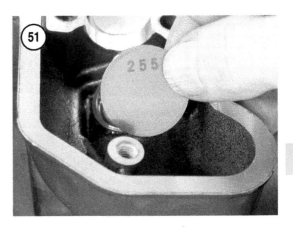

9. Install the oil pipe (A, **Figure 36**). Install new seal washers and torque the banjo bolt to 20 N•m (15 ft.-lb.).

    a. If the complete oil pipe assembly was removed from the crankcase, align and finger-tighten all banjo bolts and the retainer bolt before torquing.

    b. Torque the retainer bolt to 8 N•m (71 in.-lb.).

10. Install the valve lifter and shim sets at their appropriate locations as follows:

    a. Lubricate the valve lifters with engine oil, then insert them into their bores (**Figure 50**).

    b. Seat the shims into the valve lifters. If the shim size is visible (stamped on the shim) (**Figure 51**), place the number down, so it does not get worn away by the cam lobe.

NOTE
If the valve assembly was not reconditioned or disturbed when the cylinder head was removed, reinstall the original shim. If clearance was incorrect before the head was removed, refer to the valve adjustment procedure in Chapter Three to determine the correct size of shim to install. If the valves were reconditioned, install the original shim at this time. Recheck valve clearance after the camshafts are installed. It is possible that the camshafts will have to be removed a second time to correctly set the valves.

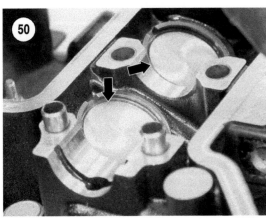

    a. Stuff the cam chain tunnel with clean shop cloths to prevent parts from entering the engine.

    b. Refer to the torque sequence in **Figure 49**.

    c. Torque the bolts with a socket and extension. Check that the socket and extension fit together tightly and cannot separate when lowered into the tunnel. Torque the bolts to 10 N•m (88 in.-lb.).

11. Install the cam chain tensioner, cam shafts and cylinder head cover as described in this chapter.

12. Install a new spark plug (Chapter Three).

13. Install the water temperature sending unit, thermostat and coolant hose (Chapter Ten).

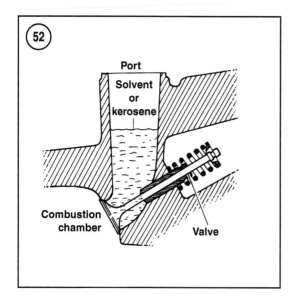

14. Install the engine coolant (Chapter Three).

15. Install the carburetor (Chapter Eight).

16. Install the exhaust pipe as described in this chapter.

VALVES

Solvent Test

Perform the solvent test with the valve assembly in the cylinder head. The test can reveal if valves are fully seating, as well as expose undetected cracks in the cylinder head.

1. Remove the cylinder head as described in this chapter.

2. Check that the combustion chamber is dry and the valves are seated.

3. Support the cylinder head so the port faces up (**Figure 52**).

4. Pour solvent or kerosene into the port.

5. Inspect the combustion chamber for leakage around the valve.

6. Repeat Steps 3-5 for the other valves.

7. If there is leakage, it can be caused by:

   a. A worn or damaged valve face.

   b. A worn or damaged valve seat (in the cylinder head).

   c. A bent valve stem.

   d. A crack in the combustion chamber.

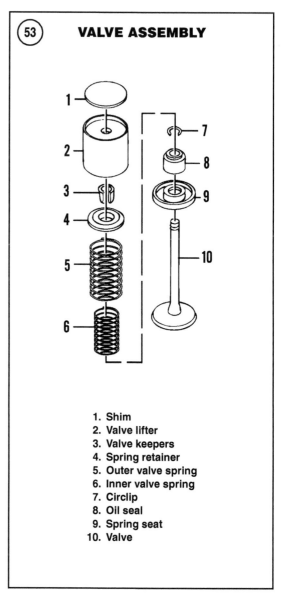

VALVE ASSEMBLY

1. Shim
2. Valve lifter
3. Valve keepers
4. Spring retainer
5. Outer valve spring
6. Inner valve spring
7. Circlip
8. Oil seal
9. Spring seat
10. Valve

Valve Removal

Refer to **Figure 53**.

1. Remove the cylinder head, shims, and valve lifters as described in this chapter.

2. Perform the solvent test on the intake and exhaust valves as described in this section.

3. Install a valve spring compressor (part No. 57001-241 or equivalent) and adapter (part No. 57001-243 or equivalent) over the valve assembly. Fit the stationary end of the tool squarely over the valve head. Fit the other end of the tool squarely on the spring retainer (**Figure 54**).

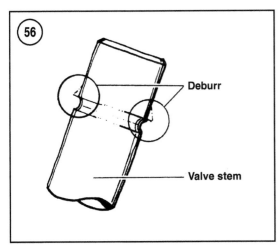

Deburr

Valve stem

4. Tighten the compressor until the spring retainer no longer holds the valve keepers in position. Lift the keepers from the valve stem. A magnetic tool works well (**Figure 55**).

CAUTION
Do not overtighten and compress the valve springs. This can result in loss of valve spring tension.

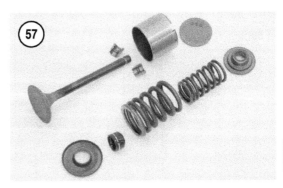

5. Slowly relieve the pressure on the valve springs and remove the compressor from the head.

6. Remove the spring retainer and valve springs.

7. Slide the circlip on the oil seal down, then remove the oil seal, circlip and spring seat.

8. Inspect the valve stem for sharp and flared metal (**Figure 56**) around the groove for the keepers. If necessary, deburr the valve stem before removing the valve from the head. Burrs on the valve stem can damage the valve guides.

9. Remove the valve from the cylinder head.

10. After removing each valve assembly, store the parts with the shim and valve lifter for that location (**Figure 57**).

CAUTION
Install valve components that are within specification in their original positions. Replace the oil seal and circlip.

11. Repeat Steps 3-10 for the remaining valves.

Valve Component Inspection

During the cleaning and inspection of the valve assemblies, do not allow the sets of parts to get intermixed. Work with one set of parts at a time, repeating the procedure until all parts are inspected. After inspecting each set of parts, return them to their storage container. Refer to **Table 2** for specifications.

NOTE
In the following procedure, whenever the valves, valve guides and valve seats must be replaced or reconditioned, it is recommended that the work be done by a dealership. These parts should be replaced or recondi-

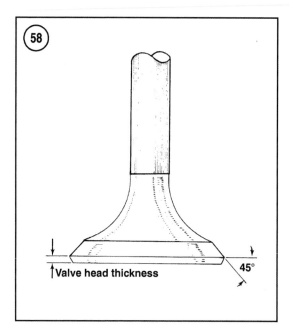

Valve head thickness   45°

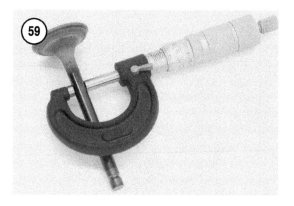

*tioned as a set. Replacing and servic-
ing of these parts require special
equipment, as well as experience in
replacing and fitting the parts.*

1. Clean the valve assembly in solvent.

> *CAUTION*
> *The valve seating surface is a critical
> surface and must not be damaged. Do
> not scrape on the seating surface or
> place the valve where it could roll off
> the work surface.*

2. Inspect the valve head as follows:
   a. Inspect the top and perimeter of each valve.
      Check for burning or other damage on the top
      and seating surface. Replace the valve if dam-
      age is evident. If the valve head appears uni-
      form, with only minor wear, the valve can be
      lapped (described in this section) and reused,
      if the other valve measurements are accept-
      able.
   b. Measure the thickness (margin) (**Figure 58**).
      Record the measurement.
3. Inspect the valve stem as follows:
   a. Inspect the stem for obvious wear and scor-
      ing. Also check the end of the valve stem for
      flare.
   b. Measure the valve stem diameter (**Figure
      59**). Record the measurement.

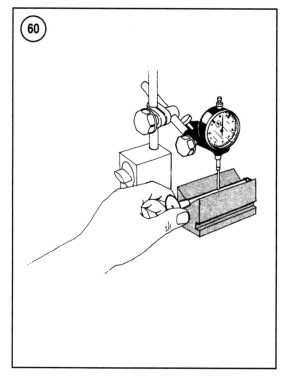

   c. Check the valve stem for runout. Place the
      valve in a V-block and measure runout with a
      dial indicator (**Figure 60**). Record the mea-
      surement.

4A. If a small hole gauge and micrometer are available, inspect the valve guides as follows:

   a. Clean the valve guides (**Figure 61**) so they
      are free of all carbon and varnish. Use solvent
      and a stiff, narrow brush.
   b. Measure each valve guide hole at the top, cen-
      ter and bottom. Use a small hole gauge and
      micrometer to make the measurements. Re-
      cord the measurements.

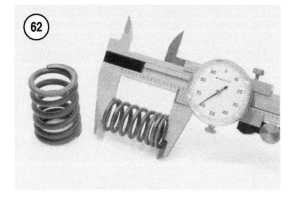

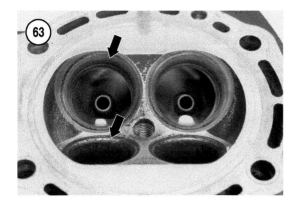

4B. If a small hole gauge and micrometer are not available, inspect the valve guides as follows:

a. Insert the appropriate valve into the guide.

b. With the valve head off the seat, move the valve stem side to side in the valve guide. Move the valve in several directions, checking for obvious perceptible play. If movement is easily detected, the valve guide and/or valve is worn. Take the valves and cylinder head to a dealership and have the parts accu-

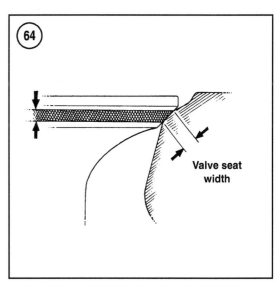

Valve seat width

rately measured to determine the extent of wear.

5. Check the inner and outer valve springs as follows:

a. Visually check the springs for damage.

b. Measure the length of each valve spring (**Figure 62**).

6. Inspect the valve spring seat, spring retainer, keepers and valve lifter for wear or damage.

7. Inspect the valve seats in the cylinder head (**Figure 63**) to determine if they must be reconditioned.

a. Clean and dry the valve seat and valve mating area with contact cleaner.

b. Coat the valve seat with machinist's marking fluid.

c. Install the appropriate valve into the guide, then *lightly* tap the valve against the seat so the fluid transfers to the valve contact area. Do not rotate the valve.

d. Remove the valve from the guide and measure the imprinted valve seat width (**Figure 64**) at several locations.

e. Clean all marking fluid from the valves and seats.

Valve Installation

Perform the following procedure for each set of valve components. All components should be clean and dry. Refer to **Figure 53** as needed.

1. Coat the valve stem and interior of the oil seal with molydisulfide grease.

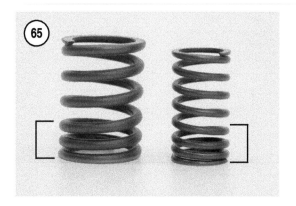

2. Install the spring seat into the head.

3. Install a new oil seal onto the valve guide, checking that the circlip seats onto the seal and valve guide.

4. Insert the appropriate valve into the cylinder head. Rotate the valve stem as it enters and passes through the seal. Check that the seal remains seated, then hold the valve in place.

5. Install the inner and outer valve springs with the *small* coil pitch, facing down (**Figure 65**).

6. Install the spring retainer.

7. Install a valve spring compressor over the valve assembly. Fit the tool squarely onto the spring retainer.

8. Tighten the compressor until the spring retainer is compressed enough to install the valve keepers.

> *CAUTION*
> *Do not overtighten and compress the valve springs. This can result in loss of valve spring tension.*

9. Insert the keepers around the groove in the valve stem (**Figure 66**).

10. Slowly relieve the pressure on the spring retainer, then remove the compressor from the head.

11. Tap the end of the valve stem with a soft mallet to ensure that the keepers are seated in the valve stem groove.

12. After all valves are installed, perform the solvent test as described in this chapter.

13. Install the cylinder head as described in this chapter.

Valve Lapping

Valve lapping restores accurate sealing between the valve seat and valve contact area, without ma-

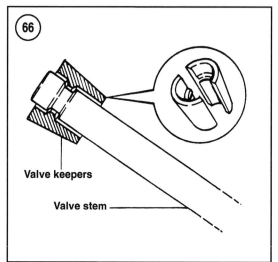

Valve keepers
Valve stem

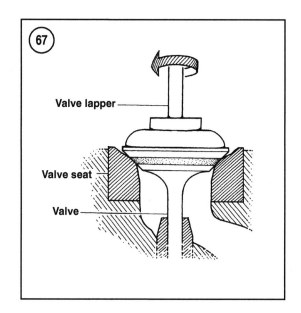

Valve lapper
Valve seat
Valve

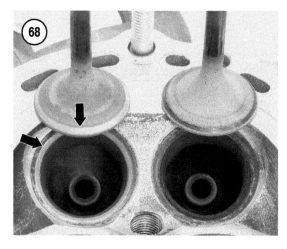

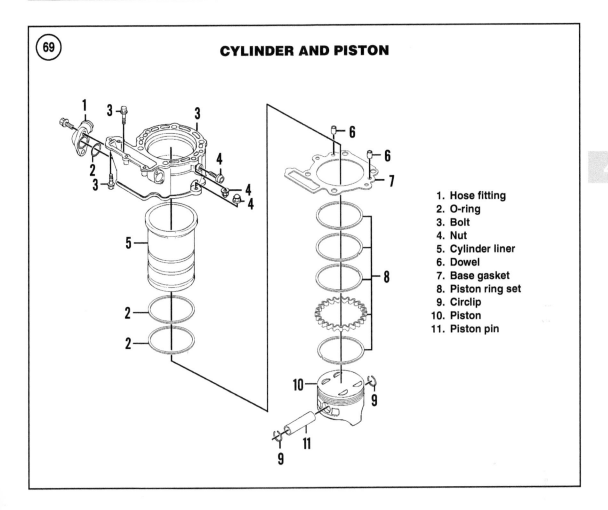

CYLINDER AND PISTON

1. Hose fitting
2. O-ring
3. Bolt
4. Nut
5. Cylinder liner
6. Dowel
7. Base gasket
8. Piston ring set
9. Circlip
10. Piston
11. Piston pin

chining. Lapping should be performed on valves and valve seats that have been inspected and are within specifications. Lapping should also be performed on valves and valve seats that have been reconditioned.

1. Lightly coat the valve face with fine-grade lapping compound.

2. Lubricate the valve stem, then insert the valve into the head.

3. Wet the suction cup on the lapping tool and press it onto the head of the valve (**Figure 67**).

4. Spin the tool back and forth between your hands to lap the valve to the seat. Every 5 to 10 seconds, rotate the valve 180° and continue to lap the valve into the seat.

5. Frequently inspect the valve seat. Stop lapping the valve when the valve seat is smooth, even and polished (**Figure 68**). Keep each lapped valve identified so it can be installed in the correct seat during assembly.

6. Clean the valves and cylinder head in solvent and remove all lapping compound. Any abrasive allowed to remain in the head will cause premature wear and damage to other engine parts.

7. After the valves are installed in the head, perform a *Solvent Test* as described in this section. If there is leakage, remove that valve and repeat the lapping process.

CYLINDER

Removal

The cylinder and piston can be removed with the engine mounted in the frame. Read all procedures completely before attempting a repair. Refer to **Figure 69**.

1. Remove the cylinder head as described in this chapter.

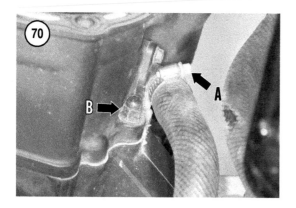

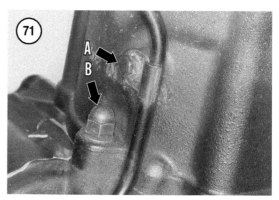

2. Remove the coolant hose at the front of the cylinder (A, **Figure 70**).

3. Remove the oil pipe retainer and bolt (A, **Figure 71**) at the rear of the cylinder.

4. If the engine crankcases will be split, remove the remaining banjo bolts and seal washers (**Figure 72**) securing the oil pipe.

5. Remove the cylinder mounting bolt, located in the cam chain tunnel (**Figure 73**). Stuff clean shop cloths into the tunnel to prevent parts from entering the engine.

6. Remove the cylinder mounting nuts at the front (B, **Figure 70**) and rear (B, **Figure 71**) of the cylinder.

7. Loosen the cylinder by tapping around the base. If necessary, apply penetrating oil to the joint.

8. *Slowly* lift the cylinder from the crankcase.

   a. Account for the two dowels under the cylinder (**Figure 74**). If loose, remove the dowels to prevent them from possibly falling into the engine.

   b. Route and secure the cam chain out of the cylinder.

9. Remove the base gasket.

10. Stuff clean shop cloths into the cam chain tunnel and around the piston. Support the piston and rod so it does not contact the crankcase or studs (**Figure 75**).

11. Inspect the cylinder as described in this chapter.

Inspection

1. Remove all gasket residue from the top and bottom cylinder block surfaces.

2. Wash the cylinder in solvent and dry with compressed air.

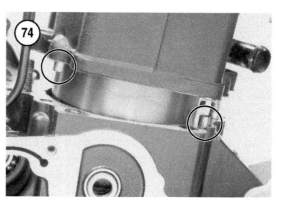

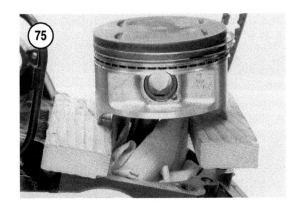

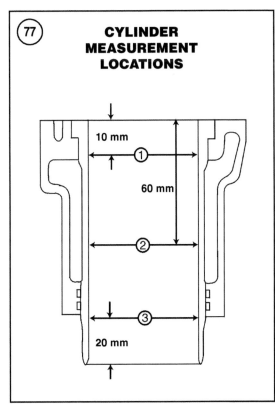

CYLINDER MEASUREMENT LOCATIONS

10 mm

60 mm

20 mm

3. Inspect the overall condition of the cylinder (**Figure 76**) for obvious wear or damage.

   a. Inspect the cylinder bore for scoring or gouges. If damaged, overbore the cylinder.

   b. Inspect the water jackets for deposits.

   c. Inspect all threads for condition and cleanliness.

4. Measure and check the cylinder for wear. Measure the inside diameter of the cylinder with a bore gauge or inside micrometer as follows:

   a. Measure the cylinder at three points along the bore axis (**Figure 77**). At each point, measure the cylinder front to back (measurement X) and side to side (measurement Y). Record the measurements for each location.

   b. Compare the largest X or Y measurement recorded to the specifications and service limit listed in **Table 2**. If the cylinder bore is not within the service limit, rebore and hone the cylinder. The rebored cylinder diameter should not vary more than 0.01 mm (0.0004 in.) at any point.

NOTE
Oversize pistons and ring sets are available in 0.5 mm (0.020 in.) and 1.0 mm (0.040 in.). The new service limit for the cylinder is 0.1 mm (0.004 in.) more than the actual diameter of the cylinder after boring. The new service limit for the piston is 0.15 mm (0.006 in.) less than the actual diameter of the piston. If wear exceeds the replacement piston sizes, replace the cylinder liner. Use a standard piston and ring set.

5. If cylinder boring is necessary, take the cylinder to a dealership or machine shop to have any machine work performed. If the cylinder is to be overbored and fitted with the next size piston and rings, take the new piston to the shop so the cylinder can be accurately bored and honed to accommodate the actual piston size.

6. If the cylinder is within all service limits, and a new piston and rings will be installed (see *Piston Inspection* in this chapter), deglaze the cylinder, preferably with a 240-grit hone. Cylinder glaze appears as a hard, shiny surface (**Figure 78**). After deglazing, the crosshatching in the cylinder provides a uniform surface, capable of retaining oil and mating with the rings (**Figure 79**).

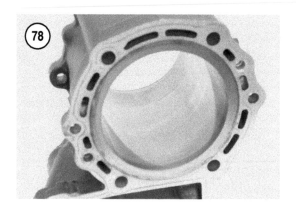

7. Thoroughly wash and scrub the cylinder in hot, soapy water after inspection and service, to remove all fine grit and material/residue left from machine operations. Check cleanliness by rubbing a clean, white cloth over the bore. No residue should be evident. When the cylinder is thoroughly clean and dry, immediately coat the cylinder bore with oil to prevent corrosion. Wrap the cylinder until engine reassembly.

> *CAUTION*
> *Wash the cylinder in hot, soapy water. Solvents do not remove the fine grit left in the cylinder. This grit causes premature wear of the rings and cylinder.*

8. Perform any service to the piston assembly before installing the cylinder.

Installation

Refer to **Figure 69**.
1. Check that all gasket residue is removed from all mating surfaces.
2. Install the dowels and a new base gasket onto the crankcase. **Figure 80** shows the gasket and dowels installed before the piston is installed. Whenever the piston is removed, it is easier to install the gasket before the piston, since the cam chain must also be handled and routed through the gasket. Gasket bending or other damage is less likely to occur.
3. Lubricate the following components with engine oil:
   a. Piston and rings.
   b. Piston pin and connecting rod.
   c. Cylinder bore.

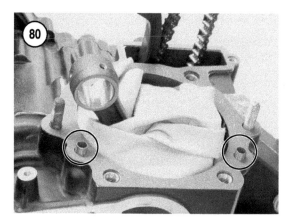

4. Support the piston so the cylinder can be lowered into place.
5. Stagger the piston ring gaps onto the piston as shown in **Figure 81**. Note that the top and second ring gaps are opposite one another, pointing to the front and back of the piston. The oil ring rails are about 30° apart, pointing to the sides.
6. Lower the cylinder onto the crankcase.
   a. Route the cam chain and guide through the chain tunnel. Secure the cam chain so it cannot fall into the engine.
   b. As the piston enters the cylinder, compress each ring so it can enter the cylinder. A ring compressor can also be used. When the bottom ring is in the cylinder, remove any holding fixture and shop cloths from the crankcase.
7. Install and torque the cylinder mounting nuts at the front (B, **Figure 70**) and rear (B, **Figure 71**) of the cylinder.
   a. If necessary, remove the coolant hose fitting in order to torque the front nut.
   b. Torque the nuts equally in several passes.

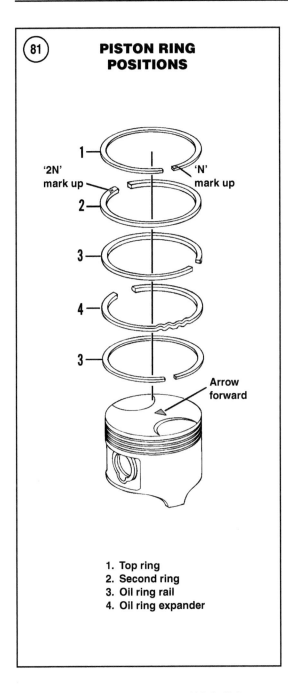

PISTON RING POSITIONS

'2N' mark up

'N' mark up

1
2
3
4
3

Arrow forward

1. Top ring
2. Second ring
3. Oil ring rail
4. Oil ring expander

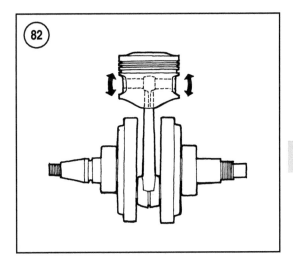

a. If the entire oil pipe was removed, install new seal washers on the banjo bolts, then finger-tighten the bolts (**Figure 72**).

b. Torque all the oil pipe banjo bolts and retainer bolt after installing the cylinder head.

10. Install the coolant hose at the front of the cylinder (A, **Figure 70**).

11. Install the cylinder head as described in this chapter.

PISTON AND PISTON RINGS

The piston is made of aluminum alloy and fitted with three rings. The piston is held on the small end of the connecting rod by a chrome-plated, steel piston pin. The pin is a precision fit in the piston and rod, and is held in place by circlips.

As each component of the piston assembly is cleaned and measured, record and identify all measurements. Refer to the measurements when checking dimensions and wear limits.

Piston Removal

1. Remove the cylinder as described in this chapter.

2. Before removing the piston, check the piston and piston pin for obvious play. Hold the rod and try to tilt the piston side to side (**Figure 82**). If a tilting (not sliding) motion is detected, there is wear on either the piston pin, pin bore or connecting rod. Wear could be on any combination of the three parts. Careful inspection is required to determine which parts should be replaced.

c. Torque the nuts to 25 N•m (18 ft.-lb.).

8. Install and torque the cylinder mounting bolt, located in the cam chain tunnel (**Figure 73**).

   a. Stuff clean shop cloths into the tunnel to prevent parts from entering the engine.

   b. Torque the bolt to 10 N•m (88 in.-lb.).

9. Install the oil pipe retainer and bolt (A, **Figure 71**) at the rear of the cylinder. Torque the bolt to 8 N•m (71 in.-lb.).

3. Stuff clean shop cloths around the connecting rod and in the cam chain tunnel to prevent parts from entering the crankcase.

4. Rotate the ends of the circlips to the removal gaps, then remove the circlips from the piston pin bore (**Figure 83**). Discard the circlips.

NOTE
New circlips must be installed during assembly.

5. Press the piston pin out of the piston by hand (**Figure 84**). If the pin is tight, a pin puller (part No. 57001-910) and adapter (part No. 57001-1211) are available from a Kawasaki dealership. A simple removal tool can be made as shown in **Figure 85**. The end of the padded pipe rests against the piston, not the piston pin. The hole in the pipe must be larger than the diameter of the piston pin. As the nut on the end of the rod is tightened, the nut and washer at the opposite end drive the piston pin into the pipe.

CAUTION
Do not attempt to drive the pin out with a hammer and drift. The piston and connecting rod assembly will likely be damaged.

6. Lift the piston off the connecting rod.

7. Inspect the piston and piston pin as described in this section.

Piston Inspection

1. Remove the piston rings as described in this section.

2. Clean the piston.
   a. Clean the carbon from the piston crown. Use a soft scraper, brushes and solvent. Do not use

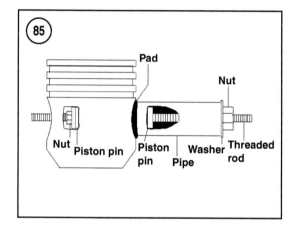

tools that can gouge or scratch the surface. This type of damage can cause hot spots on the piston during engine operation.
   b. Clean the piston pin bore, ring grooves and piston skirt. Clean the ring grooves with a soft brush, or use a broken piston ring to remove carbon and oil residue (**Figure 86**). Mild galling or discoloration can be polished off the piston skirt with fine emery cloth and oil.

3. Inspect the piston. Replace the piston if damage is evident.
   a. Inspect the piston crown (A, **Figure 87**) for wear or damage. If the piston is pitted, overheating is likely occurring. This can be caused by a lean fuel mixture and/or preignition.
   b. Inspect the ring grooves (B, **Figure 87**) for dents, nicks, cracks or other damage. The grooves should be square and uniform for the circumference of the piston. Particularly inspect the top compression ring groove. It is lubricated the least and is nearest the combustion temperatures. If the oil ring appears

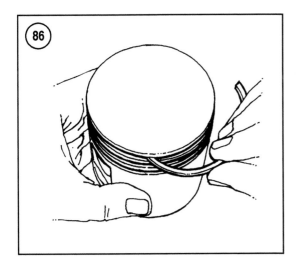

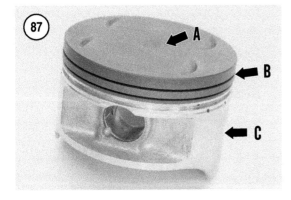

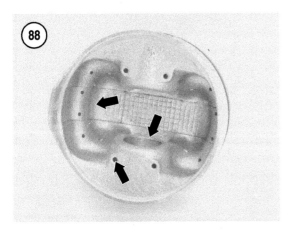

the oil ring was difficult to remove, the piston has likely overheated and distorted.

c. Inspect the piston skirt (C, **Figure 87**). If the skirt shows signs of severe galling or partial seizure (bits of metal embedded in the skirt), replace the piston.

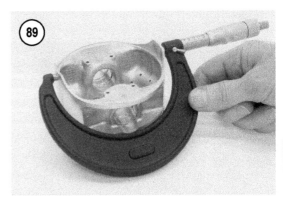

d. Inspect the interior of the piston (**Figure 88**). Check the crown, skirt, piston pin bores and bosses for cracks or other damage. Check the oil holes and circlip grooves for cleanliness and damage.

4. Measure the width of all ring grooves. Replace the piston if any measurement exceeds the service limit listed **Table 2**.

5. Inspect the piston ring-to-ring groove clearance as described in *Piston Ring Inspection and Removal* in this section.

Piston-to-Cylinder Clearance Check

Determine the clearance between the piston and cylinder to determine if the parts can be reused. If parts do not fall within specification, the cylinder should be bored oversize to match an oversize piston assembly. Clean and dry the piston and cylinder before measuring.

1. Measure the outside diameter of the piston. Measure 5 mm (0.2 in.) up from the bottom edge of the piston skirt and 90° to the direction of the piston pin (**Figure 89**). Record the measurement.

2. Determine clearance by subtracting the piston measurement from the largest cylinder measurement. If cylinder measurements are not yet known, the procedure is described under *Cylinder Inspection* in this chapter. If the clearance exceeds the specifications in **Table 2**, the cylinder must be overbored and fitted with an oversize piston assembly.

Piston Pin Inspection

1. Clean the piston pin.

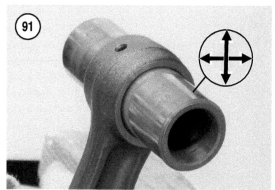

2. Inspect the pin for chrome flaking, wear or discoloration from overheating.

3. Inspect the bore in the small end of the connecting rod (**Figure 90**). Check for scoring, uneven wear, and discoloration from overheating.

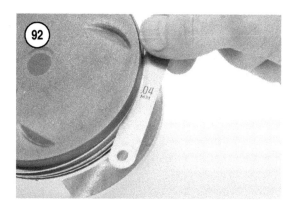

4. Lubricate the piston pin and slide it into the connecting rod. Slowly rotate the pin and check for radial play (**Figure 91**). If play is detectable, one or both of the parts are worn. Kawasaki does not provide specifications for the pin or connecting rod bore. Therefore, if neither part has obvious wear, have the pin diameter compared to the diameter of a new pin. If the pin is worn, replace the pin and recheck for play in the connecting rod. If play still exists, replace the rod.

5. Lubricate the piston pin and slide it into the piston bores. Slowly rotate the pin and check for radial play. If play is detectable, one or both of the parts are worn. Kawasaki does not provide specifications for the pin or piston pin bore. Therefore, if neither part has obvious wear, have the pin and piston pin bore dimensions compared to the dimensions of new parts. Replace worn parts.

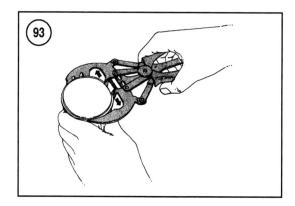

Piston Ring Inspection and Removal

The piston is fitted with two compression rings and an oil control ring assembly. The oil ring assembly consists of two side rails and an expander ring.

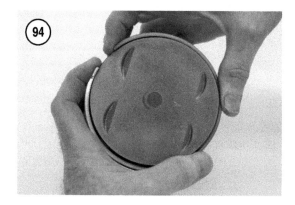

1. Check the piston ring–to–ring groove clearance as follows:

    a. Clean the rings and grooves so accurate measurements can be made with a flat feeler gauge.

    b. Press the top ring into the piston groove.

    c. Insert a flat feeler gauge between the ring and groove (**Figure 92**). Record the measure-

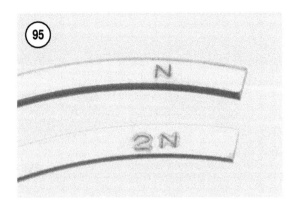

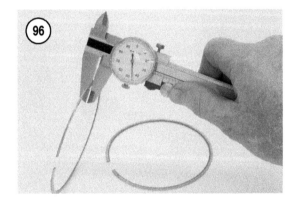

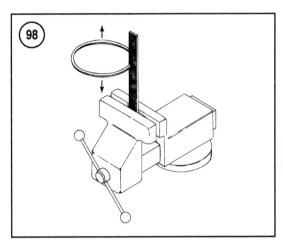

a. Spread the rings only enough to clear the piston.

b. As each ring is removed, check that the *top* mark is visible. The top ring should be marked with *N* and the second ring marked with *2N* (**Figure 95**). If necessary, mark the rings on their top surface if they will be reinstalled. The oil control ring rails are not marked.

3. Remove the oil ring assembly by first removing the top rail, followed by the bottom rail. Remove (by hand) the expander ring last.

4. Clean and inspect the piston as described under *Piston Inspection* in this section.

5. Measure the thickness of the top and second rings (**Figure 96**). Replace all the rings if any measurement exceeds the service limit listed in **Table 2**.

6. Inspect the end gap of the top and second rings as follows:

a. Insert a ring into the bottom of the cylinder. Use the piston to square the ring to the cylinder wall. Push the ring about 10 mm (0.4 in.) into the cylinder.

b. Measure the end gap with a feeler gauge (**Figure 97**). Replace all the rings if any measurement exceeds the service limit listed in **Table 2**. Always replace rings as a set. If new rings are to be installed, gap the new rings after the cylinder has been serviced. If the new ring gap is too narrow, carefully widen the gap using a fine-cut file as shown in **Figure 98**. Work slow and measure often.

7. Roll each ring around its piston groove and check for binding or snags (**Figure 99**). Repair minor damage with a fine-cut file.

ment. Repeat this step at other points around the piston. Replace the rings if any measurement exceeds the service limit in **Table 2**. If excessive clearance remains after new rings are installed, replace the piston.

d. Repeat substeps b and c for the second compression ring. The oil control ring is not measured.

2. Remove the top and second rings with a ring expander (**Figure 93**) or by hand (**Figure 94**).

Piston Ring Installation

If new piston rings will be installed, the cylinder must be honed. This is necessary to deglaze and crosshatch the cylinder surface. The newly honed surface is important in providing lubrication pockets for the new rings, helping them seat and seal against the cylinder. A dealership or machine shop can hone the cylinder for a minimal cost. Refer to *Cylinder Inspection* in this chapter to determine if the cylinder should be honed and reused. If the cylinder needs to be overbored, a larger piston and rings have to be installed.

1. Check that the piston and rings are clean and dry. When installing, spread the rings only enough to clear the piston.
2. Install the rings as follows:
   a. Install the oil ring expander in the bottom groove, followed by the bottom rail and top rail. The ends of the expander must *not* overlap. The rails can be installed in either position and direction.
   b. Install the second ring. Check that the *2N* mark faces up (**Figure 95**).
   c. Install the top ring. Check that the *N* mark faces up (**Figure 95**).
3. Check that all rings rotate freely in their grooves.

Piston Installation

1. Install the piston rings onto the piston as described in this section.
2. Check that all parts are clean and ready to be installed. Use new circlips when installing the piston.

CAUTION
Never install used circlips. Severe engine damage could occur. Circlips fatigue and distort when they are removed, even though they appear reusable.

3. Install a new circlip into the left piston pin boss. Rotate the ends of the circlip away from the gap. By installing the circlip in the left boss the remaining circlip can be installed at the right side of the engine, where work space is better.
4. Lubricate the following components with engine oil:
   a. Piston pin.
   b. Piston pin bores.

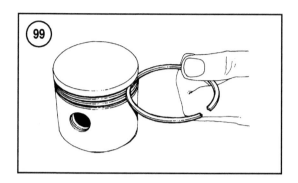

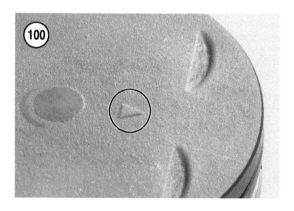

   c. Connecting rod bore.

5. Start the piston pin into the open pin bore, then place the piston over the connecting rod. The arrow stamped on the piston crown (**Figure 100**) must point forward.

CAUTION
The piston must be installed correctly. Failure to install the piston correctly can lead to severe engine damage.

6. Align the piston with the rod, then slide the pin through the rod and into the other piston bore.

7. Install a new circlip into the right piston pin boss. Rotate the ends of the circlip away from the gap.

8. Stagger the piston ring gaps on the piston as shown in **Figure 81**. Note that the top and second ring gaps are opposite one another, pointing to the front and back of the piston. The oil ring rails are about 30° apart, pointing to the sides.

9. Install the cylinder as described in this chapter.

10. Refer to Chapter Three for break-in procedures.

Table 1 GENERAL ENGINE SPECIFICATIONS

Engine type	Four-stroke, single cylinder,
Valve system	Chain-driven DOHC 4-valve
Cooling system	Liquid cooled
Lubrication system	Forced pressure, wet sump
Horsepower	47.3 at 6500 rpm (U.S. model)
Torque	40.5 ft.-lb. (U.S. model)
Engine displacement	651 cc (39.7 cu. in.)
Bore and stroke	100.0 × 83.0 mm (3.94 × 3.27 in.)
Compression ratio	9.5:1
Valve timing	
Intake valve opens	19° BTDC
Intake valve closes	69° ABDC
Intake valve duration	268°
Exhaust valve opens	57° BBDC
Exhaust valve closes	31° ATDC
Exhaust valve duration	268°

Table 2 ENGINE TOP END SPECIFICATIONS

	New mm (in.)	Service limit mm (in.)
Camshaft		
Chain length (20 links/21pins)	127.0-127.4	128.9
	(5.0-5.02)	(5.07)
Bearing oil clearance	0.030-0.064	0.15
	(0.0012-0.0025)	(0.0059)
Bearing inside diameter	23.000-23.013	23.07
	(0.9055-0.9060)	(0.9083)
Journal outside diameter	22.949-22.970	22.92
	(0.9035-0.9043)	(0.9024)
Lobe height		
Intake	36.75-36.85	36.65
	(1.4468-1.4507)	(1.4429)
Exhaust	36.25-36.35	36.15
	(1.4271-1.4311)	(1.4232)
Cylinder compression	530-855 kPa (77-124 psi)	–
Cylinder head warp limit	–	0.05
		(0.002)
Cylinder		
Inside diameter	100.000-100.012	100.10
	(3.9370-3.9374)	(3.9409)
Piston–to–cylinder clearance	0.043-0.070	–
	(0.0017-0.0028)	–
Piston		
Mark direction	Arrow facing forward	
Outside diameter		
measurement point	5.0 (0.2) from bottom	–
Outside diameter	99.942-99.957	99.80
	(3.9347-3.9353)	(3.9291)
Ring groove width		
Top and second ring	1.21-1.22	1.31
	(0.0476-0.0480)	(0.0516)
Oil ring	2.81-2.83	2.91
	(0.1106-0.1114)	(0.1146)
Piston rings		
Thickness (top and second)	1.17-1.19	1.10
	(0.0461-0.0469)	(0.0433)
(continued)		

Table 2 ENGINE TOP END SPECIFICATIONS (continued)

	New mm (in.)	Service limit mm (in.)
Ring–to–groove clearance		
Top and second ring	0.020-0.050 (0.0008-0.0020)	0.16 (0.006)
End gap		
Top and second ring	0.20-0.40 (0.008-0.016)	0.70 (0.03)
Top mark identification		
Top ring	N	
Second ring	2N	
Valve clearance		
Intake	0.10-0.20 (0.004-0.008)	–
Exhaust	0.15-0.25 (0.006-0.010)	–
Valve head thickness (margin above seat)		
Intake	1.0 (0.040)	0.5 (0.020)
Exhaust	1.0 (0.040)	0.7 (0.028)
Valve seat		
Angle	45°	
Width	0.8-1.2 (0.031-0.047)	
Intake outside diameter	36.9-37.1 (1.453-1.461)	–
Exhaust outside diameter	31.9-32.1 (1.256-1.264)	–
Valve stem		
Intake outside diameter	6.965-6.980 (0.2742-0.2748)	6.95 (0.2736)
Exhaust outside diameter	6.955-6.970 (0.2738-0.2744)	6.94 (0.2732)
Runout (total indicator reading)	0.0-0.01 (0.0-0.0004)	0.05 (0.0020)
Valve guide inside diameter		
Intake and exhaust	7.000-7.015 (0.2756-0.2762)	7.08 (0.2787)
Valve spring free length		
Inner intake and exhaust	37.6 (1.48)	36.2 (1.43)
Outer intake and exhaust	40.5 (1.59)	39.0 (1.54)

Table 3 ENGINE TOP END TORQUE SPECIFICATIONS

	N•m	in.-lb.	ft.-lb.
Camshaft cap bolts	12	106	–
Camshaft sprocket bolts	49	–	36
Cylinder head bolts*			
6 mm	10	88	–
8 mm	18	–	13
10 mm			
First stage	20	–	15
Second stage	65	–	48
Cylinder head cover bolts	8	71	–
Cylinder head nuts	25	–	18
Cylinder mounting bolt	10	88	–
	(continued)		

Table 3 ENGINE TOP END TORQUE SPECIFICATIONS (continued)

	N•m	in.-lb.	ft.-lb.
Cylinder mounting nuts	25	–	18
Oil pipe			
Banjo bolts	20	–	15
Retainer bolt	8	72	–
Spark plug	14	–	10
Upper chain guide			
Outer bolts	10	88	–
Inner bolt	8	71	–
Upper engine mounting bolts	25	–	18
Water temperature sending unit	15	–	11

*Refer to text for sequence

CHAPTER FIVE

ENGINE LOWER END

This chapter provides procedures for servicing or removing the following lower end components:

1. Cam chain and lower guide.
2. Left crankcase cover.
3. Engine balancer.
4. Crankcase, bearings and seals.
5. Crankshaft and connecting rod.

To access and service some of these components, the engine must be removed from the frame, then dismantled of all assemblies attached to the crankcase.

Read this chapter before repairing the engine lower end. Become familiar with the procedures, photos and illustrations to understand the skill and equipment required. Refer to Chapter One for tool usage and techniques.

SHOP CLEANLINESS

Prior to removing and disassembling the engine, clean the engine and frame with degreaser. The disassembly job goes easier and there is less chance of dirt entering the assemblies. Keep the work environment as clean as possible. Store parts and assemblies in well-marked plastic bags and containers. Keep reconditioned parts wrapped and lubricated until they are installed.

ENGINE

The following removal and installation procedure outlines the basic steps necessary to remove the engine from the frame. Depending on the planned level of disassembly, consider removing top end components, and those located in the crankcase covers, while the engine remains in the frame. Since the frame keeps the engine stabilized, tight nuts and bolts are easier to remove if the engine is held steady. Also, if the actual engine problem is unknown, it may be discovered in another assembly, other than the crankcase.

During engine removal, make note of mounting bolt directions and how cables and wire harnesses are routed. Refer to the appropriate chapters for removal, inspection and installation procedures for the components in the engine top end and crankcase covers.

Removal and Installation

1. Support the motorcycle so it is stable, level and the rear wheel is off the ground.
2. If possible, perform a compression test (Chapter Three) and leakdown test (Chapter Two) before dismantling the engine.

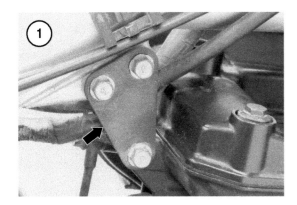

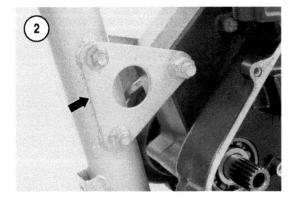

11. Disconnect the clutch cable at the engine (Chapter Six).

12. Remove the exhaust pipe (Chapter Four).

13. Disconnect the starter motor leads (Chapter Nine).

14. Remove the carburetor (Chapter Eight).

15. Remove the spark plug lead and water temperature sending unit lead.

16. Remove the coil.

17. Remove the sprocket guard and drive chain (Chapter Eleven).

18. Disconnect the stator, pickup coil and neutral switch wiring harness connectors (Chapter Nine).

19. Remove the crankcase breather hose.

20. Inspect the engine and verify that it is ready for removal. If desired, remove any additional components that make engine removal and handling easier.

21. Remove the engine mounting bolts and brackets as follows:

> *WARNING*
> *When removing the bolts, be aware that the engine may shift in the frame. Keep hands protected and check the stability of the motorcycle and engine after removing each set of bolts. Get assistance when removing the engine from the frame.*

   a. Remove the bolts and upper mounting brackets (**Figure 1**).

   b. Remove the bolts and lower mounting brackets (**Figure 2**).

   c. Remove the rear mounting bolt (A, **Figure 3**).

   d. Remove the swing arm pivot nut (B, **Figure 3**). Do not remove the pivot bolt at this time.

> *WARNING*
> *In the following step, do not completely remove the swing arm pivot bolt from the frame. Pull the bolt out only far enough to remove the engine (**Figure 4**). Complete removal allows the swing arm to fall.*

   e. Pull the swing arm pivot bolt out so it clears the engine, then remove the engine from the frame.

   f. Push the swing arm pivot bolt back into the frame.

3. Drain the engine oil (Chapter Three).

4. Remove the skid plate, side covers, seat, fuel tank and radiator covers (Chapter Fifteen).

5. Disconnect the battery leads (Chapter Three).

6. Drain the engine coolant (Chapter Three).

7. Disconnect the coolant hoses at the engine.

8. Remove the footpegs.

9. Remove the shift lever.

10. Remove the rear brake pedal (Chapter Fourteen).

g. Clean and inspect the frame in the engine bay. Check for cracks and damage, particularly at welded joints.

22. Refer to the procedures in this chapter for servicing the crankcase assembly.

23. Reverse this procedure to install the engine. Note the following:

a. If the chain is endless (no master link) check that it is routed over the swing arm pivot before installing the swing arm pivot bolt.

b. Install the engine mounting brackets, bolts and nuts. Finger-tighten all nuts before torquing. Note the original direction of the bolts during installation. Torque all 8 mm mounting bolts, followed by the 10 mm mounting bolts. Lastly, torque the swing arm pivot bolt. Refer to **Table 2** for the torque specifications.

> *NOTE*
> *If the engine top end is not installed when the engine is mounted in the frame, do not tighten any mounting nuts until the upper mounting brackets have been installed and aligned with their mounting points.*

c. Carefully route electrical wires so they are not pinched or in contact with surfaces that get hot.

d. Apply dielectric grease to electrical connections before reconnecting.

e. If assemblies have been removed from the top end or crankcase covers, install those components. Refer to the appropriate chapters for inspection and installation procedures.

f. Fill the engine with engine oil (Chapter Three).

g. Fill the cooling system with coolant (Chapter Three).

h. Adjust the clutch free play (Chapter Three).

i. Check throttle cable adjustment (Chapter Three).

j. Check rear brake pedal height. (Chapter Three).

k. Check chain adjustment (Chapter Three).

l. Start the engine and check for leaks.

m. Check throttle and clutch operation.

n. If the engine top end has been rebuilt, perform a compression check. Record the result and compare it to future checks.

o. Read the *Engine Break-In* procedure in Chapter Three.

CAM CHAIN AND LOWER GUIDE

If the engine balancer assembly will be removed, the cam chain and lower guide can be removed during that procedure. Use the following procedure when only the cam chain and lower guide need to be removed.

Removal, Inspection and Installation

1. Remove the cam shafts, then remove the cam chain from the camshaft sprockets (Chapter Four).

2. Remove the alternator cover and rotor (Chapter Nine).

3. Remove the cam chain guide and the balancer chain guide (**Figure 5**). If necessary, rotate the crankshaft to access the bolt behind the front left balancer weight.

4. Remove the cam chain from the sprocket.

5. Inspect the cam chain and guide (**Figure 6**).

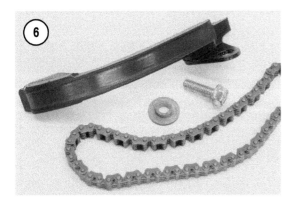

a. Pull the chain tight and measure a 20-link span (21 pins). Refer to **Table 1** for specifications. Repeat the check on several lengths of the chain.

b. Inspect the chain guide for wear on its face.

6. Reverse this procedure to install the cam chain and guides. Note the following:

a. Check that the shouldered spacer seats in the cam chain guide. The guide should pivot on the spacer.

b. Apply threadlocking compound to the bolt threads.

c. Torque the bolts to the specifications listed in **Table 2**.

LEFT CRANKCASE COVER

Removal and Installation

The left crankcase cover is located between the alternator cover and left crankcase.

1. Remove the alternator cover, starter drive gears and rotor (Chapter Nine). If the crankcase halves

will be disassembled, also remove the starter (Chapter Nine).

2. Remove the tensioner bolt and shaft lever from the balancer tensioner shaft (**Figure 7**).

3. Remove the nine bolts from the perimeter of the crankcase cover (**Figure 8**).

4. Remove the gasket and 2 dowels behind the cover.

5. Inspect the parts for damage. To inspect the tensioner bolt and shaft lever, refer to the inspection procedures in *Engine Balancer* in this chapter.

6. Reverse this procedure to install the cover, tensioner bolt and and shaft lever. Note the following:

a. Install the lever so the weld joint is facing in.

b. Tighten the cover bolts in several passes and in a crossing pattern.

ENGINE BALANCER

The engine uses a rotating balancer assembly to dampen vibration that is inherent to a single-cylinder engine. The three balancer weights are shaft-mounted and synchronized with the crankshaft. The balancer shaft at the front of the engine has a weight at each end of the shaft, while the rear balancer shaft has a single weight, located between the crankcase halves. Both balancer shafts are chain-driven by a sprocket on the left end of the crankshaft.

Early and Late Model Differences

In 1996, from engine serial number KL650AE032206-on, the left and rear balancer

ENGINE BALANCER

Early: 1987-1996 (to KL650AE032205)
Late: 1996-on (from KL650AE032206)

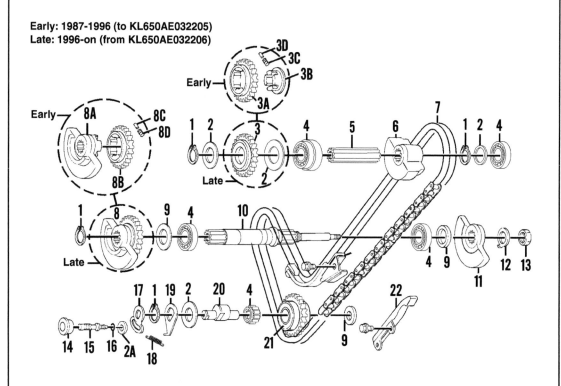

1. Snap ring	8C. Roller
2. Washer	8D. Spring
2A. Washer (all 1996-on models)	9. Spacer
3. Rear balancer sprocket	10. Front balancer shaft
3A. Rear balancer sprocket	11. Right balancer weight
3B. Coupling	12. Lockwasher
3C. Roller	13. Nut
3D. Spring	14. Plug
4. Bearing	15. Tensioner bolt
5. Rear balancer shaft	16. O-ring
6. Rear balancer weight	17. Shaft lever
7. Balancer chain	18. Tensioner spring
8. Left balancer weight/sprocket	19. Spring lever
	20. Tensioner shaft
8A. Left balancer weight	21. Tensioner sprocket
8B. Left balancer sprocket	22. Chain guide

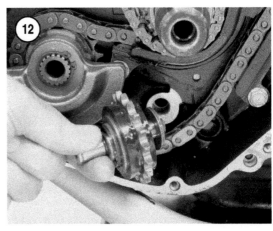

sprockets (**Figure 9**) underwent design changes. The left sprocket became rubber-mounted in its balancer weight, making the components a single part number. The earlier design used springs and rollers between the sprocket and weight. Each of these components is available as a separate part number.

These same changes occurred at the rear sprocket, eliminating the springs and rollers between the sprocket and shaft coupling. Additionally, the later design places a washer between the crankcase bearing and rear sprocket.

On *all* 1996-on models, a washer is used on the tensioner bolt. Although some components do not appear exactly as those shown in the following procedures, the component changes do not affect the function or servicing of the balancer assembly.

Removal

The following procedure details the complete removal of the balancer components from the engine crankcase. Refer to **Figure 9**.

1. Remove the right crankcase cover (Chapter Six).
2. Remove the alternator cover (Chapter Nine).

3. Remove the right balancer weight as follows:
   a. Flatten the lockwasher so the nut can be removed from the balancer shaft.
   b. Hold the rotor with the rotor holder tool. The tool can be braced against the bottom of the left footpeg. Remove the nut and lockwasher securing the right balancer weight (**Figure 10**). Remove the weight and spacer from the shaft.
4. Remove the left crankcase cover, tensioner bolt and shaft lever as described in this chapter.
5. Remove the balancer chain tensioner assembly as follows:
   a. Remove the tensioner spring (**Figure 11**). If the spring must be used again, do not score, scratch or bend the spring ends. Fracturing can occur.
   b. Rotate the spring lever counterclockwise to relieve chain tension, then carefully pull the tensioner shaft assembly from the crankcase (**Figure 12**). Keep the opposite end of the shaft tilted up, to prevent dropping the spacer, located behind the sprocket.

CAUTION
If the spacer falls, it can roll into the crankcase opening located below the tensioner shaft bore. If this occurs, use a magnetic tool or wire to retrieve the part.

6. Remove the snap ring from the left end of the front balancer shaft (**Figure 13**), then remove the weight/sprocket and spacer.
7. Remove the five bolts (**Figure 14**) securing the lower cam chain guide and balancer chain guides.

8. Remove the cam chain (A, **Figure 15**).

9. Remove the snap ring, washer(s), rear balancer sprocket and balancer chain (B, **Figure 15**).

10. Refer to *Crankcase* in this chapter to remove the balancer shafts (**Figure 16**), located between the crankcase halves.

11. Inspect the complete balancer assembly as described in this section.

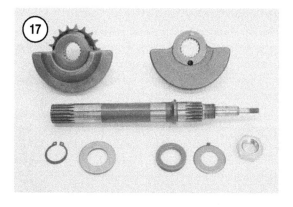

Inspection

During inspection, replace any parts that are obviously damaged or worn. Refer to **Figure 9**.

1. Inspect the front balancer shaft assembly (**Figure 17**).

   a. Inspect the spacers and locknut.

   b. Inspect the splines, threads, snap ring groove and polished surfaces on the shaft and right weight (**Figure 18**).

   c. Inspect the splines, sprocket teeth and rubber mounting on the left sprocket (**Figure 19**). On early models, inspect the condition of the springs and rollers in the sprocket. Replace parts that are fatigued, worn or missing.

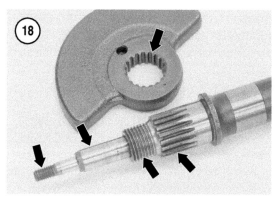

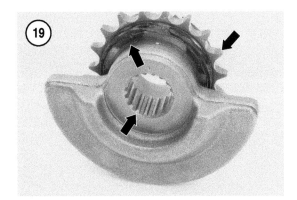

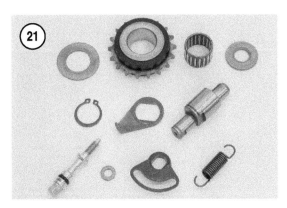

d. Install a new snap ring and lockwasher during assembly.

2. Inspect the rear balancer shaft assembly (**Figure 20**).

    a. Inspect the spacers and locknut.

    b. Inspect the splines, snap ring grooves and polished surface on the shaft.

    c. Inspect the splines in the weight.

    d. Inspect the splines, sprocket teeth and rubber mounting on the sprocket. On early models, inspect the condition of the springs and rollers in the sprocket. Replace parts that are fatigued, worn or missing.

    e. Install a new snap ring at the left end of the shaft during assembly.

    f. Install a new snap ring on the right end of the shaft at this time. Install the snap ring so the sharp edge faces in (toward balancer weight).

3. Inspect the tensioner assembly (**Figure 21**).

    a. Inspect the spacer and washer(s).

    b. Install a new O-ring on the tensioner bolt.

    c. Inspect the fit of the bearing in the sprocket and on the tensioner shaft (**Figure 22**). The shaft and sprocket bore should be smooth, and the bearing should fit firmly between the parts. The sprocket teeth and rubber mounting must be in good condition.

    d. Assemble the tensioner shaft, bearing, sprocket and spring lever (**Figure 23**). The lever should fit flat and firmly to the shaft. The shaft should rotate freely and smoothly when the lever is operated. The sprocket should spin freely and smoothly on the shaft.

4. Inspect the spring lever, spring and shaft lever (**Figure 24**) for the following specific damage. If other damage is evident, replace the damaged part(s).

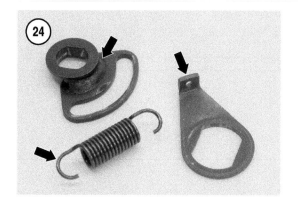

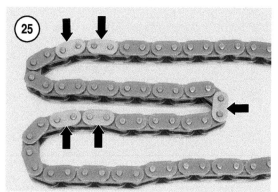

a. Inspect the weld on the shaft lever for breakage.

b. Inspect the spring hole in the spring lever. The hole should be smooth with no sharp edges that could damage the spring ends.

c. Inspect the spring ends and coils for damage. Any scratches, scoring or abnormal bends of the spring wire could weaken the spring, causing it to break. Preferably, install a new spring whenever the tensioner assembly is removed.

5. Inspect the balancer chain.

a. Pull the chain tight and measure a 20-link span (21 pins). Refer to **Table 1** for specifications. Repeat the check on several lengths of the chain.

b. Note the positions of the silver chain links (**Figure 25**). These links are used to position and synchronize the balancer shaft assemblies with the crankshaft. The silver links must face out when the chain is installed.

6. Inspect the chain guides (**Figure 26**) for wear.

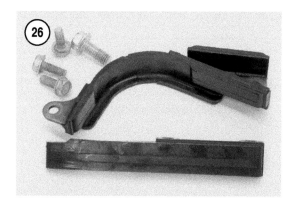

Installation and Synchronization

During assembly the sprockets and weights must be properly aligned on their respective shafts (**Figure 27**). In addition, the crankshaft and sprocket/shaft assemblies must be aligned and engaged with the proper silver links of the balancer chain (**Figure 28**).

Follow the procedure closely and check all alignments often as the balancer is assembled. If necessary, refer to the *Early and Late Model Differences* section and understand the differences in the components. In this procedure, early models are 1987-1996, to engine serial number

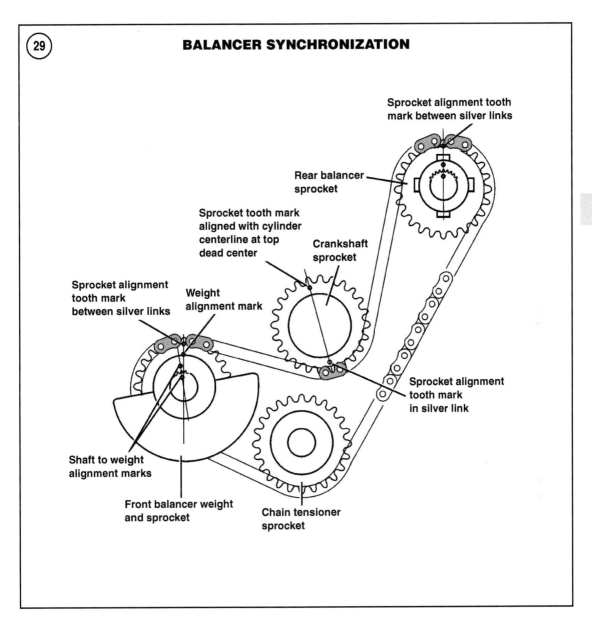

BALANCER SYNCHRONIZATION 29

Sprocket alignment tooth mark between silver links

Rear balancer sprocket

Sprocket tooth mark aligned with cylinder centerline at top dead center

Crankshaft sprocket

Sprocket alignment tooth mark between silver links

Weight alignment mark

Sprocket alignment tooth mark in silver link

Shaft to weight alignment marks

Front balancer weight and sprocket

Chain tensioner sprocket

KL650AE032205. Late models are 1996-on, from engine serial number KL650AE032206-on. During assembly, lubricate all components with engine oil. Refer to **Figure 9** as needed.

1. Assemble the crankcase halves as described in *Crankcase* of this chapter.

2. Install the following component(s):

    a. Install the spacer on the left end of the front shaft.

    b. On late models, install the washer on the rear shaft.

    c. On early models, check that the marks on the left and rear sprockets are aligned with the marks on the mating balancer weight and coupling.

3. Position the crankshaft at top dead center. The alignment mark(s) on the crankshaft sprocket should be parallel to the cylinder centerline.

4. Orient the balancer chain and identify the silver links (**Figure 25**). There are two pairs of silver links, with a single silver link placed between the pairs. The silver links are used to position and synchronize the balancer shaft assemblies with the

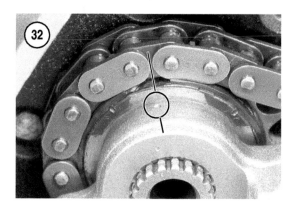

crankshaft. The silver links must face out when the chain is installed.

5. Position and align the balancer chain on the sprockets as follows. Refer to **Figure 29** to verify proper installation.

    a. Identify the pair of silver links to the *right* of the single silver link. These links will be engaged with the rear sprocket.

    b. Identify the alignment tooth mark on the rear sprocket.

    c. Align and engage the rear sprocket alignment tooth mark with the chain, then install the sprocket onto the shaft. The alignment mark must be *between* the silver links. If installed correctly, the shaft to sprocket alignment marks will also be aligned between the silver links (**Figure 30**).

    d. Route the chain under the crankshaft, engaging the single silver link with the alignment tooth mark on the crankshaft sprocket (**Figure 31**).

> *NOTE*
> *During chain routing and alignment with the crankshaft and front balancer shaft, the shafts may rotate slightly. As long as the chain remains seated in the aligned sprocket(s), overall alignment is not affected.*

    e. Align and engage the front sprocket alignment tooth mark with the chain. The mark must be *between* the silver links. If installed correctly, the weight alignment mark on *top* of the weight will also be aligned between the silver links (**Figure 32**). *Do not* refer to the marks aligning the shaft and weight.

6. If necessary, slowly rotate the crankshaft and balancer shafts back into the top dead center position.

7. Check that all sprocket alignments are correct. If necessary, reposition any sprockets that are not aligned at top dead center.

8. Install the washer and a new snap ring onto the rear balancer shaft. Install the snap ring with the sharp edge facing out.

9. Install a new snap ring onto the front balancer shaft. Install the snap ring with the sharp edge facing in.

10. Engage the cam chain with the crankshaft sprocket, then route the chain toward the top of the engine. Use wire to secure the chain and to minimize slack.

11. Install the lower cam chain guide and balancer chain guides. Note the following:

    a. Check that the shouldered spacer seats in the cam chain guide. The guide should pivot on the spacer.

    b. If the front balancer weight must be rotated to install the chain guide and bolt at the crankshaft, apply tension to the lower part of the

12. Install the balancer chain tensioner assembly (**Figure 33**). Note the following:

    a. Install the sprocket with the wide shoulder facing out.

    b. Install the spring lever with the tensioner spring hole facing out.

    c. If possible, install the spring by hand. If pliers must be used, prevent excessive stretching, scoring or damage to the ends of the spring.

    d. Recheck balancer alignments after installing the tensioner.

13. Install the right balancer weight as follows:

    a. Temporarily install the rotor and finger-tighten the rotor nut.

    b. Install the spacer and weight onto the shaft. Check that the marks on the weight and shaft are aligned (**Figure 34**).

    c. Install a new lockwasher, engaging the tab on the washer with the weight.

    d. Hold the rotor with the rotor holder tool. The holder tool can be braced against the top of the left footpeg. Install and torque the locknut to 44 N•m (32 ft.-lb.).

    e. Lock one side of the washer against the nut (**Figure 35**).

    f. Remove the rotor and continue assembly.

14. Install the left crankcase cover, tensioner bolt and shaft lever as described in this chapter.

15. Install the rotor, starter drive gears and alternator cover (Chapter Nine). If the crankcase halves were disassembled, also install the starter (Chapter Nine).

16. Install the right crankcase cover (Chapter Six).

CRANKCASE

The following procedures detail the disassembly and reassembly of the crankcase. When the two halves of the crankcase are disassembled–or split–the crankshaft, balancer shafts and transmission assemblies can be removed for inspection and repair. Before proceeding with this procedure, remove the engine from the frame as described in this chapter. All assemblies located in the crankcase covers must be removed. It may be easier to remove these assemblies with the engine in the frame. The engine will remain steady during the disassembly process.

The crankcase halves are made of cast aluminum alloy. Do not hammer or pry on the cases. The cases

balancer chain. If the chain does not remain seated on the sprockets, balancer alignments may be lost. Recheck balancer alignments after installing the guides.

    c. Apply threadlocking compound to the bolt threads.

    d. Torque the bolts to the values listed in **Table 2**.

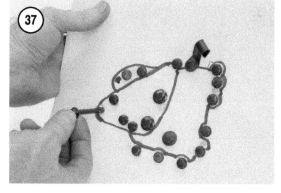

will fracture or break. The cases are aligned at the joint by dowels, and joined with liquid sealant.

The crankshaft is made of two full-circle fly-wheels, press-fitted to the crankpin. The assembly is supported at the left end by a roller bearing, and at the right end by a ball bearing. The connecting rod has a needle bearing fitted at the large end.

Disassembly

Any reference to the *left* or *right* side of the engine refers to the side of the engine as it is mounted in the frame–not on the workbench. As components are removed, keep the parts organized and clean. Leave the right case facing down until all assemblies have been removed from the case.

1. Place the engine on wooden blocks with the left side facing up (**Figure 36**).

2. Loosen the 13 bolts at the perimeter of the crankcase and the six bolts within the left crankcase cover area. Loosen each bolt 1/4 turn, working in a crossing pattern. Loosen the bolts until they can be removed by hand.

NOTE
*Since the bolt lengths vary, make a drawing of the crankcase shape on a piece of cardboard. Punch holes in the cardboard at the bolt locations. After removing each bolt from the crankcase, put the bolt in its respective hole in the template (**Figure 37**). If desired, mark each bolt hole with a reference number, indicating which of the three bolt sizes is used at that location. After the template is marked, the bolts can be removed and cleaned. Af-*

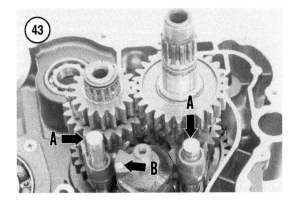

ter cleaning, return the bolts to the template.

3. Separate the crankcase halves as follows:

  a. With the left side of the engine facing up, lightly pry the cases at the notch located at the back of the engine. The notch is between the bores for the swing arm bolt and rear engine mounting bolt. Pry the case slowly and only enough to break the seal between the cases. If necessary, use a heat gun to soften the joint sealant, making separation easier.

CAUTION
Pry the cases only at the notch. Do not pry the cases on the mating surfaces. Damage to the cases will likely occur.

  b. When the case seal breaks, raise and lower the left case until it is fully released. Evenly lift the left case from the right case (**Figure 38**).

  c. Account for the two dowels between the cases when the cases are separated (**Figure 39**).

4. Remove the front balancer shaft (**Figure 40**).

5. Remove the rear balancer shaft and weight (**Figure 41**). Account for the shaft washer between the bearing and weight. Inspect the complete balancer assembly as described in this chapter.

6. Remove the crankshaft assembly (**Figure 42**) from the case. The crankshaft should free itself from the right bearing without the use of tools. Do not lift the crankshaft by the connecting rod. Inspect the crankshaft assembly and cases as described in this chapter.

CAUTION
Take extreme care when removing, handling and storing the crankshaft. Do not expose the assembly to dirt or place it in an area where it could roll and fall to the floor. If the crankshaft is dropped, the crankpin will likely be knocked out of alignment. The connecting rod may also be damaged.

7. Remove the transmission assembly from the case as follows:

  a. Remove the two shift fork shafts (A, **Figure 43**).

  b. Remove the shift drum (**Figure 44**), disengaging it from the shift forks. If necessary, raise the transmission shafts slightly to disengage the forks from the shift drum.

c. Remove the No. 1, No. 2 and No. 3 shift forks (**Figure 45**).

d. Remove the input shaft and output shaft assemblies (**Figure 46**) as a unit, keeping the gears meshed and the shafts upright.

e. Disassemble and inspect the transmission (Chapter Seven).

> *CAUTION*
> *Take extreme care when removing, handling and storing the transmission. Wrap and store the assembly until it is inspected. Do not expose the assembly to dirt or place it in an area where it could roll and fall to the floor.*

Assembly

1. Read the entire procedure to ensure that all tools, parts and supplies are on hand, as well as proper preparation of the case halves. Follow these practices when assembling the crankcase:

a. Check that all mating surfaces are smooth, clean and dry. Minor irregularities can be repaired with an oil stone. After thorough removal of the old sealant, clean all mating surfaces with a highly evaporative solvent, such as brake cleaner or electrical contact cleaner. The new sealant will not adhere to oily surfaces.

b. Lubricate the crankshaft, bearings and transmission assembly with engine oil.

c. Lubricate seal lips with grease.

d. To install the crankshaft, use the crankshaft installation jig (part No. 57001-1174). The tool can be ordered from a Kawasaki dealership. The tool (**Figure 47**) is adjusted to fit snugly between the flywheels, opposite the crankpin. As the crankcase halves are seated together, the tool prevents the flywheels from flexing or binding, causing misalignment of the flywheels.

e. A *liquid gasket* sealant is required to seal the crankcase halves. Use Kawabond 5 or Yamabond 4, available at dealerships. These sealants are durable and can be submerged in oil. Depending on experience and preference, Kawabond 5 is thinner than Yamabond 4, and may be considered more difficult to evenly apply.

> *CAUTION*
> *Do not attempt to seal the cases with RTV silicone sealant, commonly found at auto parts stores. The sealant is inadequate for this application. Oil leakage and possible engine damage is likely to occur.*

2. Place the right crankcase on wooden blocks, with the open side of the case facing up.

3. Install the transmission as follows:

a. Mesh the input shaft and output shaft assemblies. Keeping the gears meshed and the shafts upright, insert the assembly into the case (**Figure 46**).

b. Identify the shift forks. (**Figure 45**). Shift forks No. 1 and No. 2 can be identified by the casting around the guide pins (**Figure 48**). These forks mate with the output shaft gears. Shift fork No. 1 mates with fourth gear and shift fork No. 2 mates with fifth gear. Shift fork No. 3 has a smaller claw diameter and only mates with third gear on the input shaft.

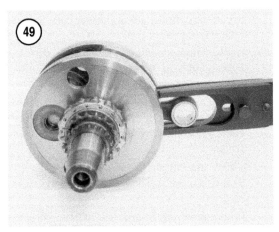

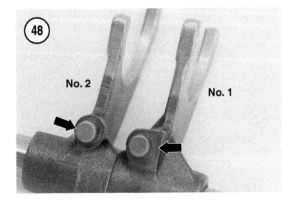

c. Install the shift forks, engaging them with the appropriate gears. Check that the guide pin on each fork points toward the shift drum area.

d. Install the shift drum (**Figure 44**), engaging it with the shift forks. If necessary, raise the transmission shafts slightly to engage the forks with the shift drum. Note that shift fork No. 1 and No. 3 share the same groove in the shift drum.

e. Align the bores in the shift forks, then insert the shift fork shafts (A, **Figure 43**) through the forks and into the case. Check that the shafts are fully seated in the case.

f. Check transmission operation. While turning the shafts and shift drum, observe the action of the shift drum and forks. The shift drum should move through its complete rotation and the forks should operate smoothly. Also check that both shafts turn freely when the shift drum is in the neutral position. The neutral position can be identified by the neutral switch indentation on the shift drum (B, **Figure 43**). When the indentation is at the 8 o'clock position, the transmission should be

in neutral. Leave the transmission in neutral during engine assembly.

4. Install the crankshaft as follows:

a. Insert the crankshaft installation jig between the crankshaft flywheels (**Figure 49**). Install the jig so it is opposite the crankpin and straddles the rod.

b. Lubricate the crankshaft and case bearing.

c. Position the crankshaft so the installation tool is aligned between the cylinder studs.

d. Insert the crankshaft into the case bearing and hand-press the crankshaft into place. If the crankshaft does not fully seat, place supports under the case and around the bearing before applying additional pressure.

e. After the crankshaft is seated, check that it moves freely. Temporarily remove the installation jig, then hold the connecting rod while rotating the crankshaft. After proper installation is verified, install the jig.

5. Install the rear balancer shaft and weight (**Figure 41**). Install the shaft with the splined end facing out. At the smooth end of the shaft, check that a new snap ring is installed and the washer is on the shaft. The washer is placed between the snap ring and bearing. If necessary, refer to **Figure 9** for proper orientation of the parts.

6. Install the front balancer shaft (**Figure 40**).

7. Install the left crankcase onto the right crankcase as follows:

a. Check that all mating surfaces are clean and dry.

b. On the right crankcase, insert the two dowels and secure the crankshaft installation jig between the cylinder studs (**Figure 50**). Strong rubber bands work well.

c. Apply liquid gasket sealant, such as Kawabond 5 or Yamabond 4, to all mating surfaces. This includes applying the sealant around all bolt holes. Use enough sealant to fill voids and provide a continuous seal on the entire joint. Do not use excessive amounts of sealant.

d. Check that all shafts are aligned vertically, then fit the left case squarely onto the right case.

e. If necessary *tap* the left case with a mallet to evenly seat the cases. Do not force the cases. If the cases are not seating, a shaft is probably misaligned with its bore. Lift the case and slightly move it side to side until the shaft(s) are properly guided.

8. Remove each crankcase bolt from the template and insert the bolts into the appropriate holes. Finger-tighten the bolts.

9. Tighten the bolts equally in several passes and in a crossing pattern. Torque the 6 mm bolts to 8 N•m (71 in.-lb.) and the 8 mm bolts to 19 N•m (14 ft.-lb.).

10. Rotate the crankshaft and transmission shafts and check for smooth operation. If binding is evident, separate the cases and correct the problem.

11. Allow the sealant to set for at least an hour before handling the crankcase.

12. Remove the installation jig from the crankshaft. If desired, leave the connecting rod supported by the rubber bands until they must be removed. Cover the cylinder opening to prevent the entry of parts or debris.

13. The crankcase assembly is ready for installation into the frame. If desired, top end components and those located in the side covers can be installed at this time. Refer to the appropriate chapters for inspection and installation procedures for the components in the engine top end and crankcase covers.

Inspection

1. In the left crankcase, remove the oil seals from the output shaft bearing (A, **Figure 51**) and shift shaft bore (B). Remove the seals as described in this chapter.

2. Remove the neutral switch (C, **Figure 51**).

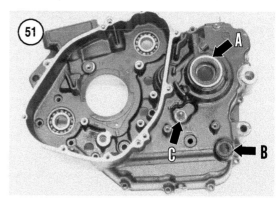

3. In the right crankcase, remove the oil pickup tube and breather tube (**Figure 52**). Do not remove the plugs from the oil passages (**Figure 53**). The plugs should remain seated in the case.

4. Remove all sealant from the gasket surfaces. Avoid gouging or scratching the surfaces.

5. Clean the crankcase halves with solvent. Flush all bearings last, using clean solvent.

6. Dry the cases with compressed air.

> *WARNING*
> *When drying the bearings with compressed air, minimize spinning the bearings. Do not spin them at high speed. Since the bearings are not lubricated, damage could occur.*

7. Flush all passages with compressed air.

8. Oil the engine bearings before inspecting their condition.

9. Inspect the bearings for roughness, pitting, galling and play. Replace any bearing that is not in good condition, or is loose in the crankcase bore. Always replace the opposing bearing at the same time.

10. Inspect the cases for fractures around all mounting and bearing bosses, stiffening ribs and threaded holes. If repair is required, the case should be taken to a dealership or machine shop that repairs precision aluminum castings.

11. Check all threaded holes for damage or sealant buildup. Clean threads with the correct size metric tap. Lubricate the tap with oil or aluminum tap fluid.

12. After the seals and any bearings are replaced:

a. Install the oil pickup tube. Install a new O-ring onto the tube. Apply threadlocking compound to the bolt threads before installing.

b. Install the breather tube.

c. Install a new seal washer onto the neutral switch, then install and torque the switch to 15 N•m (11 ft.-lb.).

SEAL REPLACEMENT

Output Shaft

1. Pry out the old seal (A, **Figure 51**). If necessary, place a block of wood on the case to improve leverage and protect the case from damage.

> *CAUTION*
> *When prying, do not allow the end of the tool to touch the seal bore or snag the oil hole in the bore. Scratches in the bore will cause leakage and heavy-handed prying can break the casting.*

2. If a new bearing will be installed, replace the bearing before installing the new seal.

3. Clean the oil seal bore.

4. Apply grease to the lip and sides of the new seal.

5. To help prevent the new seal from being driven too deep, pass a wire tie, or similar object, through the oil hole and leave it there until the new seal is installed (**Figure 54**).

6. Place the seal in the bore, with the closed side of the seal facing out. The seal must be square to the bore.

7. Drive or press the seal into place. Use a driver that fits at the perimeter of the seal.

8. Move the object placed in the oil hole and verify that the hole is uncovered. Remove the object from the hole.

Shift Shaft

1. Pry out the old seal (B, **Figure 51**). If necessary, place a block of wood on the case to improve leverage and protect the case from damage.

> *CAUTION*
> *When prying, do not allow the end of the tool to touch the seal bore.*

Scratches in the bore will cause leakage.

2. Clean the oil seal bore.

3. Apply grease to the lip and sides of the new seal.

NOTE
This seal is commonly replaced when the engine is assembled. If the shift shaft is passing through the bore, cover the shaft splines with plastic wrap, to prevent tearing of the seal lip when it passes over the shaft.

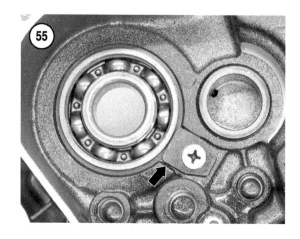

4. Place the seal in the bore, with the closed side of the seal facing out. The seal must be square to the bore.

5. Press the seal into place by hand. If necessary, use a driver that fits at the perimeter of the seal.

CRANKCASE BEARING REPLACEMENT

Refer to Chapter One for additional removal and installation techniques for bearings. Also refer to *Interference Fit* if using heat is preferred for the removal and installation of the bearings in the housings.

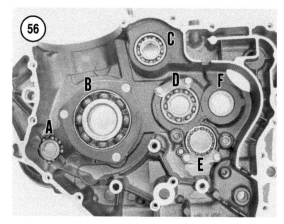

Crankcase Bearing Drivers and Puller

Preferably, remove crankcase bearings with a press. Pressure can be controlled and applied more evenly. If bearing drivers and pullers must be used, Kawasaki recommends the following tools.

1. Bearing driver set (part No. 570011-1129), or an equivalent. The set can be ordered from a dealership.

2. Blind bearing puller. Used for transmission shaft bushings.

Crankcase Bearing Identification

1. When replacing crankcase bearings, note the following:

   a. Where used, remove bearing retainers (**Figure 55**) before attempting bearing removal. When installing retainers, clean the screw threads and install the screws using threadlocking compound.

   b. Identify and record the size code of each bearing before removing it from the case. This

eliminates confusion when installing the bearings in their correct bores.

   c. Record the orientation of each bearing in its bore. Note if the size code faces toward the inside or outside of the case. Commonly, the markings should *face up* when installing the bearing.

   d. Use a hydraulic press or a set of bearing drivers to remove and install bearings. All bearings in the crankcases are an interference-fit. Removing and installing the bearings is eased by using heat, as described in *Interference Fit* in Chapter One.

   e. Bearings that are only accessible from one side of the case can be removed with a blind bearing puller. The puller is fitted through the bearing, then expanded to grip the back side of the bearing. A sliding weight on the tool is quickly pulled back to impact and dislodge the bearing.

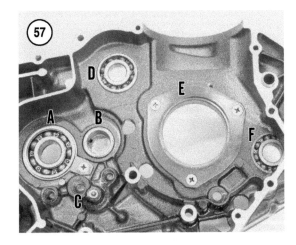

2. The following substeps identify the *right* crankcase bearings (**Figure 56**). The direction of the manufacturer's marks on each bearing is also indicated. When the marks are facing *in*, this indicates they face the *inside* of the case. When the marks are facing *out*, this indicates they face the *outside* of the case.

    a. Front balancer shaft bearing (A). Manufacturer's marks facing out.

    b. Crankshaft bearing (B). Manufacturer's marks facing in.

    c. Rear balancer shaft bearing (C). Manufacturer's marks facing in.

    d. Input shaft bearing (D). Manufacturer's marks facing in.

    e. Shift drum bearing (E). Manufacturer's marks facing in.

    f. Output shaft bushing (F).

3. The following substeps identify the *left* crankcase bearings (**Figure 57**). The direction of the manufacturer's marks on each bearing is also indicated. When the marks are facing *in*, this indicates they face the *inside* of the case. When the marks are facing *out*, this indicates they face the *outside* of the case.

    a. Output shaft bearing (A). Manufacturer's marks facing out.

    b. Input shaft bushing (B).

    c. Shift drum bushing (C).

    d. Rear balancer shaft bearing (D). Manufacturer's marks facing in.

    e. Crankshaft bearing race (E). Manufacturer's marks facing in. The roller bearing is mounted on the crankshaft.

    f. Front balancer shaft bearing (F). Manufacturer's marks facing out.

Crankcase Bearing Replacement

All crankcase bearings can be removed and installed using the following steps. Read the entire procedure before replacing any bearing.

> *CAUTION*
> *Before performing this procedure, refer to Steps 1-3 in the **Crankcase Bearings** section to determine if there is specific information related to the bearing being installed. Do not install a bearing until any specific information is known.*

1. Remove the seal from the bearing, if applicable.
2. Make note of which side of the bearing is facing up.
3. Heat the crankcase as described in *Interference Fit* in Chapter One. Observe all safety and handling procedures when the case is heated.

> *CAUTION*
> *Do not heat the housing or bearing with a propane or acetylene torch. The direct heat will destroy the case hardening of the bearing and will likely warp the housing.*

4. Support the heated crankcase on wooden blocks, allowing space for the bearing to fall from the bore.
5. Remove the damaged bearing from the bore, using a press, hand-driver set or bearing puller.
6. Clean and inspect the bore. Check that all oil holes (where applicable) are clean.
7. Place the new bearing in a freezer and chill for at least one hour.
8. When the bearing is chilled, reheat the crankcase.
9. Support the heated crankcase on wooden blocks, then lubricate the mating surface of the bore and bearing. Place the bearing squarely over the bore and check that it is properly oriented.

    a. If the correct orientation of a bearing is not known, *generally*, bearings that are not sealed on either side should be installed with the manufacturer's marks (stamped on the side of the bearing) facing up.

5

b. Generally, bearings with one side sealed should have the sealed side facing down.

10. Press the bearing into place using a driver that fits onto the outer bearing race.

> *CAUTION*
> *If a press is not available, the bearing can be seated by hand, using a driver and hammer. Place the driver squarely over the bearing, then drive the bearing into the case. Prevent using excessive force when driving the bearing. Bearing and case damage could occur.*

11. Install the seal (if applicable) as described in this chapter.

CRANKSHAFT

Inspection

Carefully handle the crankshaft assembly during inspection. Do not place the crankshaft where it could accidentally roll off the workbench. The crankshaft is an assembly-type, with its two halves joined by a crankpin. The crankpin is hydraulically pressed into the flywheels and aligned, both vertically and horizontally, with calibrated equipment.

If any part of the crankshaft assembly is worn or damaged, have a dealership evaluate all the parts to determine the practicality of repair.

1. Clean the crankshaft with *clean* solvent and dry with compressed air. Lubricate the rod bearing and shaft bearing with engine oil.

2. Inspect the right end of the crankshaft (**Figure 58**).

a. Inspect the shaft threads. The primary drive-gear nut is torqued to these threads. Light damage can be corrected with a thread file.

b. Inspect the keyway and seating surfaces for the primary drive gear and oil pump drive gear.

c. Inspect the bearing surface for scoring, heat discoloration or other damage. Burnishing can be removed with 320-grit carborundum cloth.

3. Inspect the left end of the crankshaft (**Figure 59**).

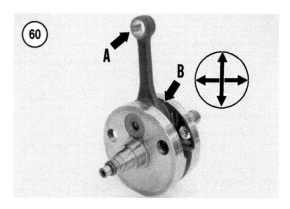

a. Inspect the keyway and seating surface for the rotor. Burnishing can be removed with 320-grit carborundum cloth.

b. Inspect the cam chain sprocket and balancer sprocket for wear and broken teeth.

c. Inspect the bearing for heat discoloration or other damage. Check that the rollers turn freely and are firmly in the bearing race.

> *NOTE*
> *On 1987-1996 models (to engine serial number KL650AE032205), the*

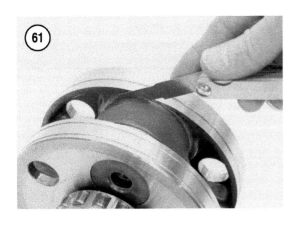

left crankshaft bearing and cam chain/balancer sprocket assembly are available as replacement parts. On later models, the parts are only available as part of the left crankshaft half.

4. Inspect the connecting rod.

   a. Inspect the rod small end (A, **Figure 60**) for scoring, galling or heat damage. Refer to Chapter Four for additional inspections of the rod bore, piston pin and piston.

   b. Inspect the rod big end and bearing for obvious scoring, galling or heat damage.

   c. Inspect the rod for radial clearance (B, **Figure 60**). Mount the crankshaft in a set of V-blocks and accurately measure play. An acceptable method is to grasp the rod and feel for radial play in all directions. There should be no perceptible play. Refer to **Table 1** for specifications.

   d. Measure the connecting side clearance (**Figure 61**). Refer to **Table 1** for specifications.

5. Place the crankshaft in a flywheel alignment jig and measure crankshaft runout with a dial indicator. The jig centers should be inserted into the ends of the crankshaft. Measure at the two points shown in **Figure 62**. The maximum difference in gauge readings is the total runout. If the runout exceeds the specifications listed in **Table 1**, have a dealership evaluate and possibly retrue the crankshaft.

NOTE
If it is known that the crankshaft was dropped or damaged, or if the engine exhibited abnormal vibration, have the crankshaft alignment checked before assembling the engine.

Table 1 ENGINE LOWER END SPECIFICATIONS

	New mm (in.)	Service limit mm (in.)
Balancer chain length (20 links/21 pins)	190.0-190.5 (7.48-7.50)	193.4 (7.61)
Camshaft Chain length (20 links/21 pins)	127.0-127.4 (5.0-5.02)	128.9 (5.07)
Connecting rod Side clearance	0.25-0.35 (0.010-0.014)	0.60 (0.024)
Radial clearance	0.008-0.020 (0.0003-0.0008)	0.07 (0.003)
Crankpin–to–flywheel tolerance (cold)	0.093-0.122 (0.0037-0.0048)	–

(continued)

Table 1 ENGINE LOWER END SPECIFICATIONS (continued)

	New mm (in.)	Service limit mm (in.)
Crankshaft runout (total indicator reading)		
Left half	0.03 (0.0012)	0.10 (0.004)
Right half	0.04 (0.0016)	0.10 (0.004)

Table 2 ENGINE LOWER END TORQUE SPECIFICATIONS

	N•m	in.-lb.	ft.-lb.
Balancer shaft nut			
(right side)	44	–	32
Balancer chain guide bolts			
6 mm	12	106	–
8 mm	25	–	18
Engine mounting bolts			
8 mm	25	–	18
10 mm	44	–	32
Cam chain guide bolt	25	–	18
Crankcase bolts			
6 mm	8	71	–
8 mm	19	–	14
Neutral switch	15	–	11
Oil drain plug	23	–	17
Oil pipe			
Banjo bolts	20	–	15
Retainer bolt	8	71	–
Oil pressure relief valve	15	–	11
Primary drive gear nut	120	–	89
Swing arm pivot bolt	98	–	72

CHAPTER SIX

CLUTCH, GEARSHIFT LINKAGE AND LUBRICATION SYSTEM

This chapter provides service procedures for the following components on the right side of the engine crankcase:

1. Right crankcase cover.
2. Clutch.
3. External gearshift linkage.
4. Oil pump.
5. Primary drive gear.
6. Clutch cable replacement.
7. Kickstarter.

Read this chapter before attempting repairs to the components in the right crankcase cover. Become familiar with the procedures, photos and illustrations to understand the skill and equipment re-

quired. Refer to Chapter One for tool usage and techniques.

RIGHT CRANKCASE COVER

Removal and Installation

1. Remove the skid plate (Chapter Fifteen).
2. Drain the engine oil and remove the oil filter (Chapter Three).
3. Drain the cooling system (Chapter Three).
4. Remove the rear brake pedal (Chapter Fourteen).
5. Remove the right footpeg.
6. Remove the clutch cable holder, then remove the clutch cable from the release lever (**Figure 1**). The cable does not have to be removed from the cable holder.
7. Remove the water pump cover and impeller assembly (Chapter Ten).
8. Remove the 15 bolts from the perimeter of the crankcase cover (**Figure 2**).
9. Turn the clutch release lever so it points to the back. This disengages it from the clutch lifter, located behind the cover.

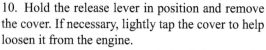

10. Hold the release lever in position and remove the cover. If necessary, lightly tap the cover to help loosen it from the engine.

11. Account for the two dowels that fit between the crankcase and cover (A, **Figure 3**).

12. Remove the O-ring at the oil filter housing (B, **Figure 3**).

13. Remove the cover gasket.

14. Remove the crankcase oil strainer (**Figure 4**).

15. Inspect the cover assembly as described in this section.

16. Reverse these steps to install the right crankcase cover assembly. Note the following:

   a. Install a new, lubricated O-ring on the oil filter housing.

   b. Clean all residue and oil from the engine gasket surface.

   c. Check that the crankcase oil strainer is installed and seated (A, **Figure 5**).

   d. Apply molybdenum disulfide grease to the end of the crankshaft (B, **Figure 5**) and impeller shaft (C). Do not plug the oil hole in the crankshaft.

   e. Install a new cover gasket on the crankcase dowels.

   f. Apply molybdenum disulfide grease to the contact points of the clutch release lever and clutch lifter.

   g. Turn the clutch release lever so it points back, then install the cover. When the cover is installed, turn the release lever forward to engage it with the clutch lifter.

   h. Install a new oil filter and fill the engine with oil (Chapter Three).

   i. Fill the engine with coolant (Chapter Three).

   j. Check clutch operation. If necessary, adjust the clutch (Chapter Three).

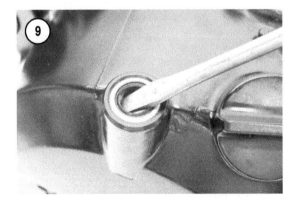

Inspection and Disassembly

1. Wipe the cover clean. Do not submerge the cover in solvent unless all bearing and seal assemblies are removed.

2. Inspect the crankshaft oil seal in the crankcase cover (**Figure 6**). This seal fits over the end of the crankshaft and must be in good condition. Oil under pressure passes from the oil filter housing and into the end of the crankshaft, where it then goes to the crankpin and connecting rod. If this seal leaks, oil pressure will be reduced to these parts. Replace the seal as follows:

    a. Remove the bolt and seal retainer plate.

    b. Pry the seal from its bore (**Figure 7**).

    c. Lubricate the new seal with grease.

    d. Place the seal over the bore, with the closed side facing up.

    e. Place a driver or socket over the seal (**Figure 8**) that is sized to fit the perimeter of the seal.

    f. Drive the seal into place.

    g. Install the seal retainer plate and bolt. Note the correct orientation of the plate (**Figure 6**).

3. Inspect the clutch release lever, seal and bearing. On 1987-1995 models, the lever is secured with a retaining bolt. Note the position of the lever before removing the bolt.

    a. Inspect the lever for shaft wear and fit it in the bearing.

    b. Inspect the seal. If oil leakage is evident, replace the seal. Pry the seal from its bore (**Figure 9**), then apply grease to the new seal. Seat the seal using a driver that fits on the perimeter of the seal. The wear ring on the seal should rest slightly above the case. The wear ring prevents the clutch release lever from contacting the case.

    c. Inspect the bearing (**Figure 10**). If damaged or worn, remove the seal and drive the bearing out of the bore. Lubricate the new bearing, then drive it into the bore. Use a driver that fits on the perimeter of the bearing. Install the bearing with the manufacturers marks facing up. Drive the bearing no deeper than flush with the inside of the case. Install a new seal.

4. Inspect the relief valve (**Figure 11**). If engine or oil pump damage has occurred, remove the valve from the crankcase cover. Remove the snap ring from the valve, then disassemble and clean the parts. Also clean the oil passage in the cover. When

installing, apply threadlocking compound to the relief valve threads.

5. Inspect and clean the crankcase oil strainer (**Figure 4**) whenever the crankcase cover is removed. Oil passes from this screen to the oil pump. A plugged, damaged or missing screen could cause engine damage. If the screen or grommet is distorted or damaged, replace the parts. Inspect and clean the passage where the strainer seats.

CLUTCH

The clutch assembly consists of an outer housing and a clutch hub. A set of clutch plates and friction plates are alternately locked to the two parts. The gear-driven clutch housing is mounted on the transmission input shaft and can rotate freely. The housing receives power from the primary drive gear mounted on the crankshaft. As the clutch is engaged, the housing and friction plates transfer the power to the clutch plates, locked to the clutch hub. The clutch hub is splined to the input shaft and powers the transmission. The plate assembly is engaged by springs and disengaged by a cable-actuated release lever and lift assembly.

The clutch operates immersed in the engine oil supply. Oil additives should not be added to the oil supply, since these can cause poor clutch operation. The friction plates can also become contaminated.

On 1987-1995 models, the clutch uses seven friction plates and six clutch plates. On 1996-on models, the clutch uses eight friction plates and seven clutch plates. The removal, inspection and installation procedures are the same for all models.

Removal

Refer to **Figure 12**.

1. Remove the right crankcase cover as described in this chapter.

2. Remove the five bolts compressing the clutch springs (**Figure 13**). Make several passes to relieve the pressure equally on the bolts.

3. Remove the bolts and springs from the clutch (**Figure 14**).

4. Remove the pressure plate. When removing, keep the plate tilted up so the lifter bearing in the back of the plate does not fall out (**Figure 15**).

5. Remove the clutch lifter (**Figure 16**).

6. Remove the plates from the clutch housing and clutch hub (**Figure 17**).

> *NOTE*
> *On 1996-on models, note that the outer friction plate seats into the shallow slots of the outer housing (**Figure 18**).*

7. Remove the clutch locknut (**Figure 19**) as follows:

  a. Temporarily install the right footpeg.

  b. Attach a clutch holder tool to the clutch hub and rest it on the footpeg (**Figure 20**).

> *CAUTION*
> *Do not jam the gears with screwdrivers or other tools. This can cause gear breakage. The Kawasaki clutch holder tool (part No. 57001-305) can be ordered from a Kawasaki dealership.*

  c. Using a 30 mm socket, loosen the locknut.

  d. Remove the clutch locknut and lockwasher.

8. Remove the clutch hub (**Figure 21**).

9. Remove the washer (**Figure 22**).

10. Remove the clutch housing (**Figure 23**).

11. Remove the clutch housing collar (**Figure 24**).

12. Remove the spacer (**Figure 25**).

13. Inspect the clutch assembly as described in this section.

Inspection

Always replace clutch plates, friction plates or springs as a set if they do not meet specifications. If any part shows signs of wear or damage, replace it, regardless of its specification. Refer to the specifi-

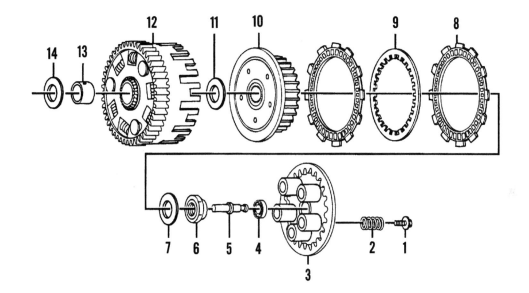

CLUTCH ASSEMBLY

1. Bolt
2. Spring
3. Pressure plate
4. Lifter bearing
5. Lifter
6. Locknut
7. Lockwasher

8. Friction plate
9. Clutch plate
10. Clutch hub
11. Washer
12. Clutch housing
13. Collar
14. Spacer

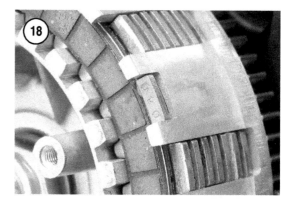

cations listed in **Table 1** for component service limits, when applicable.

1. Clean the parts in solvent and dry with compressed air. Also clean the transmission shaft.

2. Measure the thickness of each friction plate (**Figure 26**). Measure at several locations around the perimeter.

3. Inspect the tabs on the friction plates (**Figure 27**). The tabs must not be damaged. Check that each plate slides smoothly in the clutch housing.

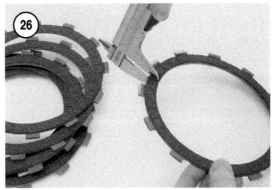

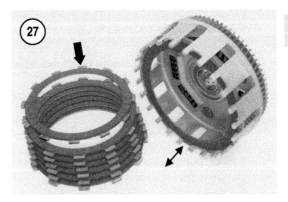

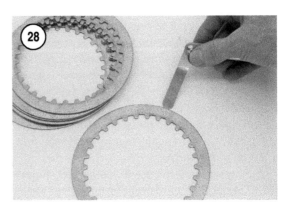

4. Measure each clutch plate (**Figure 28**) for warp. Lay each plate on a surface plate, or thick piece of glass, and measure any gap around the perimeter of the plate.

5. Inspect the inner teeth on the plates (**Figure 29**). The teeth must not be damaged. Check that each plate slides smoothly on the clutch hub.

6. Measure the free length of each clutch spring (**Figure 30**).

7. Inspect the clutch housing and collar (A, **Figure 31**).

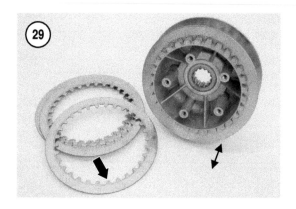

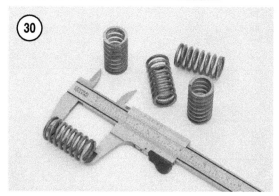

a. Inspect the surface of the collar and housing bore for wear and damage. Check the fit of the collar in the bore and on the transmission shaft. The collar should move freely, but have no obvious play.

b. Inspect the oil pockets in the housing bore for cleanliness.

c. Inspect the gear teeth (B, **Figure 31**) for wear or damage.

d. Inspect the damper springs (C, **Figure 31**) and rivets (D) on both sides of the housing for looseness or damage.

e. Inspect the slots for nicks, wear and damage (**Figure 32**). The slots must be smooth and free of defects so the friction plates smoothly engage and disengage. If *chatter* marks are evident (**Figure 33**), light damage can be smoothed using a fine-cut file or oilstone.

8. Inspect the clutch hub.

a. Inspect the shaft splines (A, **Figure 34**). The hub should fit on the transmission shaft with no obvious play.

b. Inspect the perimeter of the hub (B, **Figure 34**) for wear and damage on the contact area.

c. Inspect the bosses (C, **Figure 34**) for damage.

d. Inspect the outer splines (D, **Figure 34**) for nicks, wear and damage. The splines must be smooth and free of defects so the clutch plates smoothly engage and disengage. If *chatter* marks are evident, light damage can be smoothed using a fine-cut file or oilstone.

9. Inspect the pressure plate, lifter and bearing (**Figure 35**).

a. Inspect the pressure plate for cracks, particularly around the bosses and bearing seat.

b. Inspect the perimeter of the pressure plate for wear and damage on the contact area.

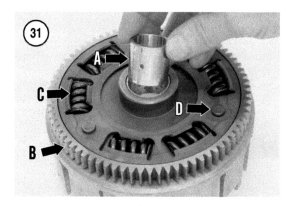

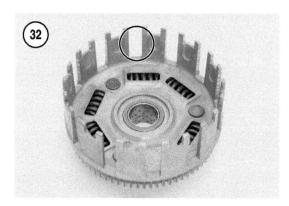

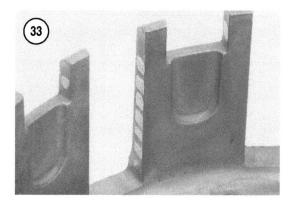

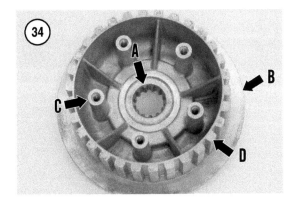

c. Inspect the bearing for radial and axial play. The bearing should fit in the pressure plate with minimal play. If the bearing is replaced, install the bearing with the manufacturer's marks facing up.

d. Inspect the lifter for wear at the contact area with the bearing. The lifter is a loose-fit in the bearing. When the lifter is installed in the transmission shaft, and the bearing is installed in the pressure plate, the parts self-center.

e. Inspect the lifter for fit with the release lever (**Figure 36**). Replace both parts if either part is damaged.

10. Inspect the bolts, washers and spacer for obvious damage.

NOTE
Do not reuse the locknut. If the lockwasher is no longer cupped, replace the lockwasher.

11. Inspect the splines, threads and polished surfaces on the transmission shaft for damage.

12. Install the clutch assembly as described in this section.

Installation

During assembly, lubricate the parts and transmission shaft with engine oil, unless specified otherwise.

1. Install the spacer.
2. Install the clutch housing collar.
3. Install the clutch housing onto the collar.
4. Install the washer.
5. Install the clutch hub.
6. Install the lockwasher, with the words OUT SIDE facing out (**Figure 37**).
7. Install a new locknut, with the wide side of the nut in contact with the lockwasher (**Figure 38**).
8. Tighten the clutch locknut as follows:

a. Temporarily install the right footpeg.

b. Attach a clutch holder tool to the clutch hub and rest it against the footpeg (**Figure 39**).

CAUTION
Do not jam the gears with screwdrivers or other tools. This can cause gear breakage. The Kawasaki clutch holder tool (part No. 57001-305) can

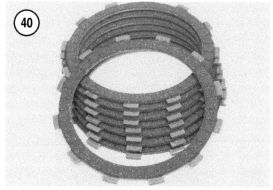

be ordered from a Kawasaki dealership.

   c. Using a 30 mm socket, tighten the locknut to
     130 N•m (96 ft.-lb.).

9. Install the plates into the clutch housing and
clutch hub as follows:

   a. Lubricate the plates with engine oil. To pre-
     vent possible seizure, particularly with new
     friction plates, it is important that the face of
     the friction plates (**Figure 40**) be completely
     coated with oil. If possible, allow the plates to
     soak in engine oil before installing.

   b. Beginning with a friction plate, alternately in-
     stall friction plates and clutch plates into the
     clutch housing and clutch hub (**Figure 41**).

NOTE
*On 1996-on models, note that the
outer friction plate seats into the shallow slots of the outer housing (**Figure
18**).*

10. Install the clutch lifter. Apply molydisulfide
grease to the lifter, particularly at the bearing contact point.

11. Install the pressure plate and bearing.

12. Lock the pressure plate to the clutch hub as follows:

   a. Install the clutch springs on the hub bosses.

   b. Finger-tighten the five bolts (**Figure 42**).

   c. Tighten the bolts, working in a crossing pat-
     tern. Make several passes so all bolts are
     tightened in equal steps to the specification in
     Table 3.

13. Install the right crankcase cover as described in
this chapter.

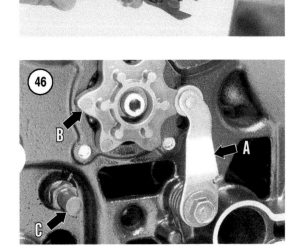

EXTERNAL GEARSHIFT LINKAGE

Removal

The external gearshift linkage is located behind the right side cover and is partially covered by the clutch (**Figure 43**). Visual inspection of the shift shaft assembly is possible without removing the clutch. This procedure details the complete removal of the gearshift linkage.

1. Remove the right crankcase cover as described in this chapter.
2. Remove the clutch as described in this chapter.
3. Put the transmission in neutral. This can be verified by the position of the lever on the shift cam. The roller should rest in the neutral detent on the shift cam (**Figure 44**).
4. Remove the shift lever from the left side of the engine. Note where the lever is located on the shaft splines, so the lever can be installed in the same position.
5. Pull the shift shaft assembly from the engine (**Figure 45**).
6. Remove the lever and torsion spring (A, **Figure 46**).
7. Remove the shift cam (B, **Figure 46**).
8. Remove the pin from the end of the shift drum (**Figure 47**).
9. If obviously damaged or loose, remove the return spring post (C, **Figure 46**).
10. Inspect the parts as described in this section.

Inspection

Inspect the components of the gearshift linkage and replace parts that are worn, damaged or fatigued. Inspect as follows:
1. Clean the parts in solvent.

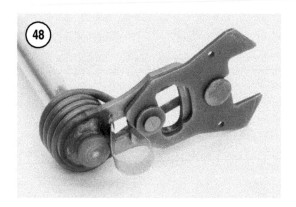

2. Inspect the shift shaft assembly (**Figure 48**).
   a. Inspect the return spring for fatigue or damage.
   b. Inspect the pawl tips for wear.
   c. Check that the pawl slides freely and returns to the extended position. If the pawl jams or drags, remove the band spring and pawl from the shift shaft (**Figure 49**), then determine which part(s) are damaged. When installing the band spring, engage the narrow end with the pawl.
   d. Inspect the condition of the splines on the shaft and in the shift lever.

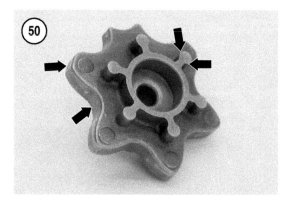

3. Inspect the shift cam (**Figure 50**).
   a. Inspect the pawl engagement points for wear.
   b. Inspect the lever engagement points for wear in the high and low areas. The neutral detent should not be worn away and should allow the lever roller to rest in the detent.
   c. Inspect the shift cam bolt, pin and pin bore (**Figure 51**). The pin and bore should not be worn. It is normal for the pin bore to be elongated.
   d. Inspect the fit of the pin in the shift drum. The pin should fit with minimal perceptible play.
   e. Inspect the bolt thread condition and cleanliness. All threadlocking compound must be removed before torquing the bolt.

4. Inspect the lever, bolt and spring (**Figure 52**).
   a. The lever roller should be round, turn freely and have no play in its pivot.
   b. The shoulder on the bolt should fit into the lever with minimal, or no play. If either part is loose, replace both parts.
   c. Inspect the ends of the spring for wear.
5. Inspect the return spring post for wear and tightness (C, **Figure 46**).

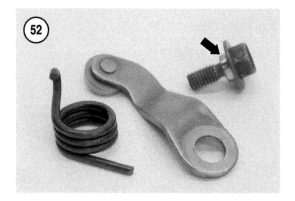

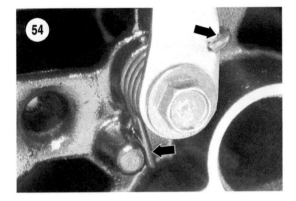

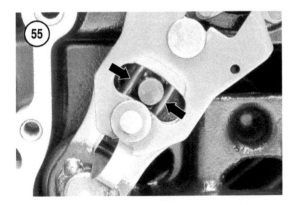

Installation

1. If removed, install the return spring post (C, **Figure 46**). Apply threadlocking compound to the threads before tightening the post.

2. Install the pin in the end of the shift drum.

3. Install the shift cam, engaging the bore in the shift cam with the pin (**Figure 53**).

4. Apply threadlocking compound to the threads of the shift cam bolt, then install and tighten the bolt to 12 N•m (106 in.-lb.).

5. Install the lever and torsion spring (A, **Figure 46**).

    a. Check that the spring is correctly installed with the lever and against the crankcase (**Figure 54**).

    b. Check that the bolt shoulder is seated into the lever before tightening.

6. Insert the shift shaft assembly through the engine (**Figure 45**). Seat the return spring around the post (**Figure 55**).

7. Install the shift lever onto the left side of the engine. If necessary, get assistance to hold the shaft in place on the right side while the lever is installed.

8. Check the shifting action of the linkage.

    a. Turn the transmission shaft to aid in shifting.

    b. The lever roller should lock into each dent on the shift cam as the transmission is shifted. In neutral, the lever roller should rest in the neutral detent (**Figure 44**).

9. Install the clutch as described in this chapter.

10. Install the right crankcase cover as described in this chapter.

OIL PUMP

The engine is a *wet-sump* engine, and therefore stores the engine oil in the crankcase. The oil supply flows from the crankcase oil strainer (A, **Figure 56**) to the oil pump (B), where it is pumped under pressure through the oil filter and to the engine components. After oil passes through a lubrication point, it falls to the crankcase, where it again circulates to the oil pump. The oil pump is driven by a gear on the end of the crankshaft.

Removal and Installation

1. Remove the right crankcase cover as described in this chapter.

2. With the engine in neutral, rotate the crankshaft and align the holes in the oil pump gear so the three pump mounting screws are accessible (**Figure 57**). Fit an open-end wrench on the primary drive-gear nut to turn the crankshaft.

3. Remove the three mounting screws and the oil pump.

4. Account for the two dowels that join the pump to the crankcase.

5. Remove the gasket (**Figure 58**).

6. Check that the plugs (**Figure 59**) are in place. Do not remove the plugs.

7. Disassemble, inspect and assemble the oil pump as described in this section.

8. Reverse these steps to install the oil pump. Note the following:

   a. Inspect and clean the crankcase oil strainer (**Figure 60**) and oil passage.

   b. Fill the pump with engine oil prior to mounting.

   c. Install a new oil pump gasket.

   d. Apply threadlocking compound to the threads of the mounting screws before installing.

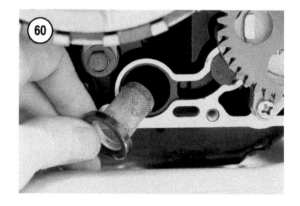

Disassembly and Assembly

1. Remove the retaining screw from the pump body (**Figure 61**), then remove the rear plate.

2. Lift out the inner rotor (A, **Figure 62**) and outer rotor (B).

3. Remove the pin (**Figure 63**), then pull the gear out of the pump body.

4. Inspect the parts (**Figure 64**) as described in this section.

5. Reverse these steps to assemble the oil pump. Note the following:

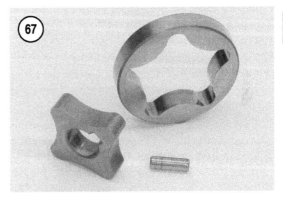

a. Install the rotors into the pump body with the marks (**Figure 65**) facing *down*. The marks should not be visible when the rotors are installed.
b. Check that the rear plate is fully seated at its perimeter before screwing it into place.
c. Apply threadlocking compound to the threads of the retaining screw before installing.
d. Install the oil pump as described in this section.

Inspection

Visually inspect all parts for obvious wear or damage. Component specifications are not available for the oil pump.
1. Clean the parts in solvent.
2. Inspect the pump shaft, gear and pin (**Figure 66**) for wear or scoring. The pin should firmly fit into the shaft. The pump shaft should fit into the pump body and rear plate with minimal perceptible play.
3. Inspect the rotors (**Figure 67**) for wear or scoring. When assembled, the inner rotor and pin should firmly fit onto the shaft.

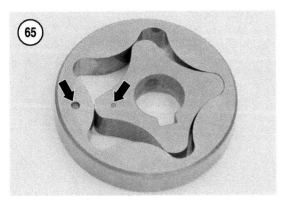

4. Inspect the pump body and rear plate (**Figure 68**) for wear or scoring.

5. Assemble the oil pump as described in this section.

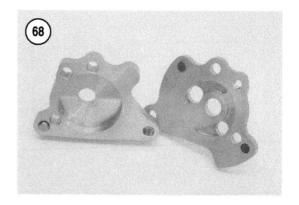

PRIMARY DRIVE GEAR

Removal and Inspection

Tools

The primary drive gear nut is very tight. The crankshaft must be held stable so the nut can be loosened. The following two methods can be used:

1. Use the Kawasaki rotor holder tool (part No. 57001-1184), or similar tool, to hold the rotor stable (**Figure 69**). The tool is placed against the top of the footpeg during removal, and against the bottom of the footpeg during installation.

2. Lock the primary drive gear and clutch housing gear together using a *jam* gear. This is typically a discarded drive gear. Place the gear above the primary and driven gears during removal (A, **Figure 70**), and below the gears during installation (B). The gear *must* completely mesh with both sets of gear teeth. Check with a dealership or motorcycle repair shop for a jam gear.

Procedure

1. Remove the right crankcase cover as described in this chapter.

2. Remove the oil pump as described in this chapter.

3A. If a rotor holder tool is used:

    a. Remove the alternator cover as described in Chapter Nine.

    b. Remove the clutch as described in this chapter.

    c. Flatten the perimeter of the lockwasher (**Figure 71**) so a socket fully seats on the nut.

    d. Fit the rotor holder tool onto the rotor.

    e. Loosen the nut.

3B. If a jam gear will be used:

    a. Flatten the perimeter of the lockwasher (**Figure 71**) so a socket fully seats on the nut.

    b. Mesh the jam gear with the clutch housing gear and primary drive gear.

CAUTION
Do not jam the gears with screwdrivers or other tools. This could cause gear tooth breakage.

   c. Loosen the nut.
   d. Remove the clutch as described in this chapter.
4. Remove the primary drive gear nut and washer.
5. Remove the oil pump drive gear, primary drive gear, shaft key and spacer (**Figure 72**).
6. Inspect the parts (**Figure 73**).
   a. Inspect the gears for worn or broken teeth.
   b. Inspect the fit of the key in the gears and crankshaft (**Figure 74**). The key must fit in all parts with no play.
   c. Inspect the spacer for obvious damage.

Installation

1. Lubricate the crankshaft with engine oil, then install the spacer, shaft key, primary drive gear and oil pump drive gear.
2. Install a new lockwasher with the cupped side facing out.
3. Apply engine oil to the crankshaft threads and primary drive gear nut. Finger-tighten the nut.
4A. If a rotor holder tool is used:
   a. Fit the rotor holder tool onto the rotor.
   b. Tighten the nut to 120 N•m (89 ft.-lb.).
   c. Flatten one side of the lockwasher (**Figure 75**) against the nut.
   d. Install the alternator cover as described in Chapter Nine.
   e. Install the clutch as described in this chapter.
4B. If a jam gear is used:
   a. Install the clutch as described in this chapter.
   b. Mesh the jam gear with the clutch housing gear and primary drive gear.

CAUTION
Do not jam the gears with screwdrivers or other tools. This could cause gear tooth breakage.

   c. Tighten the nut to 120 N•m (89 ft.-lb.).
   d. Flatten one side of the lockwasher (**Figure 75**) against the nut.
5. Install the oil pump as described in this chapter.
6. Install the right crankcase cover as described in this chapter.

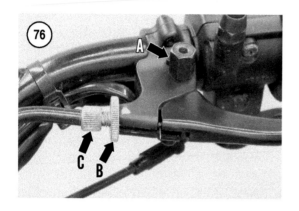

CLUTCH CABLE REPLACEMENT

1. Remove the left hand guard. Carefully flex the guard to remove it from the clutch lever assembly.

2. Remove the pivot bolt (A, **Figure 76**) and lower nut from the lever.

3. Loosen the clutch cable locknut (B, **Figure 76**), then turn the adjuster (C) in until the cable and lever can be removed from the slots in the adjuster assembly (**Figure 77**). Remove the cable from the lever.

4. Remove the cable from the release lever and cable holder (**Figure 78**).

5. Remove the cable from the bike, noting the routing of the cable.

6. Clean the levers, pivot bolt and clutch lever housing.

7. Lubricate the new cable with an aerosol cable lubricant. Lubricate the lever, pivot bolt and cable ends with lithium grease.

8. Route the cable from the handlebar to the release lever.

9. Route the cable into the cable holder, then attach the cable to the release lever. Center the adjuster and locknuts in the cable holder.

10. At the handlebar, attach the cable to the lever, then install the lever assembly.

11. Adjust the clutch cable (Chapter Three).

KICKSTARTER

The KLR650 does not come standard with a kickstarter. A kickstarter kit is available from a Kawasaki dealership (part No. 99995-1007). The engine cases are already bored and threaded to accept the components of the kit. The kit includes the lever assembly and kickstarter idle gear.

On 1987-1996 models (to engine serial number KL650AE032209), the kickstarter idle gear is already installed behind the right crankcase cover, on the end of the transmission output shaft. If the idle gear is removed from the shaft, install the gear with the shallow side facing out (gear teeth away from the crankcase). Install the snap ring with the sharp inner edge facing out.

Table 1 CLUTCH SPECIFICATIONS

	New mm (in.)	Service limit mm (in.)
Friction plate quantities		
1987-1995	7	
1996-on	8	
Clutch plate (steel) quantities		
1987-1995	6	
1996-on	7	
Friction plate thickness	2.80-3.1 (0.110-0.122)	2.60 (0.102)
Clutch plate (steel) warp	0.0-0.2 (0.0-0.008)	0.3 (0.012)
Spring free length		
1987-1995	34.2 (1.35)	33.1 (1.30)
1996-on	38.7 (1.52)	36.4 (1.43)

Table 2 OIL PUMP SPECIFICATIONS

Lubrication system	Troichoid pump, forced pressure, wet sump
Oil pump pressure at 4000 rpm	78-147 kPa (11-21 psi)
Relief valve opening pressure	430-590 kPa (62-85 psi)

Table 3 CLUTCH TORQUE SPECIFICATIONS

	N•m	in.-lb.	ft.-lb.
Clutch hub locknut	130	–	96
Clutch spring bolts			
1987-1995	10	88	–
1996-on	9	80	–
Oil pressure relief valve	15	–	11
Primary drive gear nut	120	–	89
Shift cam bolt	12	106	–

CHAPTER SEVEN

TRANSMISSION AND INTERNAL SHIFT MECHANISM

This chapter provides service procedures for the transmission and internal shift mechanism. Read this chapter and become familiar with the procedures, photos and illustrations to understand the skill and equipment required. Refer to Chapter One for tool usage and techniques.

TRANSMISSION OPERATION

The machine is equipped with a five-speed constant-mesh transmission. The gears on the input shaft (A, **Figure 1**) are meshed with the gears on the output shaft (B). Each pair of meshed gears represents one gear ratio. For each pair of gears, one of the gears is splined to its shaft, while the other gear freewheels on its shaft.

Next to each freewheeling gear is a gear that is splined to the shaft. This locked gear can slide on the shaft and lock into the freewheeling gear, making that gear ratio active. Anytime the transmission is *in gear* one pair of meshed gears are locked to their shafts, and that gear ratio is selected. All other meshed gears have one freewheeling gear, making those ratios inoperative.

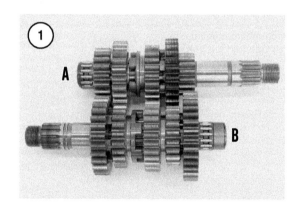

To engage and disengage the various gear ratios, the splined gears are moved by shift forks. The shift forks are guided by the shift drum, which is operated by the shift lever. As the bike is upshifted and downshifted, the shift drum rotates and guides the forks to engage and disengage pairs of gears on the transmission shafts.

SERVICE NOTES

The engine crankcase must be split to remove the transmission and shift assemblies (**Figure 2**). Re-

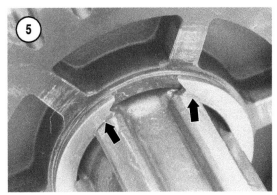

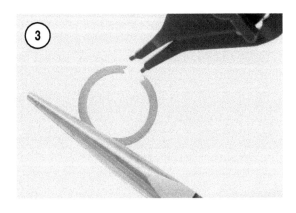

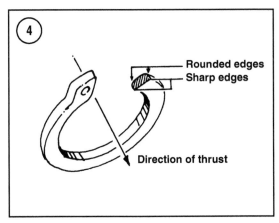

Rounded edges
Sharp edges

Direction of thrust

move and install the transmission assemblies as described in *Crankcase* in Chapter Five.

After the transmission is removed from the crankcase, disassembly, inspection and assembly can be performed. Careful inspection of the parts is required, as well as keeping the parts oriented so they can be reinstalled in the correct direction on the shafts. The gears, washers and snap rings *must* be installed in the same direction as they were before disassembly. If necessary, slide the parts onto a long

dowel or screwdriver as the parts are removed, or make an identification mark on each part to indicate position and orientation.

Always install new snap rings. The snap rings fatigue and distort when they are removed. Do not reuse them, although they appear to be in good condition. To install a new snap ring without distorting it, hold the closed side of the snap ring with a pair of pliers while the open side is spread with snap ring pliers (**Figure 3**). While holding the spread ring with both tools, slide it over the shaft and into position. This technique is particularly useful when removing the notched-type snap ring shown in **Figure 3**. During removal, this type of snap ring tends to bind and grip the shaft.

Usually, snap rings have one rounded edge, while the other side has a sharp edge (**Figure 4**). The inner sharp edge prevents the snap ring from lifting out of the shaft groove when lateral pressure is applied to the snap ring. Always look at the inner and outer edges of the snap ring. Some snap rings are manufactured with the inner and outer sharp edge on opposite sides. If a snap ring has no identifiable sharp edge, the snap ring can be installed in either direction. In all cases, new snap rings must be installed at assembly, and when applicable, must be installed in the same direction as the removed snap rings.

When installed on a splined shaft, the snap ring gap should be positioned over a groove in the shaft (**Figure 5**). When a spline washer is used next to the snap ring, the teeth of the spline washer should be offset from the ends of the snap ring (**Figure 6**).

INPUT SHAFT

Use the following procedures to disassemble and assemble the input shaft. Although the design of the

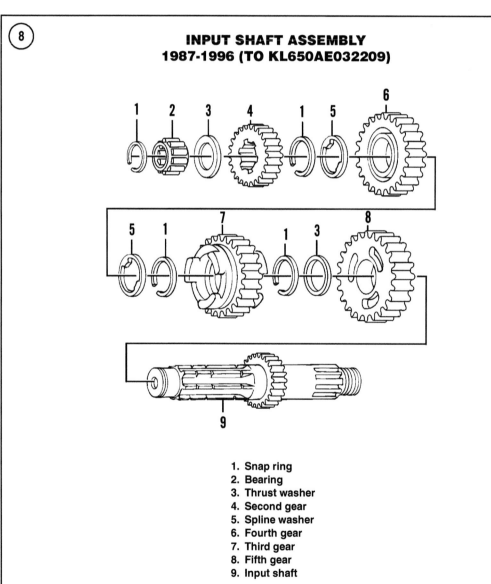

INPUT SHAFT ASSEMBLY
1987-1996 (TO KL650AE032209)

1. Snap ring
2. Bearing
3. Thrust washer
4. Second gear
5. Spline washer
6. Fourth gear
7. Third gear
8. Fifth gear
9. Input shaft

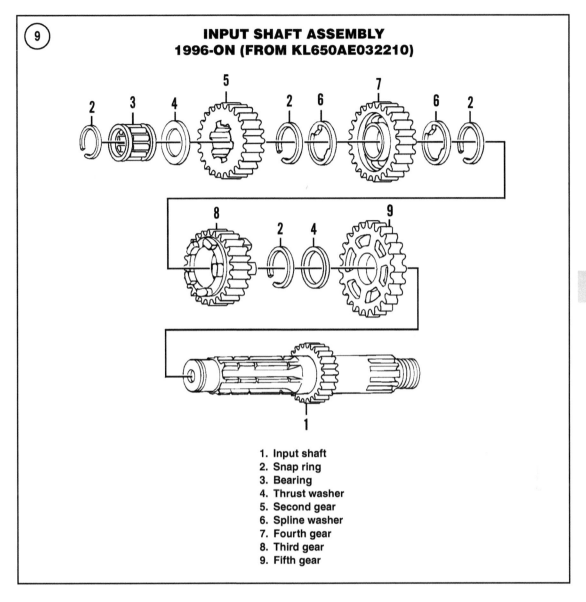

**INPUT SHAFT ASSEMBLY
1996-ON (FROM KL650AE032210)**

1. Input shaft
2. Snap ring
3. Bearing
4. Thrust washer
5. Second gear
6. Spline washer
7. Fourth gear
8. Third gear
9. Fifth gear

input shaft gears changed in 1996 (after engine serial number KL650AE032209), the service procedures can be used for all models. Figures of both input shaft assemblies are provided. The photographs in the procedures show the later input shaft assembly.

**Disassembly
All Models**

Remove the parts from the input shaft (**Figure 7**) in the following order. Clean and organize the parts

as they are removed. Refer to **Figure 8** or **Figure 9** as needed.

CAUTION
Keep parts oriented and note their direction and original location on the shaft. Preferably, mark the outer face of each part as it is removed from the shaft. Check all snap rings and note which direction the inner, sharp edge is facing. For the model shown in this procedure, the notched snap rings were identical on both inner edges, and therefore could be installed in either direction.

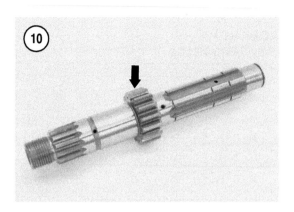

1. Disassemble the input shaft in the following order:

   a. Snap ring.
   b. Bearing.
   c. Thrust washer.
   d. Second gear.
   e. Snap ring.
   f. Spline washer.
   g. Fourth gear.
   h. Spline washer.
   i. Snap ring.
   j. Third gear.
   k. Snap ring.
   l. Thrust washer.
   m. Fifth gear.

2. Inspect each part as described in this chapter, then return it to its place until assembly.

Assembly
All Models

Before beginning assembly, have new snap rings on hand. Throughout the procedure, the orientation of many parts is made in relation to first gear (**Figure 10**), on the input shaft. If desired, lock the lower portion of the shaft (below first gear) in a *padded* vise. With the shaft held stable and vertical, installation of the snap rings will be easier. Do not allow the vise to damage the shaft.

For the model shown in this procedure, the notched snap rings were identical on both inner edges, and therefore could be installed in either direction. Refer to **Figure 8** or **Figure 9** as needed.

> *CAUTION*
> *Refer to **Service Notes** in this chapter*
> *for the snap ring installation tech-*
> *nique. This technique is recom-*

mended for installing the notched snap rings. The installed snap rings must grip the shaft tightly. If necessary, practice the technique using discarded snap rings.

1. Clean and dry all parts before assembly. Lubricate all parts with engine oil.

2. Install fifth gear (**Figure 11**). The gear dog recesses must face *out* (away from first gear).

3. Install the thrust washer and snap ring (**Figure 12**). The snap ring must seat in the shaft groove.

4. Install third gear (**Figure 13**). The side of the gear with the shift fork groove must face *out* (away from first gear).

5. Install the snap ring and spline washer (**Figure 14**). The snap ring must seat in the shaft groove. The ends of the snap ring must be offset from the spline washer teeth.

6. Install fourth gear (**Figure 15**). The flat side, with the recesses, must face *in* (toward first gear).

7. Install the spline washer and snap ring (**Figure 16**). The snap ring must seat in the shaft groove. The ends of the snap ring must be offset from the spline washer teeth.

8. Install second gear (**Figure 17**). The chamfered, inner edge of the gear must face *in* (toward first gear).

9. Install the thrust washer and bearing (**Figure 18**).

10. Install the snap ring (**Figure 19**). The snap ring must seat in the shaft groove. The sharp edge of the snap ring must face *out* (away from first gear).

11. Check that all parts are secure and that the gears spin, slide and engage freely on the shaft. Wrap and store the assembly until it is ready to install into the crankcase. Install the complete transmission assembly as described in Chapter Five.

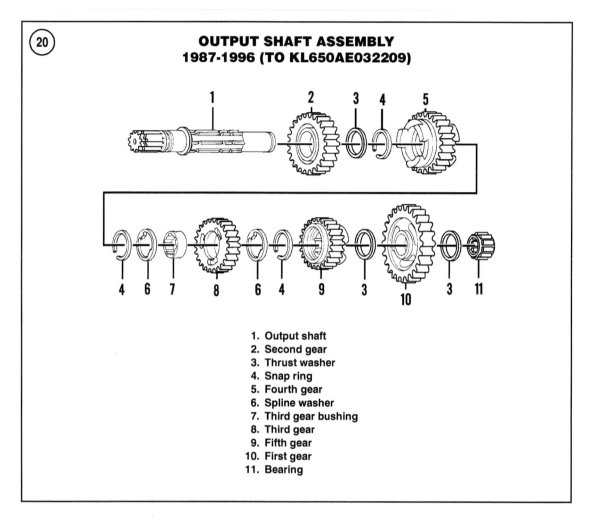

OUTPUT SHAFT ASSEMBLY
1987-1996 (TO KL650AE032209)

1. Output shaft
2. Second gear
3. Thrust washer
4. Snap ring
5. Fourth gear
6. Spline washer
7. Third gear bushing
8. Third gear
9. Fifth gear
10. First gear
11. Bearing

OUTPUT SHAFT

Use the following procedures to disassemble and assemble the output shaft. The design of the output shaft and gears changed significantly in 1996 (after engine serial number KL650AE032209). The early shaft design is shouldered to the outside of second gear and all parts are removed from one end of the shaft. The later shaft design is stepped to the inside of second gear and requires parts to be removed from both ends of the shaft. Figures of both output shaft assemblies are provided.

Disassembly
1987-1996 (To Engine Serial Number KL650AE032209)

Remove the parts from the output shaft in the following order. The parts are removed from one end

of the shaft. Clean and organize the parts as they are removed. Refer to **Figure 20**.

> *CAUTION*
> *Keep parts oriented and note their direction and original location on the*

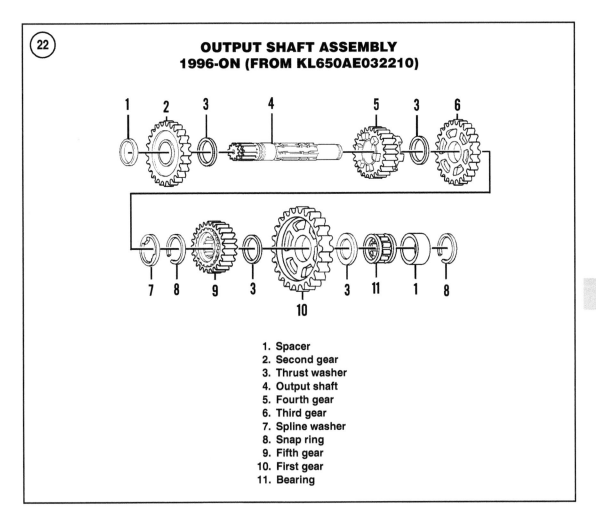

OUTPUT SHAFT ASSEMBLY
1996-ON (FROM KL650AE032210)

1. Spacer
2. Second gear
3. Thrust washer
4. Output shaft
5. Fourth gear
6. Third gear
7. Spline washer
8. Snap ring
9. Fifth gear
10. First gear
11. Bearing

shaft. Preferably, mark the outer face of each part as it is removed from the shaft. Check all snap rings and note which direction the inner, sharp edge is facing.

1. At the smooth end of the shaft, disassemble the parts in the following order:
   a. Bearing.
   b. Thrust washer.
   c. First gear.
   d. Thrust washer.
   e. Fifth gear.
   f. Snap ring.
   g. Spline washer.
   h. Third gear.
   i. Third gear bushing.
   j. Spline washer.
   k. Snap ring.
   l. Fourth gear.
   m. Snap ring.
   n. Thrust washer.
   o. Second gear.

2. Inspect each part as described in this chapter, then return it to its place until assembly.

Disassembly
1996-On (From Engine Serial Number KL650AE032210)

Remove the parts from the output shaft (**Figure 21**) in the order below. The parts are removed from both ends of the shaft. Clean and organize the parts as they are removed. Refer to **Figure 22**.

CAUTION
Keep parts oriented and note their direction and original location on the shaft. Preferably, mark the outer face

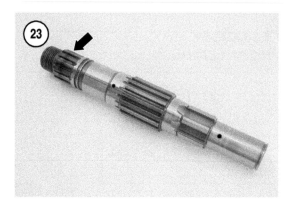

of each part as it is removed from the shaft. Check all snap rings and note which direction the inner, sharp edge is facing. For the model shown in this procedure, the notched snap rings were identical on both inner edges, and therefore could be installed in either direction.

1. Beginning at the splined end of the shaft, disassemble the parts in the following order:
   a. Spacer.
   b. Second gear.
   c. Thrust washer.
2. At the smooth end of the shaft, disassemble the parts in the following order:
   a. Snap ring.
   b. Spacer.
   c. Bearing.
   d. Thrust washer.
   e. First gear.
   f. Thrust washer.
   g. Fifth gear.
   h. Snap ring.
   i. Spline washer.
   j. Third gear.
   k. Thrust washer.
   l. Fourth gear.
3. Inspect each part as described in this chapter, then return it to its place until assembly.

Assembly
1987-1996 (To Engine Serial Number
KL650AE032209)

Before beginning assembly, have new snap rings on hand. Throughout the procedure, the orientation of the parts is made in relation to the splined end of

the shaft, where the drive sprocket is mounted. If desired, lock the splined end of the shaft in a *padded* vise. With the shaft held stable and vertical, installation of the snap rings will be easier. Do not allow the vise to damage the shaft. Refer to **Figure 20**.

CAUTION
*Refer to **Service Notes** in this chapter for the snap ring installation technique. This technique is recommended for installing the notched snap rings. The installed snap rings must grip the shaft tightly. If necessary, practice the technique using discarded snap rings.*

1. Clean and dry all parts before assembly. Lubricate all parts with engine oil.
2. Install second gear. The side of the gear with the gear dog recesses must face *out* (away from the splined end).
3. Install the thrust washer and snap ring. The snap ring must seat in the shaft groove.
4. Install fourth gear. The side of the gear with the shift fork groove must face *out* (away from the splined end).

5. Install the snap ring and spline washer. The snap ring must seat in the shaft groove. The ends of the snap ring must be offset from the spline washer teeth.

6. Align and install the third gear bushing.

CAUTION
The oil hole in the bushing must be aligned with the oil hole in the shaft.

7. Install third gear. The side of the gear with the gear dog recesses must face *in* (toward the splined end).

8. Install the spline washer and snap ring. The snap ring must seat in the shaft groove. The ends of the snap ring must be offset from the spline washer teeth.

9. Install fifth gear. The side of the gear with the shift fork groove must face *in* (toward the splined end).

10. Install the thrust washer.

11. Install first gear. The side of the gear with the gear dog recesses must face *in* (toward splined end). This is also the deep side of the gear.

12. Install the thrust washer.

13. Install the bearing.

14. With the parts in their correct positions, wrap a strong rubber band around the end of the shaft. Check that all parts are secure and that the gears spin, slide and engage freely on the shaft. Wrap and store the assembly until it is ready for installation into the crankcase. Install the complete transmission assembly as described in Chapter Five.

Assembly
1996-On (From Engine Serial Number
KL650AE032210)

Before beginning assembly, have new snap rings on hand. Throughout the procedure, the orientation of many parts is made in relation to the splined end of the shaft (**Figure 23**). If desired, lock the splined end of the shaft in a *padded* vise. With the shaft held stable and vertical, installation of the snap rings will be easier. Do not allow the vise to damage the shaft.

For the model shown in this procedure, the notched snap rings were identical on both inner edges, and therefore could be installed in either direction. Refer to **Figure 22**.

CAUTION
*Refer to **Service Notes** in this chapter for the snap ring installation technique. This technique is recommended for installing the notched snap rings. The installed snap rings must grip the shaft tightly. If necessary, practice the technique using discarded snap rings.*

1. Clean and dry all parts before assembly. Lubricate all parts with engine oil.

2. Install fourth gear (**Figure 24**). The side of the gear with the shift fork groove must face *out* (away from the splined end).

3. Install the thrust washer (**Figure 25**).

4. Install third gear (**Figure 26**). The side of the gear with the gear dog recesses must face *in* (toward the splined end).

5. Install the spline washer (**Figure 27**).

6. Install the snap ring (**Figure 28**). The snap ring must seat in the shaft groove. The ends of the snap ring must be offset from the spline washer teeth.

7. Install fifth gear (**Figure 29**). The side of the gear with the shift fork groove must face *in* (toward the splined end).

8. Install the thrust washer (**Figure 30**).

9. Install first gear (**Figure 31**). The side of the gear with the gear dog recesses must face *in* (toward the splined end).

10. Install the thrust washer (**Figure 32**).

11. Install the bearing and spacer (**Figure 33**).

12. Install the snap ring (**Figure 34**). The snap ring must seat in the shaft groove. The sharp edge of the snap ring must face *out* (away from the splined end).

13. At the splined end of the shaft, install the thrust washer (**Figure 35**).

14. Install second gear (**Figure 36**). The side of the gear with the gear dog recesses must face *in* (away from the splined end).

15. Install the spacer (**Figure 37**).

16. With the parts in their correct positions, wrap a strong rubber band around the end of the shaft.

17. Check that all parts are secure and that the gears spin, slide and engage freely on the shaft. Wrap and store the assembly until it is ready for installation into the crankcase. Install the complete transmission assembly as described in Chapter Five.

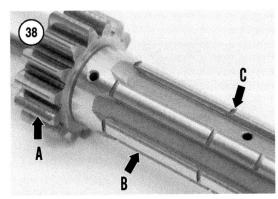

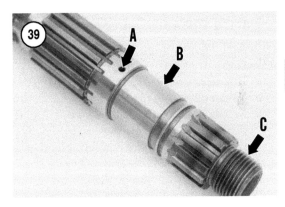

TRANSMISSION INSPECTION

Use the following procedures to inspect the shaft assemblies for all models. Because of component design changes, some inspections may not be required for the assembly being serviced.

Shaft Inspection

1. Inspect the shafts for the following:
   a. Broken or damaged gear teeth (A, **Figure 38**) on the input shaft.
   b. Worn or damaged splines (B, **Figure 38**).
   c. Rounded or damaged snap ring grooves (C, **Figure 38**).
   d. Clean oil holes (A, **Figure 39**).
   e. Wear, galling or other damage on the bearing/bushing surfaces (B, **Figure 39**). A blue discoloration indicates excessive heat.
   f. Damaged threads (C, **Figure 39**). Mildly damaged threads can be trued with a thread die.
2. Assemble the shafts as described in this chapter.

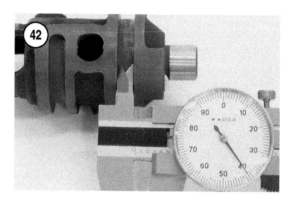

Component Inspection

1. Refer to **Figure 40** and inspect the gears for the following:
   a. Broken or damaged teeth (A).
   b. Worn or damaged splines (B).
   c. Scored, galled or fractured bore (C). A blue discoloration indicates excessive heat. For early models that use a bushing at third gear on the output shaft, check the fit of the bushing in the gear and on the shaft.
   d. Worn, damaged or rounded gear dogs and recesses (D). Any wear on the dogs and mating recesses should be uniform. If the dogs are not worn evenly, the remaining dogs will be overstressed and possibly fail. Check the engagement of the dogs by placing the gears at their appropriate positions on the countershaft, then twist the gears together. Check for positive engagement in both directions. If damage is evident, also check the condition of the shift forks, as described in this chapter.

> *NOTE*
> *The side of the gear dogs that carries the engine load will wear and eventually become rounded. The unloaded side of the dogs will remain unworn. Rounded dogs will cause the transmission to jump out of gear.*

   e. Worn or damaged shift fork groove. Measure the gear groove width (**Figure 41**). Refer to **Table 2** for specifications.
   f. Smooth gear operation on the shafts. Bored gears should fit firmly on the shaft, yet spin smoothly and freely. Splined gears should fit snugly at their position on the shaft, yet slide

smoothly and freely from side to side. If a gear is worn or damaged, also replace the gear it mates to on the other shaft.

2. Inspect the needle bearings for obvious damage or heat discoloration. The bearings should rotate freely on the shaft and in their respective bushings in the crankcase.

3. Inspect the thrust washers, spacers and spline washers. The washers should be smooth and show no signs of wear or damage. The teeth on the spline washers should not be missing or damaged.

4. Install the parts onto their shafts as described in this chapter.

SHIFT DRUM AND FORKS

As the machine is upshifted and downshifted, the shift drum and fork assembly engages and disengages pairs of gears on the transmission shafts. Gear shifting is done by the shift forks, which are guided by cam grooves in the shift drum.

It is important that the shift drum grooves, shift forks and mating gear grooves are in good condition. Excessive wear between the parts will cause unreliable and poor engagement of the gears. This

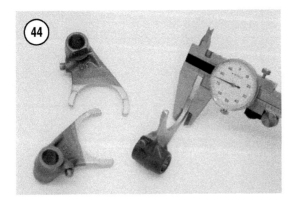

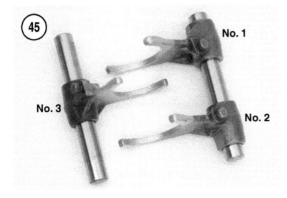

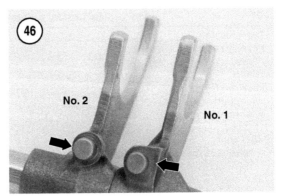

can lead to premature wear of the gear dogs and other parts.

Inspection

When inspecting the shift fork and drum assembly, replace parts that are worn or not within the specifications in **Table 2**.

1. Clean all parts in solvent and dry with compressed air.

2. Inspect the shift drum for the following:

   a. Worn shift drum grooves and *cam points*. The grooves should be a uniform width. Worn grooves can prevent complete gear engagement, which can cause rough shifting and allow the transmission to disengage. Measure the width of each groove in several locations to determine wear (**Figure 42**).

   b. Worn or damaged bearing surfaces. Besides wear, look for overheating discoloration and a lack of lubrication. Fit the shift drum into the crankcase bearings and check for play. If necessary, replace the shift drum bearings as described in Chapter Five.

3. Inspect each shift fork for wear and damage. Inspect the:

   a. Guide pin. The pin should be symmetrical and not flat on the side. Measure the diameter of the guide pin (**Figure 43**).

   b. Shift fork thickness. Measure both claws at the end (**Figure 44**).

4. Inspect the shift fork shafts for wear and obvious damage.

5. Install the No. 1, No. 2 and No. 3 shift forks onto the fork shafts (**Figure 45**). The forks should slide and pivot smoothly with no excessive play or tightness. If necessary, the forks can be identified as follows:

   a. Shift forks No. 1 and No. 2 can be identified by the casting around the guide pins (**Figure 46**). These forks mate with the output shaft gears. Shift fork No. 1 mates with fourth gear and shift fork No. 2 mates with fifth gear.

   b. Shift fork No. 3 has a smaller claw diameter and only mates with third gear on the input shaft.

6. Install the shift drum, forks and transmission assembly as described in Chapter Five.

Table 1 GENERAL TRANSMISSION SPECIFICATIONS

Transmission	Five-speed constant mesh
Primary reduction	
Type	Gear
Ratio	2.272 (75/33)
Final reduction ratio	2.866 (43/15)
Overall drive ratio	5.157 (top gear)
Gear ratios	
1st gear	2.266 (34/15)
2nd gear	
1987-1996 (to KL650 AE032209)	1.529 (26/17)
1996-on (from KL650AE032210)	1.444 (26/18)
3rd gear	
1987-1996 (to KL650AE032209)	1.181 (26/22)
1996-on (from KL650AE032210)	1.136 (25/22)
4th gear	0.954 (21/22)
5th gear	0.791 (19/24)

Table 2 TRANSMISSION SPECIFICATIONS

	Standard mm (in.)	Service limit mm (in.)
Shift drum groove width	6.05-6.20 (0.238-0.244)	6.3 (0.248)
Gear groove width	4.55-4.65 (0.179-0.183)	4.8 (0.189)
Shift fork claw thickness	4.4-4.5 (0.173-0.177)	4.3 (0.169)
Shift fork guide pin diameter	5.9-6.0 (0.232-0.236)	5.8 (0.228)

CHAPTER EIGHT

FUEL SYSTEM

This chapter provides service procedures for removing, disassembling, inspecting and repairing the carburetor. Also included is information on how the different carburetor systems operate within the carburetor. A table is provided at the end of the chapter, listing the standard carburetor specifications.

Refer to this chapter for fuel valve servicing, throttle cable replacement and air filter housing removal. Also included is a schematic of the evaporative emission control system (California models only). Refer to Chapter Three for air filter service, throttle cable adjustment and lubrication. Refer to *Safety* and *Basic Service Methods* in Chapter One.

CARBURETOR

Operation

The Keihin CVK40 is a vacuum-controlled, or constant velocity, carburetor. It uses both a throttle valve and diaphram-operated slide to regulate fuel to the engine. The throttle valve (**Figure 1**) is located on the output side of the carburetor and is connected to the throttle cables. It is not connected to any fuel-regulating device. The slide and diaphragm assembly, located at the center of the carburetor, regulates fuel by a jet needle at the bottom of the slide (**Figure 2**). The diaphragm is located at the top of the slide in the diaphragm chamber (**Figure 3**). The diaphragm divides and seals the large chamber into a lower and upper chamber.

During operation, when the throttle valve is opened, air demand and speed through the carburetor is increased. As air passes under the slide, air pressure drops in that area. This low air pressure is vented to the upper diaphragm chamber. The lower diaphragm chamber is vented to atmospheric pressure. This difference in pressure causes the slide and jet needle to rise, allowing fuel to pass into the carburetor throat. When the throttle valve is closed, the pressure differential lowers, allowing the slide and jet needle to return to a resting position.

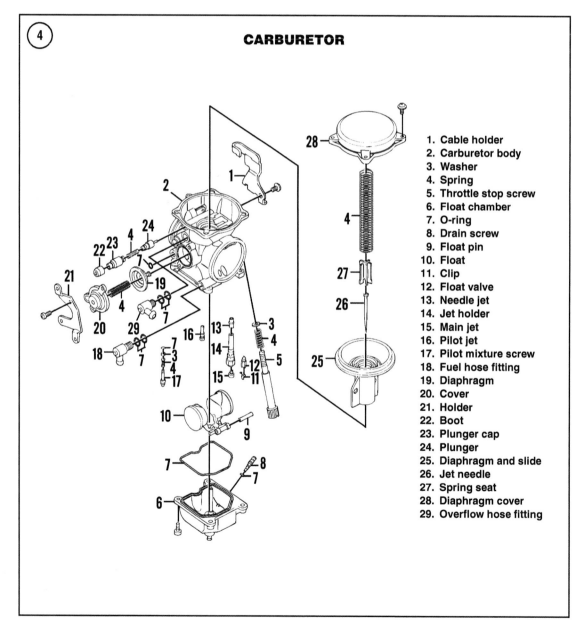

CARBURETOR

1. Cable holder
2. Carburetor body
3. Washer
4. Spring
5. Throttle stop screw
6. Float chamber
7. O-ring
8. Drain screw
9. Float pin
10. Float
11. Clip
12. Float valve
13. Needle jet
14. Jet holder
15. Main jet
16. Pilot jet
17. Pilot mixture screw
18. Fuel hose fitting
19. Diaphragm
20. Cover
21. Holder
22. Boot
23. Plunger cap
24. Plunger
25. Diaphragm and slide
26. Jet needle
27. Spring seat
28. Diaphragm cover
29. Overflow hose fitting

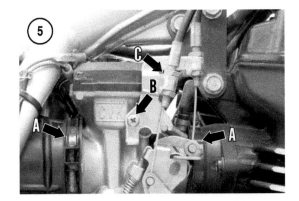

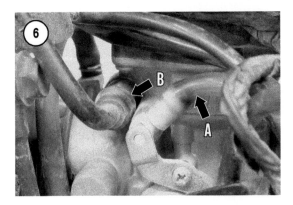

Refer to *Carburetor Systems* in this chapter for operational details of specific fuel systems within the carburetor.

Removal and Installation

Refer to **Figure 4**.

1. Support the bike so it is stable and secure.
2. Remove the fuel tank (Chapter Fifteen).
3. Remove the clamps from the air filter housing duct and intake duct (A, **Figure 5**).
4. Remove the cable holder screw (B, **Figure 5**). If the screw is too tight, disconnect the throttle cables (C, **Figure 5**) and leave the holder attached to the carburetor. If the cables are removed, mark the cables so they can be reinstalled in the correct position.
5. On the left side of the carburetor, remove the overflow hose from the fitting (A, **Figure 6**).
6. Remove the carburetor from the ducts as follows:
   a. Pry up on the air filter housing duct to break it free from the carburetor. If necessary, lightly spray penetrating lubricant under the duct to reduce its grip on the carburetor. Pry the duct off the carburetor, keeping the carburetor upright.
   b. Push the carburetor back and pry the intake duct off the carburetor.
7. Tilt the carburetor to the left to access the choke plunger. Remove the choke plunger (B, **Figure 6**).
8. Remove the carburetor out the right side of the engine (**Figure 7**).
9. Open the drain on the bottom of the float chamber and drain any remaining fuel.
10. Disconnect the choke plunger (**Figure 8**) by pushing up on the spring and removing the cable from the plunger.
11. Reverse this procedure to install the carburetor. Note the following:
   a. Clean and lightly lubricate the inside edges of the ducts so the carburetor easily seats.
   b. Align and engage the tab on the carburetor with the slot in the intake duct. Although there are slots on both sides of the duct (**Figure 9**), there is only one tab on the carburetor.
   c. Check the fuel and overflow hose routing.
   d. Check/adjust the throttle cables (Chapter Three).
   e. Check the throttle for proper operation.

8

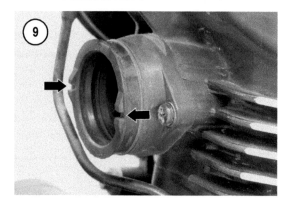

f. Check the carburetor for leaks.

g. If the carburetor was disassembled, adjust the carburetor (Chapter Three).

Disassembly and Assembly

Refer to **Figure 4**.

1. Remove the air cutoff valve assembly as follows:

    a. Remove the screws from the holder (**Figure 10**). The holder is under spring pressure. Keep pressure on the holder while removing the screws.

    b. Remove the holder, cover and spring (**Figure 11**).

    c. Remove the diaphragm and O-ring (**Figure 12**).

2. Remove the hose fittings (**Figure 13**).

3. Remove the diaphragm cover and the slide assembly as follows:

    a. Remove the cover (**Figure 14**). The cover is under slight spring pressure. Hold the cover in place while removing the four screws, then lift off the cover.

    b. Remove the spring (**Figure 15**).

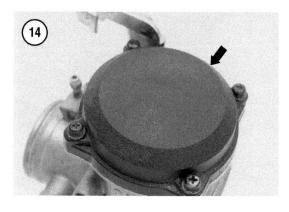

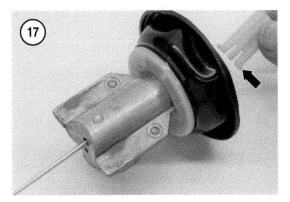

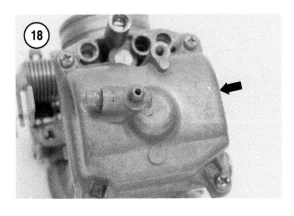

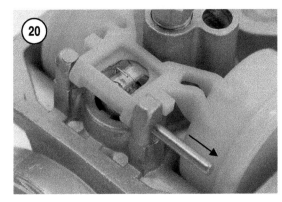

c. From the intake side, push up on the slide so it can be lifted from the carburetor (**Figure 16**).

> *CAUTION*
> *Do not lift or hold the slide by the diaphragm. Do not allow the jet needle to get damaged while raising it from the carburetor.*

d. Remove the spring seat (**Figure 17**), then tilt the slide to remove the needle.

4. Remove the float chamber (**Figure 18**). Remove the drain screw and O-ring from the chamber.

5. Remove and disassemble the float assembly (**Figure 19**) as follows:

a. Remove the pin (**Figure 20**) from the float.

b. Carefully lift out the float and float valve (**Figure 21**).

c. Slide the clip and float valve off the float tab (**Figure 22**).

6. Remove the pilot mixture screw (**Figure 23**) as follows:

a. If a plug is installed over the screw, remove the plug by drilling through it with a *small* drill bit. The hole only needs to be large

enough to insert a small sheet metal screw or pick tool. Using a large drill bit risks damaging the pilot mixture screw and the internal threads. When drilling, use a light touch so the bit does not contact the screw when the bit breaks through the plug. Carefully pry the plug after the hole is drilled.

b. Make a scratch on the edge of the bore, in line with the slot in the screw. This will be used as a reference point when installing the screw.

c. Turn the screw clockwise and accurately count the number of turns it takes to seat the screw into the carburetor.

d. Record the number of turns.

e. Remove the pilot mixture screw, spring, washer and O-ring (**Figure 24**).

7. Remove the main jet assembly as follows:

a. Remove the main jet (**Figure 25**).

b. Remove the jet holder (**Figure 26**).

c. Remove the needle jet (**Figure 27**). Unseat the jet from the carburetor venturi by hand. Note that the long end of the jet fits into the carburetor throat.

8. Remove the pilot jet (**Figure 28**).

9. Remove the throttle stop screw, spring and washer.

10. Clean and inspect the parts as described in this chapter.

11. Refer to *Carburetor Systems* in this chapter for the function of the jets and their affect on performance.

12. Reverse this procedure to assemble the carburetor. Note the following:

a. Install new, lubricated O-rings. Install the small O-ring under the air cutoff valve with the round side facing out.

b. Attach the float valve and clip to the float before installing the parts.

c. Check and adjust the float height. Refer to *Float Adjustment* in this chapter.

d. When installing the pilot mixture screw, *lightly* seat the screw, then turn it out the number of turns recorded during disassembly. Refer to the reference mark on the carburetor for the original setting. If the number of turns is not known, refer to **Table 1** for the basic setting.

e. It is easiest to install the spring, cover and holder over the air cutoff valve diaphragm as

a single assembly (**Figure 11**). Check that the spring is seated in the cover and that the hose fittings are installed before assembling.

f. When installing the slide and diaphragm, or the diaphragm in the air cutoff valve, the diaphragm must be seated at its edge before installing the cover.

g. Install the carburetor as described in this chapter.

CARBURETOR CLEANING AND INSPECTION

Use the following procedure to clean and inspect the carburetor. Refer to *Carburetor Systems* in this chapter to gain an understanding of how the carburetor passages for each system are interconnected. When cleaning, this will help in verifying whether the entire length of a passage is clean. Always replace worn or damaged parts. Refer to **Figure 4** as needed.

It is recommended to use a commercial cleaner specifically for carburetors, since the cleaner contains agents for removing fuel residues and buildup. Use a cleaner that is harmless to rubber and plastic parts. Follow the manufacturer's instructions when using the cleaner.

> *CAUTION*
> *Do not clean the jet orifices or seats with wire or drill bits. These items can scratch the surfaces and alter flow rates, or cause leaking.*

> *NOTE*
> *Because of heat and age, O-rings eventually lose there flexibility and do not seal properly. It is standard practice to replace all O-rings and gaskets when rebuilding a carburetor.*

1. Clean all the parts in carburetor cleaner. Use compressed air to clean all passages, orifices and vents in the carburetor body.

2. Inspect the main jet assembly and pilot jet (**Figure 29**). Check that all holes, in the ends and sides, are clean and undamaged.

3. Inspect the pilot mixture screw assembly and choke plunger (**Figure 30**).

   a. Inspect the screw and plunger tips for dents or wear.

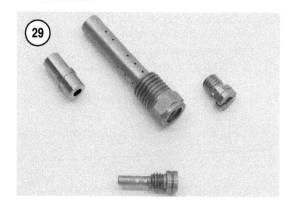

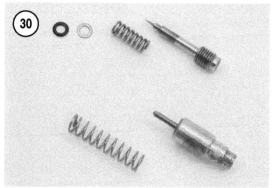

b. The spring coils should be resilient and not be crushed.

c. The plunger should move freely in its bore in the carburetor.

4. Inspect the throttle stop screw, spring and washer (**Figure 31**). Inspect the screw tip and threads for damage. The spring coils should be resilient and not be crushed.

5. Inspect the diaphragm and slide (**Figure 32**). If either part is damaged, the parts are only available together, as a single part number.

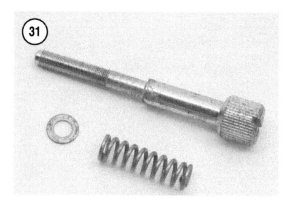

a. Inspect the slide for wear and scratches. Fit the slide into the carburetor body and check for smooth operation. The slide should vertically move freely, but have minimal front to back play.

b. Inspect the diaphragm for dryness, tears and holes. The diaphragm *must* be undamaged in order to isolate the pressure differences that are above and below the diaphragm. A leaking diaphragm prevents the slide from reaching/maintaining its normal level, for any throttle position off idle. Engine performance will be noticeably diminished.

6. Inspect the diaphragm cover and needle assembly (**Figure 33**).

a. The diaphragm cover *must* be undamaged in order to maintain low pressure in the upper chamber of the carburetor. A cracked or loose cover affects engine performance similarly to a damaged diaphragm.

b. The jet needle must be smooth and evenly tapered (**Figure 34**). If it is stepped, dented, worn or bent, replace the needle.

7. Inspect the float and float valve assembly (**Figure 35**).

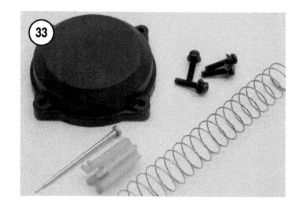

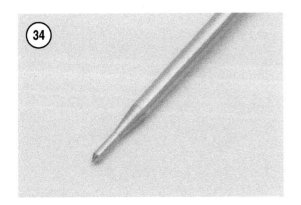

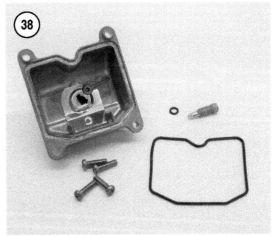

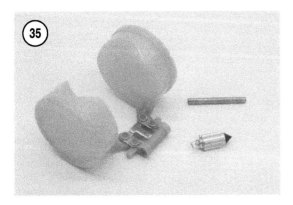

a. Inspect the tip of the float valve (**Figure 36**). If it is stepped or dented, replace the float valve.

b. Lightly press on the spring-loaded pin (**Figure 37**) in the float valve. The pin should easily move in and out of the valve. If it is varnished with fuel residue, replace the float valve.

c. Inspect the float valve seat in the carburetor body. The seat should be clean and scratch-free. If it is not, the float valve will not seat properly and the carburetor will overflow.

d. Inspect the float and pin. Submerge the float in water and check for leakage. Replace the float if water or fuel is detected inside the float. Check that the float pin is straight and smooth. It must be a slip-fit in the float.

8. Inspect the float bowl assembly (**Figure 38**).

a. Check that all residue is removed from the interior of the bowl.

b. Inspect threaded parts for damage.

c. Replace the O-ring on the bowl and drain screw.

9. Inspect the air cutoff valve assembly (**Figure 39**).

a. The diaphragm must be free of damage in order to operate properly.

b. The plunger on the end of the diaphragm stem should move freely in its bore in the carburetor.

c. Inspect the small air hole at the edge of the cover for cleanliness.

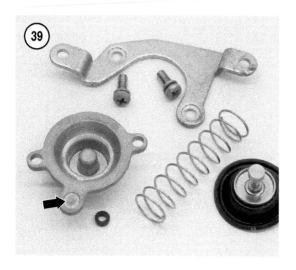

10. Inspect the hose fittings for damage (**Figure 40**). Replace the O-rings on both fittings.

FLOAT ADJUSTMENT

The float and float valve maintain a constant and measured fuel level in the float bowl. As fuel is used, the float lowers and allows more fuel past the valve. As the fuel level rises, the float closes the valve when the required fuel level is reached. If the float is out of adjustment, the fuel level will be too high or low. A low fuel level causes the engine to run as if the jetting is too lean. A high fuel level will cause the engine to run as if the jetting is too rich. It may also cause fuel overflow.

1. Remove the carburetor as described in this chapter.
2. Remove the float bowl.
3. Lightly touch the float to ensure the float valve is seated.
4. Hold the carburetor so the float valve remains seated, but the spring-loaded pin in the valve is not being compressed by the tab on the float (**Figure 41**). The tab should only *touch* the pin.
5. Measure the distance from the carburetor body gasket surface to the highest point on the float (**Figure 41**). Refer to **Table 1** for the required float height.
6. If the float height is incorrect, do the following:
   a. Remove the float assembly from the carburetor.
   b. Remove the float valve and clip.
   c. Bend the float tab in the appropriate direction to raise or lower the float. Use care when

bending the tab to prevent breaking the plastic lugs or float.
   d. Assemble the float and recheck the height. Adjust, if necessary.
7. Install the float bowl.
8. Install the carburetor as described in this chapter.

CARBURETOR SYSTEMS

Before disassembling the carburetor, understand the function of the pilot, needle and main jet systems. When evaluating or troubleshooting these systems, keep in mind that their operating ranges overlap one another during the transition from closed to fully open throttle.

Common factors that affect carburetor performance are altitude, temperature and engine load. If the engine is not running or performing up to expectations, check the following before adjusting or replacing the components in the carburetor:

1. Throttle cables. Check that the cables are not dragging and are correctly adjusted.
2. Air filter. Check that the filter is clean.

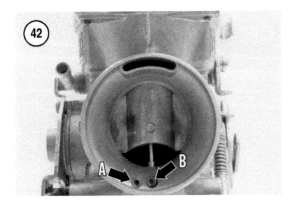

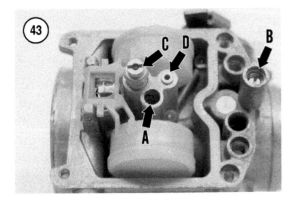

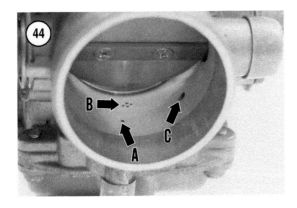

3. Fuel flow. Check that fuel is adequately flowing from the fuel tank to the carburetor. Inspect in-line filter screens for plugging.

4. Ignition timing. Check that timing is correct.

5. Choke. Check that the choke fully opens and closes.

6. Muffler. Check that the muffler is not restricting flow.

7. Brakes. Check that the brake pads are not dragging on the discs.

Pilot Jet System

The pilot system controls the air/fuel ratio from closed throttle to about 1/4 throttle. Air enters the pilot air jet (A, **Figure 42**), where it passes to the pilot jet (A, **Figure 43**). The pilot jet draws fuel from the float chamber and mixes it with the air from the pilot air jet. The atomized air/fuel mixture passes to the pilot mixture screw (B, **Figure 43**), where it is regulated into the throat of the carburetor. The mixture is discharged from the pilot hole (A, **Figure 44**). Turning the pilot mixture screw in will *lean* the air/fuel mixture entering the engine, while turning the screw out will *enrich* the mixture.

The pilot hole and mixture screw affects idle and low engine speeds. As the throttle plate is opened, it uncovers the bypass pilot holes (B, **Figure 44**), which then become effective. These holes are connected to the passage between the pilot jet and pilot mixture screw. They are not affected by the mixture screw. As engine speed increases, fuel is drawn from these passages directly from the pilot jet.

For high altitudes, the pilot jet is interchangeable with a jet that provides a leaner air/fuel mixture. Refer to **Table 1** for the recommended pilot jet sizes.

Jet Needle System

The jet needle is connected to the slide and controls the mixture from approximately 1/4 to 3/4 throttle. Air enters the main air jet (B, **Figure 42**), where it passes to the needle jet holder and needle jet. These parts are located above the main jet (C, **Figure 43**). The needle jet holder mixes fuel from the float chamber with the air from the main air jet. The atomized air/fuel mixture passes to the needle jet, where it is regulated by the jet needle into the throat of the carburetor. As the throttle is opened, the needle rises and fuel is regulated by the needle taper. The vertical position of the needle in the slide is not adjustable.

Main Jet System

The main jet (C, **Figure 43**) is screwed to the bottom of the needle jet holder and controls the mixture from approximately 3/4 to full throttle. The main jet is numbered and is interchangeable with jets that provide a leaner or richer air/fuel mixture. Replacing the standard jet with a larger numbered jet will make the mixture *richer*. Jets with a smaller number will make the mixture *leaner*. Many of these jets are for off-road use only.

For high altitudes, the main jet is interchangeable with a jet that provides a leaner air/fuel mixture. Refer to **Table 1** for the standard recommended main jet sizes.

Choke System

The choke system consists of a plunger assembly (**Figure 45**) and starter jet (D, **Figure 43**). The jet discharges fuel from the orifice in the plunger bore (A, **Figure 46**). The plunger bore also intersects an air vent (A, **Figure 47**), connecting the bottom of the slide diaphragm chamber and the throat of the carburetor.

When the choke is operated, the plunger opens the fuel and air passages. As the engine is cranked, air and fuel is drawn to the carburetor throat. When the choke plunger is closed, the air and fuel passages are closed.

This type of choke is most effective if the throttle remains closed during startup, in order to maintain high vacuum at the air and fuel passages.

Air Cutoff Valve System

The function of the air cutoff valve (**Figure 48**) is to richen the pilot jet system air/fuel mixture during compression braking, such as descending steep grades, when the engine speed is high, but the throttle is closed. Without this valve, the engine will develop a lean air/fuel mixture in the pilot system, which will cause backfiring and possible engine damage. The air cutoff valve consists of a one-piece diaphragm with cylindrical plunger, spring and cover assembly (**Figure 39**). The function of the assembly is to open and close an air passage in the plunger bore.

During acceleration and steady running speeds, the diaphragm and plunger are in the down position. This allows air to vent from a passage in the slide diaphragm chamber (B, **Figure 47**) to the plunger bore (B, **Figure 46**). The air then passes to the pilot jet system. During deceleration, when the throttle valve is closed, engine vacuum vents through a passage (C, **Figure 44**) leading to the air cutoff valve cover. The vacuum pulls the diaphragm out, causing the plunger to block the air passage in the plunger

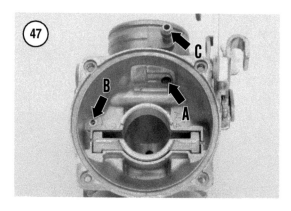

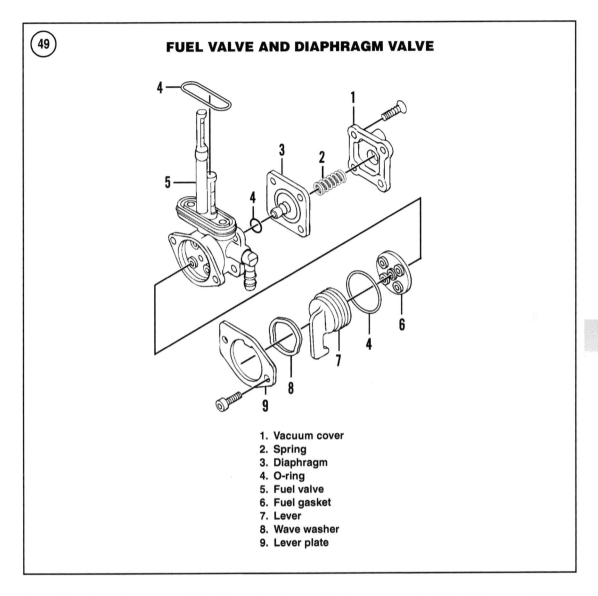

FUEL VALVE AND DIAPHRAGM VALVE

1. Vacuum cover
2. Spring
3. Diaphragm
4. O-ring
5. Fuel valve
6. Fuel gasket
7. Lever
8. Wave washer
9. Lever plate

8

bore. This reduces the amount of air going to the pilot jet system, and a rich fuel mixture is discharged from the pilot hole (A, **Figure 44**).

When acceleration resumes, the vacuum holding the diaphragm out is reduced, and the spring pushes the diaphragm and plunger down, again creating a normal fuel mixture in the pilot jet system.

FUEL VALVE

The fuel valve is equipped with a lever and diaphragm valve. The lever controls off, on and reserve fuel positions, while the diaphragm valve controls the flow of fuel out of the fuel valve. The diaphragm

valve is vacuum-actuated and is connected by a hose to a fitting on the carburetor (C, **Figure 47**). When the engine is cranked, or running, vacuum in the carburetor throat opens the diaphragm valve. In order for the diaphragm valve to operate properly, it must be kept clean and tight, in order to maintain vacuum. The vacuum hose must also be tight and in good condition.

Removal, Inspection and Installation

Refer to **Figure 49**.

1. Remove the fuel tank (Chapter Fifteen).

2. Drain the fuel from the tank. Since the fuel valve does not allow fuel to pass without vacuum applied, the tank must be siphoned or emptied through the fill opening.

> *WARNING*
> *Drain the fuel into an approved container. Perform the draining procedure a safe distance away from the work area.*

3. Place the fuel tank vertically on the workbench, then remove the bolts securing the fuel valve (**Figure 50**).

4. Remove the lever plate, wave washer, lever and fuel gasket (**Figure 51**).

> *CAUTION*
> *Before removing the cover in Step 5, check that it is separated from the thin diaphragm (**Figure 52**) that contacts the cover. If necessary, use a small tool to gently separate the cover around its perimeter.*

5. Remove the vacuum cover and spring (**Figure 53**).

6. Lift the diaphragm assembly from the fuel valve (**Figure 54**). Again, use a small tool to separate the inner diaphragm from the valve.

7. Clean the parts (**Figure 55**).

   a. Wash the valve, fuel filters and other metal parts in solvent.

   b. Gently wipe the diaphragm assembly.

8. Inspect the parts.

   a. Check for buildup in the screen filters. If buildup is evident, lightly scrub the screens with a nylon brush and solvent. Carefully

blow compressed air through the screen, from the inside to the outside.

b. Inspect the valve passages for buildup.

c. Inspect the vacuum holes (**Figure 56**) in the vacuum cover for cleanliness.

d. Inspect the diaphragm vent passage (**Figure 57**) for cleanliness. If this vent is clogged, the diaphragm will drag open when vacuum is applied.

e. Inspect the remaining parts for obvious damage.

f. Replace all O-rings. If fuel has leaked past the fuel gasket, replace the gasket. Eventually, the gasket will wear on its face, where it contacts the lever.

g. Inspect the vacuum hose connecting the fuel valve to the carburetor. The hose must be clean and tight to the fittings, in order to hold vacuum.

9. Assemble the valve. Note the following:

a. Lightly lubricate the O-rings. Also, lubricate the fuel gasket.

b. Install the diaphragm assembly so the vent (**Figure 58**) faces the fuel fitting. Check that both diaphragms lay flat before installing the vacuum cover.

c. Install the vacuum cover with the vacuum fitting facing the fuel fitting. Tighten the cover screws equally in several passes.

10. Install the fuel valve into the fuel tank and equally tighten the bolts.

11. Install the fuel tank (Chapter Fifteen).

12. When filling the tank, start with a small amount of fuel and check for leaks. Operate the lever and check that all positions are leak-free.

EVAPORATIVE EMISSIONS CONTROL SYSTEM (CALIFORNIA MODELS ONLY)

No adjustments are required for the evaporative emissions control system. Visual inspection of the hoses and connections should be made as recommended in Chapter Three. If the system is suspected of causing poor running, or obvious damage has occurred to the system, refer inspection and testing of the separator and canister to a Kawasaki dealership. Refer to **Figure 59** for parts identification and hose routing.

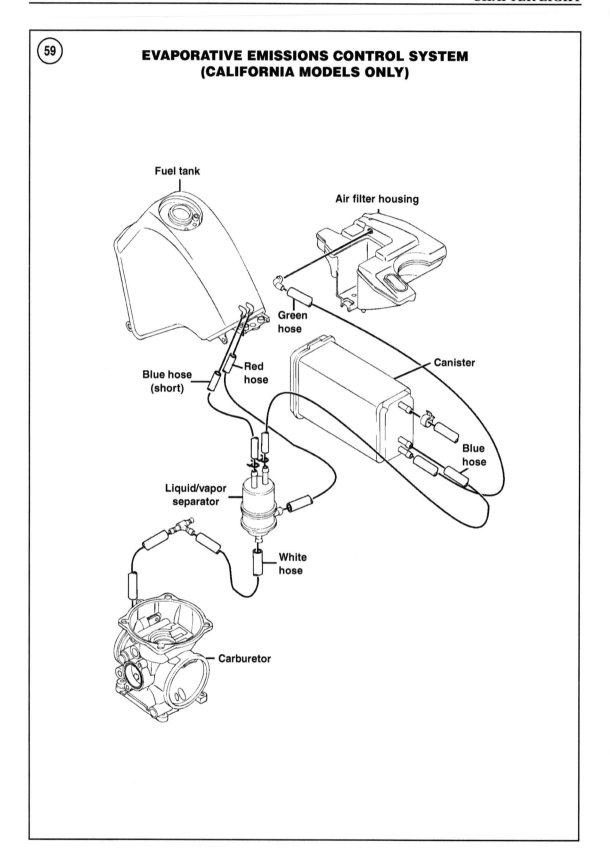

**⑲ EVAPORATIVE EMISSIONS CONTROL SYSTEM
(CALIFORNIA MODELS ONLY)**

Fuel tank

Air filter housing

Green hose

Canister

Blue hose (short)

Red hose

Blue hose

Liquid/vapor separator

White hose

Carburetor

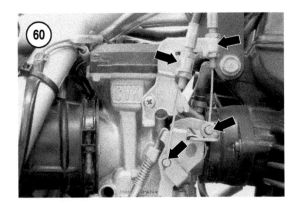

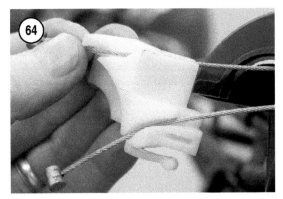

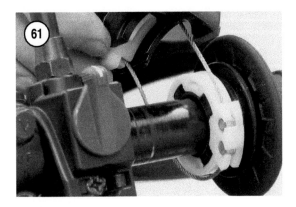

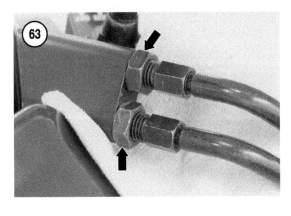

THROTTLE CABLE REPLACEMENT

The throttle uses two cables. One cable pulls the throttle open during acceleration, while the other pulls the throttle closed during deceleration. In operation, the cables always move in opposite directions to one another. Use the following procedure to replace the throttle cables.

1. Remove the fuel tank.

2. At the carburetor, loosen the locknuts and remove the cables from the holders and throttle valve (**Figure 60**). The left cable is the accelerator cable. Identify the cables so the new cables can be matched and installed in the correct position.

3. Note how the cables are routed, then pull the cables from the frame.

4. At the handlebar, remove the screws from the housing and separate the parts (**Figure 61**). Remove the complete throttle assembly from the handlebar.

5. Remove the cable ends from the throttle drum, then remove the cable guide from the upper housing (**Figure 62**).

6. Remove the rubber boots from the throttle cable adjusters. Loosen the locknuts (**Figure 63**), then unscrew the adjusters from the throttle housing. Note that the top cable is the accelerator cable.

7. Clean the throttle assembly and handlebar.

8. Lubricate the cables with an aerosol cable lubricant. Lubricate the throttle drum, cable ends and cable guide with lithium grease.

9. Install the new cables into their positions in the upper housing. Note that the angled end of the cables fits into the housing. Check that the cables are in their correct hole.

10. Fit the cables into the cable guide (**Figure 64**). The top cable is the accelerator cable.

11. Fit the cable guide into the upper housing and hold it in place.

12. Install the grip on the handlebar.

13. Route the accelerator cable over the drum and the decelerator cable under the drum (**Figure 61**). Seat the cable ends in the drum.

14. Carefully seat the assembly onto the lower housing. Hold the parts together and install the screws.

15. Route the cables through the frame and to the carburetor.

16. Install the accelerator cable in the left holder, then install the remaining cable.

17. Pull all slack out of the cables, then check that they move in the correct direction when the throttle is operated.

18. Connect the cables to the carburetor, then adjust the cables as described in Chapter Three.

19. Install the fuel tank when adjustment and proper operation is verified.

AIR FILTER HOUSING

The air filter housing is attached to the subframe. It is not necessary to remove the housing in order to service the air filter. Service the air filter as described in Chapter Three. If the air filter housing assembly must be removed, do the following:

1. Disconnect/remove the subframe (Chapter Fifteen).

2. Remove the bolts securing the housing to the subframe.

Table 1 CARBURETOR SPECIFICATIONS

Carburetor type	Keihin constant velocity
Carburetor bore diameter	40 mm
Identification number	CVK40
Idle speed	1200-1400 rpm
Main jet number	
Below 1220 m (4000 ft.)	148
Above 1220 m (4000 ft.)	145
Main air jet number	50
Needle jet number	6
Jet needle	N31R
Pilot jet number	
Below 1220 m (4000 ft.)	40
Above 1220 m (4000 ft.)	38
Pilot air jet number	70
Pilot mixture screw turns out	1 3/8
Starter jet number	52
Float height*	17.5 mm (0.69 in.)
*Refer to text.	

CHAPTER NINE

ELECTRICAL SYSTEM

This chapter provides access, service and test procedures for electronic components.

Also provided is a section that explains the basic operation of the ignition and charging systems. A basic knowledge of the systems is often helpful in logically troubleshooting and testing the components.

WIRING DIAGRAMS

Refer to the individual circuit diagrams in this chapter and the main wiring diagrams at the end of the manual.

NOTE
Scan the QR code or search for "Clymer Manuals YouTube Tech Tips" to see an overview on electrical troubleshooting with a wiring diagram.

RESISTANCE TESTING GUIDELINES

When testing component resistance, use the provided specifications as a *guide* in judging the condition of a component. Electrical test results vary, depending on ambient temperature and the quality and condition of the meter being used. Other factors include the cleanliness and overall condition of the wiring harness, terminals and connectors.

Use caution when considering the replacement of a component, particularly when it is marginally out of specification. Test the parts that interact with the component to ensure the problem is isolated. Perform a continuity test on appropriate wiring harnesses before assuming the part is faulty. Knowledge of how the system operates is also helpful in correctly determining faulty components.

Whenever in doubt of the test results, take the component to a dealership for evaluation. Comparison to a new part often reveals the condition of the component.

Refer to the wiring diagrams in this chapter and at the back of the manual to identify wires, plugs and components.

CONTINUITY
TESTING GUIDELINES

Circuits, switches, light bulbs and fuses on the motorcycle can be checked for continuity (a completed circuit), using an ohmmeter connected to the appropriate color-coded wires in the circuit. Tests can be made at the connector plug or at the part itself. Use the following procedure as a guide in performing general continuity tests.

> *CAUTION*
> *When performing general continuity checks do not turn on the ignition switch. Damage to parts and test equipment could occur. Also, verify that power from the battery is not routed directly into the test circuit, regardless of ignition switch position.*

1. Refer to the wiring diagram at the back of the manual and find the part to be checked.

2. Identify the wire colors leading to the part and determine which pairs of wires should be checked. For any check, the circuit should begin at the plug, pass through the part, then return to the plug.

3. Determine when continuity should exist.

   a. Typically, whenever a switch or button is turned on, it *closes* the circuit, and the meter should indicate continuity.

   b. When the switch or button is turned off, it *opens* the circuit, and the meter should not indicate continuity.

   c. Refer to the switch continuity diagrams within the wiring diagrams.

4. On the motorcycle, trace the wires from the part to the nearest connector plug.

5. Separate the connector plug.

6. Connect an ohmmeter to the plug half that leads to the part being checked.

7. If the test is being made at the terminals on the part, remove all other wires connected to the terminals so they do not influence the meter reading.

8. Operate the switch/button and check for continuity.

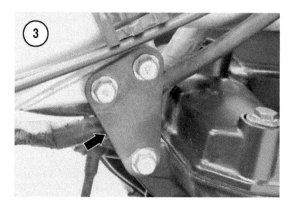

IGNITION AND CHARGING SYSTEM
OPERATION

A permanent magnet alternator is located on the left end of the crankshaft, and is the source of energy for powering the ignition system and charging the battery. A capacitor discharge ignition (CDI) system is used on this engine. When the engine is running, the current produced by the exciter coil in the alternator goes to the CDI unit and is stored in a capacitor. When the rotor is in the correct position for ignition, the ignition pickup coil signals the CDI

fully charged, the excess current is shunted and dissipated as heat by the regulator/rectifier.

ALTERNATOR COVER

Removal and Installation

The alternator cover must be removed to access the rotor, stator and pickup coil. The stator and pickup coil are mounted on the inside of the alternator cover.

1. Remove the skid plate (Chapter Fifteen).

2. Remove the chain sprocket guard (Chapter Eleven). The sprocket does not have to be removed.

3. Disconnect the neutral switch wire and remove the shift lever (**Figure 1**).

4. If the stator will be tested, remove the fuel tank (Chapter Fifteen) and disconnect the stator and pickup coil connectors, located on the right side of the frame (**Figure 2**).

5. If the stator and cover must be completely removed from the motorcycle, remove the upper engine mounting brackets (**Figure 3**) and route the wires out of the frame and engine.

6. Remove the 10 bolts from the perimeter of the cover (**Figure 4**).

7. Pull the cover away from the engine. Magnetic resistance will be felt as the cover is unseated.

> *WARNING*
> *Keep fingers away from the edge of*
> *the cover to avoid possible pinching.*

8. Remove the cover gasket, then account for the two cover dowels and the washer on each starter gear shaft (**Figure 5**).

9. Remove and/or test the stator and pickup coil (**Figure 6**) as described in this chapter.

10. Reverse this procedure to install the stator and alternator cover. Note the following:

    a. Install a new cover gasket.

    b. Check that all wires are routed and secured.

    c. Clean electrical connections, then apply dielectric grease when assembling.

    d. Tighten the upper engine mounting bolts to 25 N•m (18 ft.-lb.).

unit to release the stored charge to the coil. The charge of current into the coil primary windings induces a much higher voltage in the secondary windings, which fires the spark plug. This system is essentially maintenance free, with the spark plug being the only part requiring routine replacement.

The battery is kept charged by separate coils in the alternator. When the engine is running, the current produced by the coils is sent to the regulator/rectifier. The power is converted to direct current and regulated to the battery. If the battery is

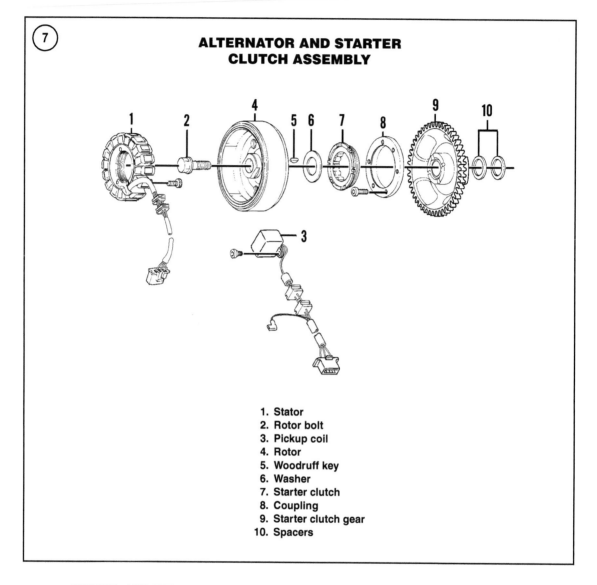

ALTERNATOR AND STARTER CLUTCH ASSEMBLY

1. Stator
2. Rotor bolt
3. Pickup coil
4. Rotor
5. Woodruff key
6. Washer
7. Starter clutch
8. Coupling
9. Starter clutch gear
10. Spacers

STATOR AND PICKUP COIL

Removal and Installation

The stator and pickup coil are mounted on the inside of the alternator cover. The stator and pickup coil can be tested without removing them from the cover. To access the wire connectors, remove the fuel tank (Chapter Fifteen) and disconnect the stator and pickup coil connectors, located on the right side of the frame (**Figure 2**). Refer to *Charging System* to test the stator charging coil resistance and *Ignition System* to test the pickup coil and stator exciter coil resistance.

Refer to **Figure 7**.

1. Remove the alternator cover as described in this chapter.

2. Remove the three bolts from the stator (A, **Figure 8**).

3. Remove the two screws from the pickup coil (B, **Figure 8**).

4. Remove the wiring harness clamps from the perimeter of the cover.

5. Remove the wire grommet from the cover, then remove the parts.

6. Reverse this procedure to install the stator and pickup coil.

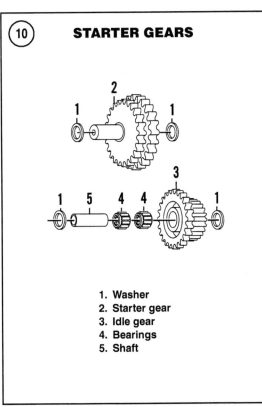

STARTER GEARS

1. Washer
2. Starter gear
3. Idle gear
4. Bearings
5. Shaft

ROTOR AND STARTER CLUTCH

The starter clutch is mounted on the back of the rotor. The starter gears and rotor must be removed to access the clutch. To remove the rotor, hold it with a rotor holder tool, then remove it with a rotor puller. The tools shown in the following procedure are the Kawasaki rotor holder (part No. 57001-1184) and rotor puller (part No. 57001-1185).

NOTE
*If troubleshooting the starter clutch, the clutch can be checked for free-wheel and lockup without removing the rotor. Turn the clutch gear clockwise (**Figure 9**). The clutch gear should turn freely and smoothly in that direction. Attempt to turn the gear counterclockwise. The gear should not turn. If the gear turns in both directions or is always locked up, disassemble and inspect the clutch assembly.*

Removal and Installation

Refer to **Figure 10**.
1. Remove the alternator cover as described in this chapter.
2. Remove the starter gears as follows:
   a. Remove the outer washer and idle gear (**Figure 11**).
   b. Remove the idle gear shaft, bearings and inner washer (**Figure 12**).
   c. Remove the starter gear and washers (**Figure 13**).
3. Remove the rotor bolt as follows:

a. Hold the rotor stationary with the rotor holder tool (**Figure 14**). Place the tool under the footpeg when loosening the rotor bolt.

b. Remove the rotor bolt.

4. Loosen the rotor and starter clutch as follows:

a. To aid in removal, spray penetrating lubricant into the rotor bore and Woodruff key area. Apply grease to the end and threads of the rotor puller.

b. Thread the puller into the crankshaft and hold the rotor stationary with the rotor holder tool (**Figure 15**). Place the tool on top of the footpeg when tightening the rotor puller.

c. Tighten the rotor puller until the rotor breaks free from the crankshaft taper. The rotor is tight and requires a wrench that provides high leverage.

d. Remove the rotor puller from the crankshaft.

5. Remove the rotor and starter clutch from the crankshaft. Hold the starter clutch to prevent it from falling from the rotor (**Figure 16**).

6. Remove the Woodruff key and spacers from the crankshaft (**Figure 17**).

7. Inspect and lubricate the parts as described in this section.

8. Reverse this procedure to install the rotor/starter clutch assembly and the starter gears. Note the following:

a. Install a new rotor bolt.

b. Tighten the rotor bolt to 120 N•m (88 ft.-lb.) to seat the rotor, then loosen the bolt.

c. Tighten the rotor bolt to 175 N•m (129 ft.-lb.).

d. Lubricate the starter clutch gear bearing, starter gear shafts and bores with molydisulfide grease. Lubricate all other parts with engine oil.

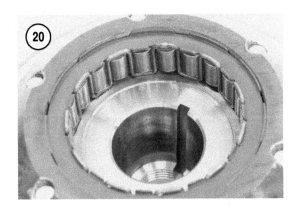

e. Install the alternator cover as described in this chapter.

Inspection

1. Inspect the clutch for proper operation as follows:

    a. With the clutch gear facing up, turn the gear *counterclockwise*. The gear should turn freely and smoothly in that direction (**Figure 18**).

    b. Attempt to turn the gear clockwise. The gear should not turn.

    c. If the gear turns in both directions or is always locked up, disassemble and inspect the clutch assembly.

2. Remove the clutch sprocket from the rotor. Turn the sprocket counterclockwise and twist it squarely away from the rotor. Remove the washer from the clutch.

3. Clean and inspect the clutch assembly (**Figure 19**).

    a. Inspect the clutch gear teeth for wear or damage.

    b. Inspect the clutch gear bearing for wear or scoring.

    c. Inspect the washer for damage.

    d. Inspect the clutch rollers (**Figure 20**). The rollers should be undamaged and operate smoothly.

4. If the clutch is damaged, remove the clutch from the rotor as follows:

    a. Remove the six bolts (**Figure 21**), then remove the clutch and coupling from the rotor.

    b. Install a new clutch and the coupling.

    c. Apply threadlocking compound to the bolts, then tighten the bolts equally in several passes to 34 N•m (25 ft.-lb.).

STARTER

22

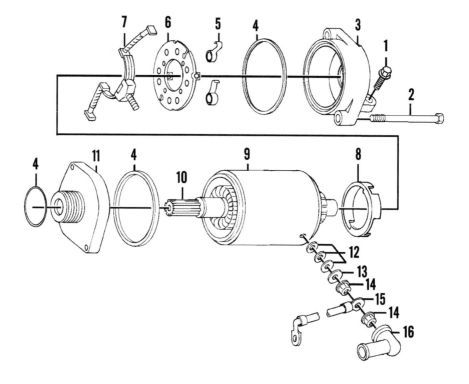

1. Bolt
2. Screw
3. Rear end cover
4. O-ring
5. Spring
6. Brush plate
7. Positive brush assembly
8. Insulator
9. Housing
10. Armature
11. Front end cover
12. Fiber washers
13. Metal washer
14. Nut
15. Cable
16. Cap

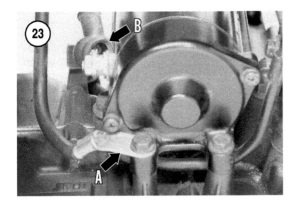

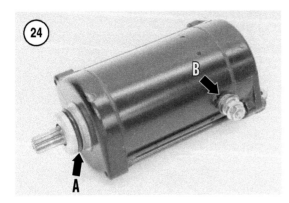

5. Lubricate the clutch rollers and washer with engine oil.

6. Install the washer and clutch sprocket into the rotor. Turn the sprocket counterclockwise and twist it squarely into the rotor. When the sprocket is fully seated, check that it turns only in the counterclockwise direction.

7. Clean and inspect the rotor components.

    a. Inspect the rotor for cracks and damage.

> *WARNING*
> *Replace the rotor if it is damaged. The rotor can fly apart at high crankshaft speeds, causing severe injury and damage to the engine.*

    b. Inspect the taper in the bore of the rotor and on the crankshaft for damage.

    c. Inspect the Woodruff key, slot and spacers for damage (**Figure 17**).

> *NOTE*
> *If the Woodruff key is bent or sheared, the rotor will not properly align on the crankshaft, which causes the engine to be out of time.*

8. Install the parts as described in this section.

STARTER

Removal and Installation

Refer to **Figure 22**.

1. Remove the exhaust pipe (Chapter Four).

2. Disconnect the ground (negative) cable from the battery and from the starter (A, **Figure 23**).

3. Disconnect the positive cable from the starter (B, **Figure 23**).

4. Remove the remaining mounting bolt, then twist the starter out of the left crankcase cover. The starter is sealed to the cover by an O-ring, which causes resistance during removal.

5. Disassemble, inspect and test the starter as described in this section.

6. Reverse this procedure to install the starter. Note the following:

    a. Lubricate the O-ring on the front end cover (A, **Figure 24**) before inserting it into the crankcase cover.

    b. Check that the fiber washers (B, **Figure 24**) on the cable post are in good condition. The washers must insulate the cable from the housing.

    c. Clean all cable connections, then apply dielectric grease to fittings and connectors before tightening.

    d. Tighten the starter mounting bolts to 10 N•m (88 in.-lb.).

Disassembly and Assembly

Refer to **Figure 22**.

1. Remove the two housing screws, rear end cover and O-ring.

> *NOTE*
> *If disassembling the starter to only check brush condition, remove only the rear end cover. The brushes can be inspected and the cover reinstalled if further disassembly is not required.*

2. Remove the front end cover and O-ring.

3. Remove the armature, brush plate assembly and insulator from the housing (**Figure 25**). Make note of any shims and their location on the armature shaft.

4. Inspect and test the starter components as described in this section.

5. Assemble the starter as follows:

a. Install the insulator and positive brush plate. Check that the small projection on the insulator faces up and that the O-ring fits between the terminal and housing (**Figure 26**). The terminal must be insulated from the housing.

b. Fit the positive brushes into the insulated brush plate slots, then seat the plate into the housing. Lock the positive brushes into the slots. The tabs on the brush plate are not spaced symmetrically. Therefore, the plate only fits correctly in one position (**Figure 27**).

c. Install the armature. Use small plastic ties, or another method, to hold back the brushes so the commutator can pass under the brushes.

d. Check that the alignment mark on the housing and the vertical tab on the brush plate are oriented as shown in **Figure 28**.

e. Install a new O-ring on the housing, then align the vertical tab on the brush plate with

the slot in the rear end cover (**Figure 29**). Seat the cover onto the housing. The alignment marks on the exterior of the housing and cover should be aligned (**Figure 30**).

f. Install a new O-ring on the housing, then align and install the front end cover. Again, the alignment marks on the exterior of the housing and cover should be aligned.

g. Install and tighten the housing screws.

h. Install the fiber washers onto the cable terminal, followed by the metal washer and nut.

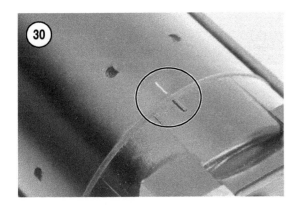

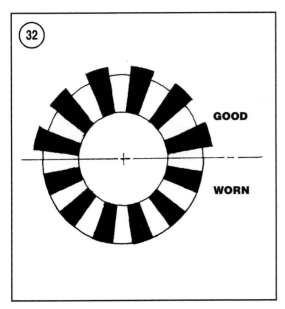

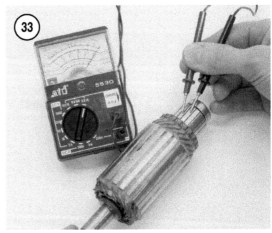

Inspection and Testing

Use an ohmmeter for all electrical tests in this procedure. If the result for any test is incorrect, the part is either shorted, or there is an open circuit between the test points. Replace or recondition parts that are worn or damaged. Refer to **Figure 22**.

1. Clean the parts as required. Use a solvent specifically for electric motors to remove buildup and contamination, particularly between the commutator bars.

2. Inspect the condition of the housing and end covers. The armature should fit in the covers with little or no play.

3. Inspect the armature.

   a. The bearing should turn smoothly and freely.

   b. Inspect any shims for cracks or damage.

   c. Inspect the windings for obvious shorts or damage.

4. Inspect and test the commutator.

   a. Measure the outside diameter (**Figure 31**). Refer to **Table 2** for specifications.

   b. Inspect the bar height. The commutator bars should be taller than the insulation between the bars (**Figure 32**).

   c. Inspect the bars for discoloration. If a pair of bars is discolored, this indicates grounded armature coils.

   d. Inspect the bars for scoring. Mild scoring can be repaired with fine emery cloth.

   e. Check for continuity across all adjacent pairs of commutator bars (**Figure 33**). There should be continuity across all pairs of bars.

The fiber washers must be in good condition so they insulate the cable from the starter housing.

  i. Install a new, lubricated O-ring onto the front end cover.

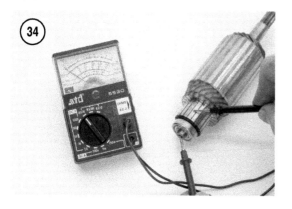

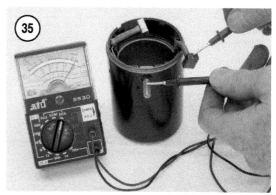

f. Check for continuity between each commutator bar and the armature shaft (**Figure 34**). There should be no continuity.

5. Inspect and test the brush assembly. The positive brushes have insulated leads.

a. Check for continuity between the positive brush and the cable terminal (**Figure 35**). There should be continuity between the parts.

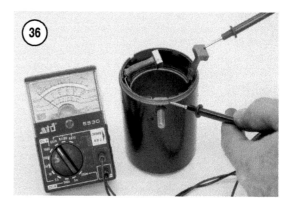

b. Check for continuity between the cable terminal and the starter housing. There should be no continuity between the parts.

c. Check for continuity between the positive brush and the starter housing (**Figure 36**). There should be no continuity between the parts.

d. Check for continuity between the grounded brush and the brush plate (**Figure 37**). There should be continuity between the parts.

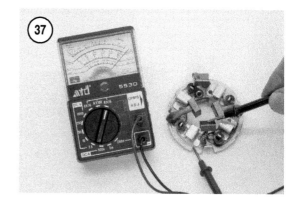

e. Check for continuity between the brush plate and the positive brush holders. There should be no continuity between the parts.

f. Measure the length of each brush (**Figure 38**). Refer to **Table 2** for specifications.

g. Inspect the condition of the brush springs.

6. Assemble the starter as described in this section.

STARTING SYSTEM

This section includes the relays and switches required to safely start and stop the engine. This includes the safety switches that prevent the engine from starting when certain startup conditions are not met. Since some components have a function in the ignition and/or charging system, refer to those sections if the component is not found in this section. Refer to **Figure 39**.

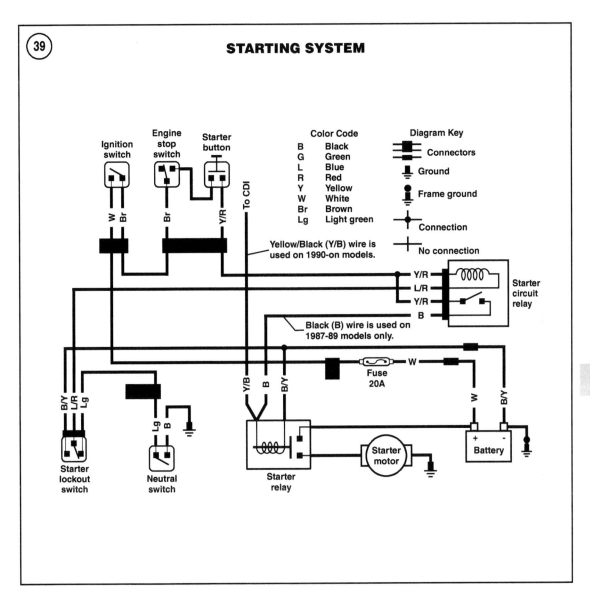

STARTING SYSTEM

Ignition switch

Engine stop switch

Starter button

Color Code

B	Black
G	Green
L	Blue
R	Red
Y	Yellow
W	White
Br	Brown
Lg	Light green

Diagram Key

Connectors

Ground

Frame ground

Connection

No connection

To CDI

Yellow/Black (Y/B) wire is used on 1990-on models.

Y/R

L/R

Y/R

B

Starter circuit relay

Black (B) wire is used on 1987-89 models only.

W

Fuse 20A

W

B/Y

Starter lockout switch

Neutral switch

Starter relay

Starter motor

Battery

Starter Relay
Testing

The starter relay (A, **Figure 40**) is located at the left side of the motorcycle, behind the plastic cover. The starter relay connects the battery to the starter. The relay is designed to temporarily carry the high electrical load between the parts during startup. The relay is activated by a small amount of current from the starter circuit relay when the starter button is pressed.

1. Remove the cover to access the switch. If necessary, remove the fuel tank to improve access.

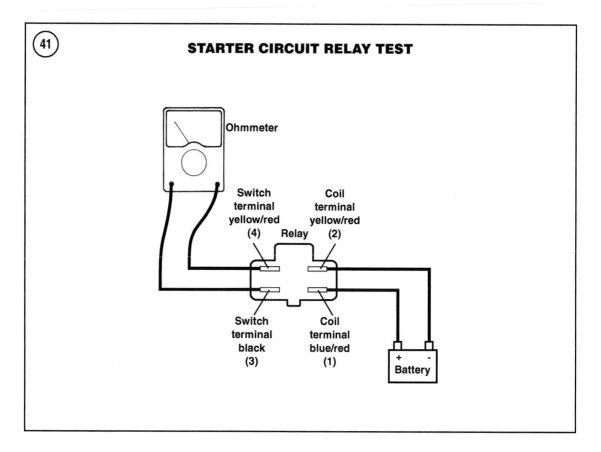

(41) STARTER CIRCUIT RELAY TEST

Ohmmeter

Switch terminal yellow/red (4) Relay Coil terminal yellow/red (2)

Switch terminal black (3) Coil terminal blue/red (1)

+ − Battery

2. Disconnect the cables leading to the starter and to the battery.

> *CAUTION*
> *The cable leading to the battery is connected to the positive terminal on the battery. Do not allow the cable end to touch the frame, engine or any other part that is grounded. Battery shorting will occur. Keep the cable end protected until it is reconnected.*

3. Connect an ohmmeter to the relay terminals. Set the meter to the R × 1 scale.
4. Check the following before testing:
   a. The neutral switch is in the on position.
   b. The engine stop switch is in the run position.
5. Turn on the ignition switch and press the starter button. Listen to the relay and observe the meter.
6. Turn off the ignition switch and note the following:
   a. If the relay *clicks* when the button is pressed and the meter reads 0 (no resistance in relay), the relay is in good condition. Check the starter and cables for damage.

   b. If the relay *clicks* when the button is pressed and the meter does not read 0 (resistance in relay), the relay is faulty.
   c. If the relay does not *click* when the button is pressed, the relay is faulty. Before replacing the relay, verify that power is passing to the relay wires when the the starter button is pressed. If not, inspect and test the starter circuit relay.

Starter Circuit Relay Testing

The starter circuit relay (B, **Figure 40**) is located at the left side of the motorcycle, behind the plastic cover. The starter circuit relay passes current to the starter relay, which activates the starter. In order for the starter circuit relay to pass current, it requires that the starter lockout switch or neutral switch be in its safe startup position. When either or both of these switches are in their safe positions, the switch in the starter circuit relay closes and completes the circuit. Refer to **Figure 39**.

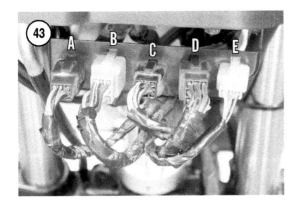

1. Remove the cover to access the relay.

2. Remove the relay from the mount, then unplug the relay.

3. Connect a 12-volt battery and ohmmeter to the relay terminals as shown in **Figure 41**.

    a. Set the ohmmeter to R × 1 ohm.

    b. Note that terminals 1 (blue/red) and 2 (yellow/red) are connected to the relay coil and terminals 3 (black) and 4 (yellow/red) are connected to the circuit relay switch.

4. Make the following checks:

    a. With the battery and ohmmeter connected as shown in **Figure 41**, check for continuity through the circuit relay switch (terminals 3 and 4). There should be continuity.

    b. Disconnect the battery and check for continuity through the circuit relay switch (terminals 3 and 4). There should be no continuity.

    c. With the battery disconnected measure the circuit relay coil resistance (terminals 1 and 2). The meter should read 65-85 ohms.

5. Replace the relay if it fails any of the checks.

Starter Lockout Switch
Testing

The starter lockout switch (**Figure 42**) is located in the clutch lever assembly. Check the switch for continuity, using an ohmmeter connected to the appropriate color-coded wires in the connector plug (E, **Figure 43**). The plug is located in front of the handlebar gauges, behind the cover. Remove the cover to access the wiring. If necessary, remove the fairing. When checking wires at the plug, connect the ohmmeter to the plug half that leads to the switch. Refer to **Figure 39**.

1. Separate the connector plug.

2. Connect an ohmmeter to the blue/red wire and the light green wire.

    a. With the clutch lever out there should be continuity.

    b. With the clutch lever pulled in there should be no continuity.

3. Connect the ohmmeter to the blue/red wire and the black/yellow wire.

    a. With the clutch lever out there should be no continuity.

    b. With the clutch lever pulled in there should be continuity.

4. Replace or repair the assembly if it fails any of the checks.

Neutral Switch
Testing

The neutral switch is located on the left side of the engine, next to the drive sprocket. Check the switch for continuity, using an ohmmeter. Refer to **Figure 39**.

1. Remove the chain sprocket guard. The sprocket does not have to be removed.

2. Put the transmission in neutral.

3. Remove the wire from the switch terminal (**Figure 44**).

4. Connect an ohmmeter to the switch terminal and to ground (an unpainted part of the engine).

    a. There should be continuity when the transmission is in neutral.

    b. There should be no continuity when the transmission is in any gear.

5. If the switch fails either test, remove the switch and test again as follows:

a. Connect the ohmmeter to the terminal and the threads on the switch.

b. Press the button on the bottom of the switch. If there is no continuity, the switch is faulty. If there is continuity, the switch button may be worn. The transmission shift drum may not be able to fully activate the switch, even though the switch operates correctly when removed from the engine. Replace the switch.

Sidestand Stop Switch
Testing

The sidestand stop switch (C, **Figure 40**) is located at the left side of the motorcycle, behind the plastic cover. The sidestand switch does not prevent the engine from being started with the sidestand either up or down. If the sidestand is down when the engine is running and the transmission is in gear, the switch will stop the engine when the clutch is released. Check the switch for continuity, using an ohmmeter. Refer to **Figure 45** as needed.

1. Remove the cover to access the switch. If necessary, remove the fuel tank to improve access.

2. Trace the wire from the switch, then separate the connector plug.

3. Connect an ohmmeter to the brown wire and the green/white wire.

a. With the sidestand down there should be no continuity.

b. With the sidestand up there should be continuity.

4. Replace or repair the assembly if it fails any of the checks.

Starter Button and Engine Stop Switch
Testing

Check the starter button and engine stop switch for continuity. Use an ohmmeter, connected to the appropriate color-coded wires in the connector plug (A, **Figure 43**). The plug is located in front of the handlebar gauges, behind the cover. Remove the cover to access the wiring. If necessary, remove the fairing. When checking wires at the plug, connect the ohmmeter to the plug half that leads to the switch and button.

If necessary, refer to the wiring diagram at the back of the manual to see how all wires at the starter button and engine stop switch assembly are con-

nected. **Figure 39** and **Figure 45** isolate the wires as they function in the starting system and ignition system.

1. Separate the connector plug.

2. Connect an ohmmeter to the brown wire and the yellow/red wire.

a. With the engine stop switch in the off position there should be no continuity.

b. With the engine stop switch in the off position, press the starter button. There should be no continuity.

c. With the engine stop switch in the run position there should be no continuity.

d. With the engine stop switch in the run position, press the starter button. There should be continuity.

3. Connect the ohmmeter to the black/white wire and the black/yellow wire.

a. With the engine stop switch in the off position there should be continuity.

b. With the engine stop switch in the off position, press the starter button. There should be no continuity.

4. Replace or repair the assembly if it fails any of the checks.

Ignition Switch
Testing

Check the ignition switch for continuity. Use an ohmmeter, connected to the appropriate color-coded wires in the connector plug (C, **Figure 43**). The plug is located in front of the handlebar gauges, behind the cover. If necessary, remove the fairing. Remove the cover to access the wiring. When checking wires at the plug, connect the ohmmeter to the plug half that leads to the switch.

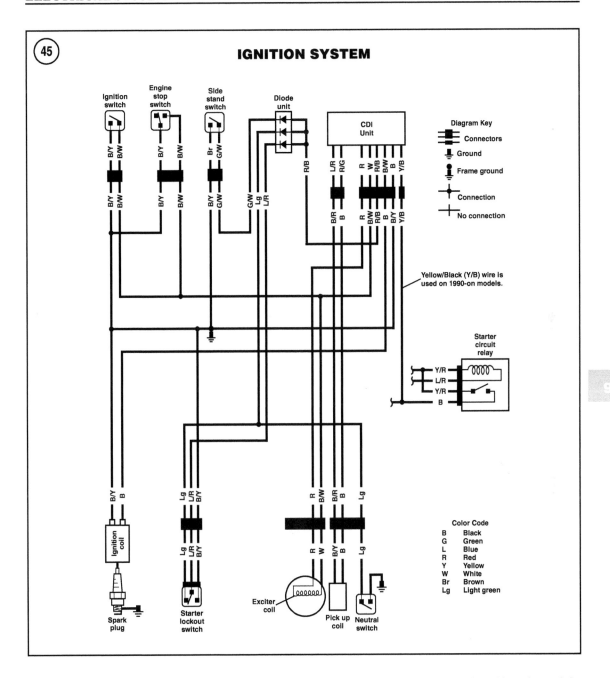

IGNITION SYSTEM

If necessary, refer to the wiring diagram at the back of the manual to see how all wires at the ignition switch assembly are connected. **Figure 39** and **Figure 45** isolate the wires as they function in the starting system and ignition system.

1. Separate the connector plug.

2. Connect an ohmmeter to the black/yellow wire and the black/white wire. When the ignition switch is in the lock, off or park position there should be continuity.

3. Connect the ohmmeter to the white wire and the brown wire. When the ignition switch is in the on position there should be continuity.

4. Connect the ohmmeter to the red wire and the red/white wire. This is the light circuit. If necessary, refer to the wiring diagram at the back of the manual. When the ignition switch is in the on position there should be continuity.

5. Replace or repair the assembly if it fails any of the checks.

IGNITION SYSTEM

This section includes many of the components required for the engine to run. Since some components have a function in the starting and/or charging system, refer to those sections if the component is not found in this section. Refer to **Figure 45**.

Precautions and Inspections

The CDI unit and other components of the ignition system can be damaged if precautions are not taken during testing and troubleshooting. Note the following:

1. Never disconnect electrical connections while the engine is running or cranking.
2. Turn off the ignition switch before disconnecting/connecting electrical components.
3. Handle the parts with care.
4. Check wiring for poor connections, corrosion and shorts before replacing components connected to the wiring.
5. Before testing components, check that the battery and spark plug are in good condition, and are not the cause of poor performance. Refer to Chapter Three for battery replacement, charging and electrolyte checks.
6. Work slowly, methodically and use test equipment that is in good condition. Record all measurements.

CDI Unit
Testing

The CDI unit is located at the right side of the motorcycle, under the fuel tank (A, **Figure 46**). Check the CDI unit for resistance and continuity by measuring the resistance between pairs of wires. If the checks indicate the unit is faulty, replace the unit.

Refer to **Figure 47** or **Figure 48** for measurement points and specifications.

Note the following when performing all checks:

> *NOTE*
> *Ohmmeter measurements vary slightly from one meter to another. Measurements that are marginally out of specification do not necessarily indicate the CDI unit is faulty.*

1. Remove the fuel tank.

2. Separate the connectors leading from the CDI unit (**Figure 49**). On 1990-on models, also disconnect the single yellow/black wire leading from the CDI unit.
3. Set the ohmmeter to the R × 1K scale. Check resistance and continuity as follows:
   a. Connect the positive meter probe and the negative meter probe to the specified wire colors, shown in the appropriate figure.
   b. Measure and record the meter reading.
   c. Repeat the steps until all wire combinations have been checked.

Diode Unit
Testing

The diode unit is located at the right side of the motorcycle, under the fuel tank (B, **Figure 46**). The three diode circuits act as one-way valves and allow power to enter one side of the unit and exit the opposite side of the unit. The purpose of the test is to check for low resistance in one direction and infinite resistance in the other direction.

1. Remove the fuel tank.
2. Remove the unit from the mount, then unplug the unit. Note the row of terminals to which the three red/black wires are connected.
3. Test the unit as follows:
   a. Set an ohmmeter to R × 1 ohm.
   b. Connect a meter probe to a red/black terminal and a meter probe to the opposite terminal on the plug.
   c. Measure and record the meter reading.
   d. Reverse the probes on the terminals and repeat the check.
   e. Measure the remaining pairs of terminals.

CDI RESISTANCE TEST (1987-1989)

Wire colors and abbreviations:

B: black
B/W: black with white tracer
L/R: blue with red tracer
R: red
R/B: red with black tracer
R/G: red with green tracer
W: white
∞: infinity

Approximate kΩ

Negative probe	Positive probe						
	W	R	R/G	L/R	B/W	B	R/B
W		∞	∞	∞	∞	∞	∞
R	10-40		4-20	10-45	∞	4-20	3-15
R/G	2-10	2-10		4-20	∞	0	1-6
L/R	60-240	60-240	30-150		∞	30-150	40-160
B/W	∞	∞	∞	∞		∞	∞
B	2-10	2-10	0	4-16	∞		1-6
R/B	4-20	4-20	1-6	5-25	∞	1-6	

f. The resistance should be infinite in one direction and low in the other direction. If any pair of terminals has low or high resistance in both directions, the unit is faulty.

Ignition Coil
Testing

The ignition coil is located at the right side of the motorcycle, under the fuel tank. Check the coil for resistance in the primary and secondary coils.

1. Remove the fuel tank.

2. Disconnect the black/yellow wire (A, **Figure 50**) and black wire (B) at the coil.

3. Remove the spark plug cap from the plug.

4. Check primary coil resistance as follows:

a. Connect the meter probes to the terminals on the coil.

b. Measure the resistance. Refer to **Table 2** for the specification.

5. Check secondary coil resistance as follows:

a. Remove the spark plug cap from the spark plug lead.

b. Connect one meter probe to the spark plug lead and the other meter probe to the terminal for the black/yellow wire.

c. Measure the resistance. Refer to **Table 2** for the specification.

NOTE
If the coil and spark plug lead pass all tests, the spark plug cap may be faulty. Although no resistance measurement is specified by Kawasaki,

CDI RESISTANCE TEST (1990-ON)

Wire colors and abbreviations:

B: black
B/W: black with white tracer
L/R: blue with red tracer
R: red
R/G: red with green tracer
W: white
Y/B: yellow with black tracer
∞: infinity

Approximate kΩ

		Positive probe					
	W	**R**	**R/G**	**L/R**	**B/W**	**B**	**Y/B**
W		∞	∞	∞	∞	∞	∞
R	10-55		5-25	5-35	∞	5-25	20-90
R/G	2-10	2-10		1-6	∞	0	10-50
L/R	4-20	4-20	1-6		∞	1-6	10-55
B/W	∞	∞	∞	∞		∞	∞
B	2-10	2-10	0	1-6	∞		10-50
Y/B	15-80	15-80	10-50	10-55	∞	10-50	

(Negative probe — left axis)

the cap on the model shown in this manual measures 5.5K ohms. If necessary, use this measurement as a guide in determining the condition of the spark plug cap.

Exciter Coil Testing

The exciter coil is located in the alternator cover. The exciter coil can be checked at the wire connector, located under the fuel tank. Check the exciter coil for resistance. Refer to **Figure 45**.

1. Remove the fuel tank.

2. Separate the connector, located on the right side of the frame (A, **Figure 51**).

3. Identify the half of the connector that leads to the stator.

4. Measure the coil resistance as follows:

a. Set the ohmmeter to R × 10 ohms.

b. Insert the meter probes into the terminals of the white wire and the red wire. Refer to **Table 2** for the specification.

c. If the exciter coil is not within specification, check the plug and wiring harness for obvious damage and shorting.

d. If necessary, remove the alternator cover and recheck the wiring harness and coil (A, **Figure 52**). Individually check the full length of the harness wires for continuity. There should be near zero resistance in all wires. Flex the harness during the check to detect erratic continuity.

e. If the harness is not shorted, check the coil at the white and red wire connections on the

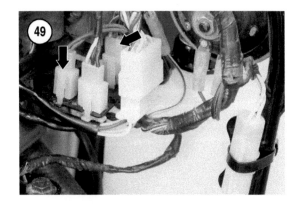

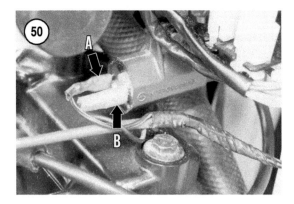

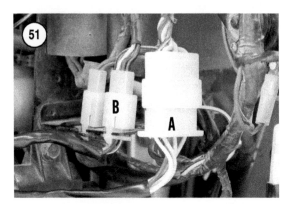

stator. If the coil fails the check, unsolder the harness connections and recheck the coil. If the coil is within specification, a solder joint was poor, or the wiring harness is actually shorted. If the coil fails the check, replace the stator.

5. Measure the resistance between the coil and ground as follows:

    a. Set the ohmmeter to R × 1 ohm.

    b. Ground one of the meter probes to the engine (or the alternator cover, if removed). Touch the other probe to the white wire, then to the red wire. The resistance should be infinity. Any other reading indicates a short, and the stator should be replaced.

Pickup Coil
Testing

The pickup coil is located in the alternator cover. The pickup coil can be checked at the wire connector, located under the fuel tank. Check the pickup coil for resistance. Refer to **Figure 45**.

1. Remove the fuel tank.

2. Separate the connector, located on the right side of the frame (B, **Figure 51**).

3. Identify the half of the connector that leads to the pickup coil.

4. Measure the coil resistance as follows:

    a. Set the ohmmeter to R × 10 ohms.

    b. Insert the meter probes into the terminals of the black/yellow wire and the black wire. Refer to **Table 2** for the specification.

    c. If the pickup coil is not within specification, check the plug and wiring harness for obvious damage and shorting.

    d. If necessary, remove the alternator cover and recheck the wiring harness and coil (B, **Figure 52**). Flex the harness during the check to detect erratic continuity. If the coil fails the check, replace the pickup coil.

5. Measure the resistance between the coil and ground as follows:

    a. Set the ohmmeter to R × 1 ohm.

    b. Ground one of the meter probes to the engine (or the alternator cover, if removed). Touch the other probe to the black/yellow wire, then the black wire. The resistance should be infinity. Any other reading indicates a short. Check for a pinched wire under the pickup

coil, or damage that allows it to short out. If no damage can be detected, the pickup coil should be replaced.

Ignition Timing

The ignition timing is electronically controlled by the CDI unit. No adjustment is possible to the ignition timing. The timing is checked to verify that the CDI unit is functioning properly.

1. Warm up the engine to operating temperature.

2. Remove the cap from the timing hole (**Figure 53**).

3. Connect a timing light following the manufacturer's instructions.

4. Start the engine and allow it to idle at 1300 rpm.

5. Direct the timing light into the timing hole and observe the timing mark. The F mark should be aligned with the index mark in the hole (**Figure 54**).

6. Raise the engine speed and check that the F mark moves clockwise. When the engine speed reaches 3300 rpm, the timing is advanced. A single line should be visible in the timing hole.

7. If the timing check is incorrect:

   a. Test the CDI unit.

   b. If the CDI unit passes all checks, inspect the Woodruff key (**Figure 55**), securing the rotor to the crankshaft. If the key is bent or sheared, the rotor will not be properly aligned on the crankshaft, causing the engine to be out of time. This slight change may not be obvious when checking ignition timing; however, the timing will be off enough to cause poor engine performance.

8. Turn off the engine and disconnect the test equipment.

9. Lubricate the O-ring on the cap, then screw the cap into the timing hole.

CHARGING SYSTEM

This section includes the components that charge and regulate power to the battery. Refer to **Figure 56**.

Precautions and Inspections

The components of the charging system can be damaged if precautions are not taken during testing and troubleshooting. Note the following:

1. Never disconnect electrical connections while the engine is running or cranking.

2. Turn off the ignition switch before disconnecting/connecting electrical components.

3. Handle the parts with care.

CHARGING SYSTEM

Diagram Key

Connectors

Ground

Frame ground

Connection

No connection

NC Not connected

Color Code
B	Black
G	Green
L	Blue
R	Red
Y	Yellow
W	White
Br	Brown
Lg	Light green

W

Fuse
20A

Br/W

W

B/Y

B/Y

B/Y

W

Br

B/Y

W

Battery

Charging
coils

Regulator/
rectifier

System
loads

Ignition
switch

4. Check wiring for poor connections, corrosion and shorts before replacing components connected to the wiring.

5. Before testing components, check that the battery is in good condition, and is not causing poor performance.

6. Work slowly, methodically and use test equipment that is in good condition. Record all measurements.

Fuse Holder

The fuse holder is located under the seat, on top of the battery case (**Figure 57**). There are two fuses in

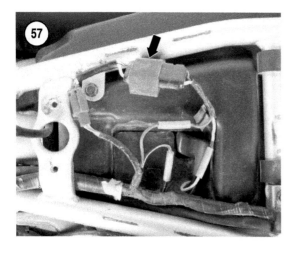

the holder. The 10 amp fuse protects the headlight, taillight and instrument gauge bulbs. The 20 amp *main circuit* fuse protects the remainder of the motorcycle. On 1990-on models, the fan circuit has its own fuse, located under the fuel tank. The fan circuit of all models, however, receives power from the main circuit, containing the 20 amp fuse.

Use an ohmmeter to check fuses for continuity. Even though a fuse may appear to be in good condition, a fine break in the fuse element is not always seen. A fine break in the element can also indicate continuity when tested cold, then break continuity when it is under load and heated in the circuit.

Battery Replacement, Charging and Electrolyte Check

Refer to Chapter Three for battery replacement, charging, and the electrolyte check.

Battery Voltage Test
Unloaded

For a conventional (original equipment) battery, check the unloaded voltage using a hydrometer or voltmeter. An unloaded test indicates the basic state of charge.
1. Connect a voltmeter to the negative and positive terminals as shown in **Figure 58**. If a voltmeter is not available, use a syringe hydrometer and measure the specific gravity of the electrolyte.
2. Measure the voltage.
  a. A fully charged battery will have a minimum of 12.6 volts (1.265 hydrometer reading).
  b. A battery that is approximately 75 percent charged will have a minimum of 12.4 volts (1.210 hydrometer reading).
  c. A battery that is approximately 50 percent charged will have a minimum of 12.1 volts (1.160 hydrometer reading).
3. If battery charging or replacement is required, refer to the procedures in Chapter Three.

Battery Voltage Test
Loaded

For a conventional (original equipment) battery, check the loaded voltage using a voltmeter. A load test requires the battery to discharge current. A load

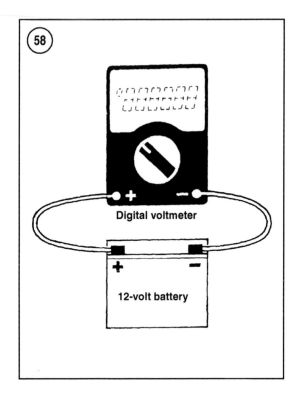

Digital voltmeter

12-volt battery

test indicates whether the battery is adequate to operate the motorcycle.
1. Connect a voltmeter to the negative and positive terminals as shown in **Figure 58**.
2. Turn on the headlight to the high beam.
3. Measure the voltage.
  a. A battery in good condition will have a minimum of 11.5 volts.
  b. If the battery needs charging or replacing, refer to the procedures in Chapter Three.

Battery Current Draw Test

If the battery is in good condition, but it discharges at a rapid rate when the motorcycle is not used, check the electrical system for a current draw. Machines that have a clock or some other type of aftermarket accessory (such as an alarm) will have a continuous parasitic current loss to operate these types of devices. This will show up as a current draw on the battery. However, on the KLR650 any current draw present when all electrical devices are shut off indicates a problem.

A short in a wire or component can allow the battery to discharge to ground. Dirt and moisture can also create a path to ground. To isolate the problem,

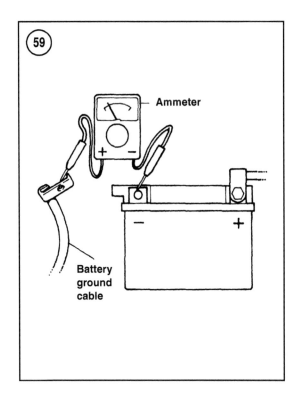

an ammeter is connected to the battery and various circuits/components disconnected while observing the current reading.

1. Remove the fuel tank.
2. Turn the ignition switch off.
3. Disconnect the negative battery cable.
4. Make sure the battery is fully charged.

CAUTION
Before connecting the ammeter in the next step, set the amperage range to the highest setting. If there is an excessive amount of current flow, the meter may be damaged.

5. Connect the ammeter to the negative battery cable and terminal as shown in **Figure 59**.
   a. The meter should indicate no current draw.
   b. If a current draw is indicated, continue the test.
6. Remove the fuses from the headlight/taillight circuit and fan circuit (1990-on models), and observe the meter.
   a. If the current draw lessens, there is a problem indicated in one of the fused circuits. Install the fuse(s) and continue to isolate the problem by separating the connectors in the circuit.
   b. If the current draw remains the same, there is a problem indicated in the main circuit.
7. Check the connector as follows:
   a. Refer to the system diagrams in this chapter, and if necessary, the wiring diagram at the back of the manual for additional circuits and part identifications.
   b. Separate the individual connectors of the appropriate parts. Work with one connector at a time, disconnecting and connecting the connectors until the meter indicates no current draw. When this occurs, the shorted circuit has been isolated.

Regulator/Rectifier

The regulator/rectifier is located under the seat (**Figure 60**). The regulator/rectifier converts the alternating current produced by the alternator into direct current to charge the battery and power any electrical loads (**Figure 56**). The unit also regulates the charging voltage to the battery. Excess voltage is dissipated as heat and radiated from the finned regulator.

Output voltage test

The following test checks for output voltage of the regulator/rectifier to charge the battery. The battery must be in good condition and charged before performing the test.

1. Start the engine and allow it to reach operating temperature, then turn off the engine.
2. Check the regulator/rectifier output voltage as follows:
   a. Set a voltmeter to DC volts. Use a scale in the 25-50 volt range.

b. Connect a voltmeter to the battery terminals as shown in **Figure 58**.

c. Start the engine and note the meter reading at idle and as the engine speed is raised.

d. Unplug the headlight and repeat the check.

e. In both checks, the meter should indicate at least 12 volts at an idle, increasing to 14-15 volts as the engine speed is raised.

f. In both checks, if the output voltage is significantly higher than 15 volts, the regulator/rectifier may not be adequately grounded, or may be faulty. If the output voltage does not rise with engine speed, the regulator/rectifier or stator coils are faulty. Before replacing parts, check the condition of the wiring harness and battery.

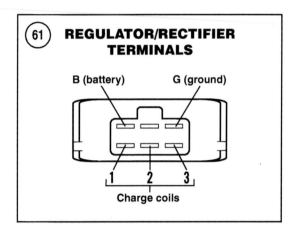

Resistance testing

The following test checks for directional resistance in the regulator/rectifier.

Use an analog ohmmeter set to the R × 10 or R × 100 scale.

For some checks, the result should be an infinity reading on the meter. For other checks, the result should be a meter reading from zero to approximately one-half of the meter scale. There is no specific resistance reading required. Each regulator/rectifier, as well as the ohmmeter used for the test, will produce slightly different results. However, the purpose of the test is to check for low resistance in one direction and infinite resistance in the other direction. For each of the following checks, the required result is indicated.

Refer to **Figure 61** for test points.

1. Remove the plug connector from the regulator/rectifier.

2. Connect the positive meter probe to the B terminal.

a. Keep the probe on the B terminal, then touch the negative probe to terminals 1, 2 and 3.

b. The meter should read zero to approximately one-half of the meter scale at all terminals.

3. Connect the negative meter probe to the B terminal.

a. Keep the probe on the B terminal, then touch the positive probe to terminals 1, 2 and 3.

b. The meter should read infinity at all terminals.

4. Connect the positive meter probe to the G terminal.

a. Keep the probe on the G terminal, then touch the negative probe to terminals 1, 2 and 3.

b. The meter should read infinity at all terminals.

5. Connect the negative meter probe to the G terminal.

a. Keep the probe on the G terminal, then touch the positive probe to terminals 1, 2 and 3.

b. The meter should read zero to approximately one-half of the meter scale at all terminals.

6. If any checks result in a reading that is high or low in both directions, the regulator/rectifier is faulty.

Stator Charging Coils
Testing

The stator coils are located in the alternator cover. Check the charging coils for resistance and continuity. Refer to **Figure 56**.

1. Remove the fuel tank.

---

63 **FAN SYSTEM**

Fan thermo switch · Fan motor · Fan relay

Fuse 10A

+ −
Battery

64

be near zero resistance in all wires. Flex the harness as the check is being made to detect erratic continuity.

f. If the harness is not shorted, check the coils at the yellow wire connections on the stator. If the coils fail the check, unsolder the harness connections and recheck the coils. If the coils are within specification, a solder joint was poor, or the wiring harness is actually shorted. If the coils fail the check, replace the stator.

5. Measure the resistance between the coils and ground as follows:

a. Set the ohmmeter to R × 1 ohm.

b. Ground one of the meter probes to the engine (or the alternator cover, if removed). Touch the other probe to the yellow wires. The resistance should be infinity. Any other reading indicates a short, and the stator should be replaced.

---

2. Separate the connector, located on the right side of the frame (A, **Figure 51**).

3. Identify the half of the connector that leads to the stator.

4. Measure the resistance as follows:

a. Set the ohmmeter to R × 1 ohm.

b. Refer to **Table 2** for the specification.

c. Insert the meter probes into the terminals of the yellow wires. Check all three combinations of the yellow wires. The resistance between all pairs of yellow wires should be within the specification.

d. If the charging coils are not within specification, check the plug and wiring harness for obvious damage and shorting.

e. If necessary, remove the alternator cover and recheck the wiring harness and coils (C, **Figure 52**). Individually check the full length of the harness wires for continuity. There should

FAN SYSTEM

When the coolant temperature reaches a set point, an electric fan (**Figure 62**) turns on to increase airflow through the radiator fins. The fan circuit consists of the fan, fan relay, fan switch and battery (**Figure 63**). On 1990-on models, a fuse is also in the circuit. The fan is turned on and off by the fan relay. The fan relay, however, is controlled by the fan switch, located at the bottom of the radiator (**Figure 64**). The fan switch is thermally sensitive and controls the electrical grounding of the fan circuit. When the engine coolant is cold, the fan switch has an open circuit to ground and the fan relay and fan are inoperative. As coolant temperature rises and begins to exceed normal operating temperature, resistance in the fan switch lowers and the switch grounds the circuit. This activates the coil in the fan relay, which closes the relay switch. When the switch is closed, current is passed to the fan and the fan turns on. As coolant temperature falls, resistance in the fan switch increases, until the switch no longer is grounded. The relay then turns off and the fan stops.

When testing or troubleshooting the fan system, it is important that all connections are clean and tight. During assembly, apply dielectric grease to connections to prevent corrosion and the entry of moisture.

NOTE
On 1990-on models, check the condition of the fan system fuse before testing individual components.

Fan
Testing

In the following steps, two tests are performed to determine if the fan is faulty.

1. If the fan does not turn on at high temperature, perform the following simple check:
    a. Remove the connector from the switch terminal (**Figure 64**).
    b. Ground the connector to an engine fin, or other unpainted ground.
    c. If the fan turns on, test the fan switch.
    d. If the fan does not turn on, continue to check the fan as described in the following steps.
2. Remove the fuel tank.
3. Separate the connector, located on the frame (A, **Figure 65**) by the fan relay (B).
4. Identify the half of the connector that leads to the fan.
5. Connect a 12-volt battery to the motor leads. Connect the positive lead to the blue wire and the negative lead to the black/yellow wire.
    a. If the fan does not turn on, replace the fan. Replacement parts are not available for the fan.
    b. If the fan turns on, test the fan relay.

Fan Switch
Testing

The following test requires that the switch be placed in heated water, to simulate actual operating conditions. Read and understand the procedure so

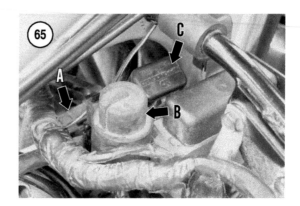

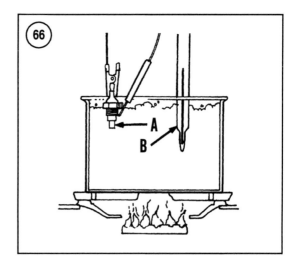

the proper equipment is on hand to safely perform the test.

1. Drain the cooling system (Chapter Three).
2. Remove the connector from the switch terminal, then remove the switch from the radiator.
3. Clean and inspect the switch for obvious damage.
4. Test the switch at ambient temperature as follows:
    a. Connect an ohmmeter to the switch terminal and to the switch threads.
    b. If the reading is anything other than infinity, the switch is faulty. Low resistance in the switch could cause the fan to come on too soon and/or not turn off.
5. Test the switch at operating temperature as follows:
    a. Connect an ohmmeter to the switch terminal and to the switch threads.
    b. Suspend the switch (A, **Figure 66**) and an accurate thermometer (B) in a container of wa-

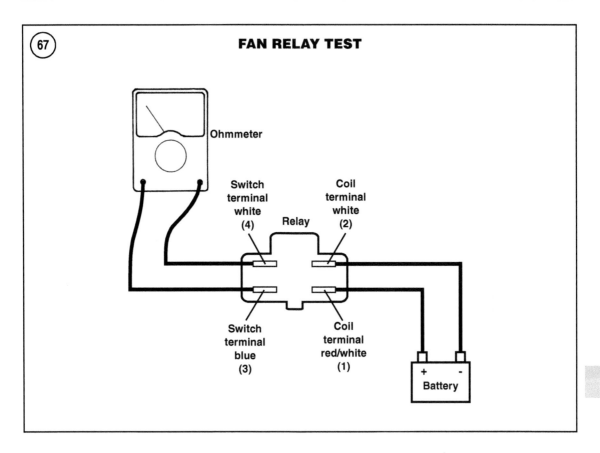

67   **FAN RELAY TEST**

Ohmmeter

Switch terminal white (4)

Coil terminal white (2)

Relay

Switch terminal blue (3)

Coil terminal red/white (1)

+ − Battery

ter. The temperature sensor and threads must be submerged. Do not allow the parts to touch the bottom or side of the container.

c. Refer to **Table 2** for the temperature and resistance specifications.

d. Slowly heat the water and observe the thermometer and ohmmeter readings. Do not excessively overheat the switch.

e. Replace the switch if it does not operate within the specifications.

6. Install the switch into the radiator. Tighten the switch to 7.4 N•m (65 in.-lb.).

7. Fill and bleed the cooling system (Chapter Three).

**Fan Relay
Testing**

The fan relay is located at the right side of the motorcycle, under the fuel tank (B, **Figure 65**). Refer to **Figure 63**.

1. Remove the fuel tank.

2. Remove the relay from the mount, then unplug the relay.

3. Connect a 12-volt battery and ohmmeter to the relay terminals as shown in **Figure 67**.

a. Set the ohmmeter to R × 1 ohm.

b. Note that terminals 1 (red/white) and 2 (white) are connected to the relay coil and terminals 3 (blue) and 4 (white) are connected to the relay switch.

4. Make the following checks:

a. With the battery and ohmmeter connected as shown in **Figure 67**, check for continuity through the relay switch (terminals 3 and 4). There should be continuity.

b. Disconnect the battery and check for continuity through the relay switch (terminals 3 and 4). There should be no continuity.

c. With the battery disconnected measure the relay coil resistance (terminals 1 and 2). The meter should read 65-85 ohms.

5. Replace the relay if it fails any of the checks.

Fan Fuse

On 1990-on models, the fan circuit fuse is located at the right side of the motorcycle, under the fuel tank (**Figure 68**). The fuse is rated at 10 amps. Use an ohmmeter to check the fuse for continuity. Even though a fuse may appear to be in good condition, a fine break in the fuse element is not always seen.

A fine break in the element can also indicate continuity when tested cold, then break continuity when it is under load and heated in the circuit.

WATER TEMPERATURE GAUGE AND SENDING UNIT

The water temperature sending unit is located on top of the cylinder head (**Figure 69**). The unit sends an electrical signal to the water temperature gauge, which reacts with a corresponding temperature reading. At low coolant temperatures, the sending unit has high resistance and the temperature gauge indicates a cold reading. As coolant temperature rises, the resistance of the sending unit lowers and the temperature gauge indicates a hotter reading.

Water Temperature Gauge Circuit Test

Use the following procedure to determine if the gauge or the sending unit should be tested.

1. Remove the fuel tank.

2. Disconnect the yellow wire from the terminal on the sending unit. Do not allow the loose wire to contact the engine or any other grounded surface.

3. Check the water temperature gauge circuit as follows:

   a. Turn on the ignition key and observe the gauge. The needle should indicate C.

   b. While observing the gauge, momentarily touch the yellow wire to the engine or another grounded (unpainted) surface on the motorcycle. The needle should should move toward H when the wire is grounded.

> *CAUTION*
> *Ground the wire only long enough to verify the action of the gauge needle. Damage to the gauge can occur with excessive grounding.*

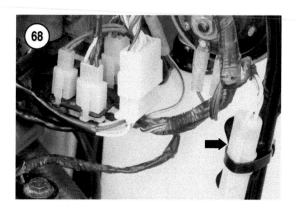

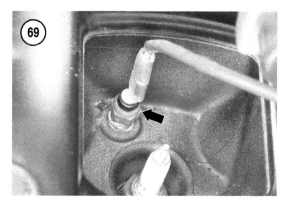

   c. If the gauge passes the check, test the water temperature sending unit as described in this section.

   d. If the gauge does not pass the check, inspect the wiring harness for damaged or poor connections. If all wiring is in good condition, test the gauge as described in this section.

Water Temperature Gauge Testing

Check the water temperature gauge for resistance. Use an ohmmeter, connected to the appropriate color-coded wires in the connector plug (B, **Figure 70**). The plug is located in front of the handlebar gauges, and behind the cover. Remove the cover to access the wiring. If necessary, remove the fairing. When checking wires at the plug, connect the ohmmeter to the plug half that leads to the gauge.

Refer to **Table 2** for resistance specifications.

1. Separate the connector plug.

2. Set the ohmmeter to R × 10 ohms.

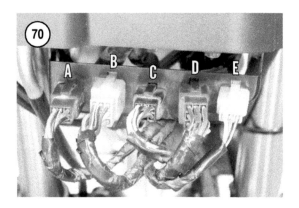

3. Connect the ohmmeter to the black/yellow wire and the yellow/white wire.

4. Connect the ohmmeter to the brown wire and the yellow/white wire.

5. Replace or repair the assembly if it fails any of the checks.

Water Temperature Sending Unit
Testing

The following test requires that the sending unit be placed in heated water, to simulate actual operating conditions. Read and understand the procedure so the proper equipment is on hand to safely perform the test.

1. Remove the sending unit from the cylinder head (Chapter Ten).

2. Clean and inspect the sending unit for obvious damage.

3. Test the sending as follows:

   a. Connect an ohmmeter to the unit terminal and to the unit threads.

   b. Suspend the unit (A, **Figure 66**) and an accurate thermometer (B) in a container of water. The temperature sensor and threads must be submerged. Do not allow the parts to touch the bottom or side of the container.

   c. Refer to **Table 2** for the temperature and resistance specifications.

   d. Slowly heat the water and observe the thermometer and ohmmeter readings. Do not excessively overheat the sending unit.

   e. Replace the unit if it does not operate within the specifications.

4. Install the sending unit (Chapter Ten).

TURN SIGNAL RELAY

Testing

The turn signal relay is located at the right side of the motorcycle, under the fuel tank (C, **Figure 65**). Use a volt/ohmmeter to test the relay and wiring.

1. Remove the fuel tank.

2. Remove the relay from the mount, then unplug the relay.

3. Check the relay resistance as follows:

   a. Set the ohmmeter to the ohms scale, then connect the meter probes to the relay terminals.

   b. There should be little or no resistance (zero ohms).

4. Check the voltage in the relay plug connector as follows:

   a. Set the meter to DC volts. Use a scale in the 25-50 volt range.

   b. Connect the negative meter probe to the orange wire and the positive meter probe to the brown wire.

   c. Turn on the ignition switch and move the turn signal switch to the left and right positions. Observe the meter.

   d. The meter should indicate battery voltage (12 volts) in both signal positions. If it does not, inspect the turn signal switch and wiring.

HEADLIGHT

Headlight Bulb Replacement

Refer to **Figure 71**.

1. Remove the cowl from the fairing (**Figure 72**). It is not necessary to remove the fairing to remove the cowl or replace the bulb.

2. Unplug the connector and remove the rubber cover from the back of the headlight.

3. Push in and turn the socket counterclockwise to remove it from the headlight lens.

4. Remove the bulb from the socket. Refer to **Table 1** for bulb specifications.

5. Clean the socket, cover and wire connector.

CAUTION
When handling the new bulb, do not touch the glass with bare hands. Handle the bulb with a clean cloth. Quartz-halogen bulbs are extremely sensitive to oil or other contaminants on their surface. Contaminants pre-

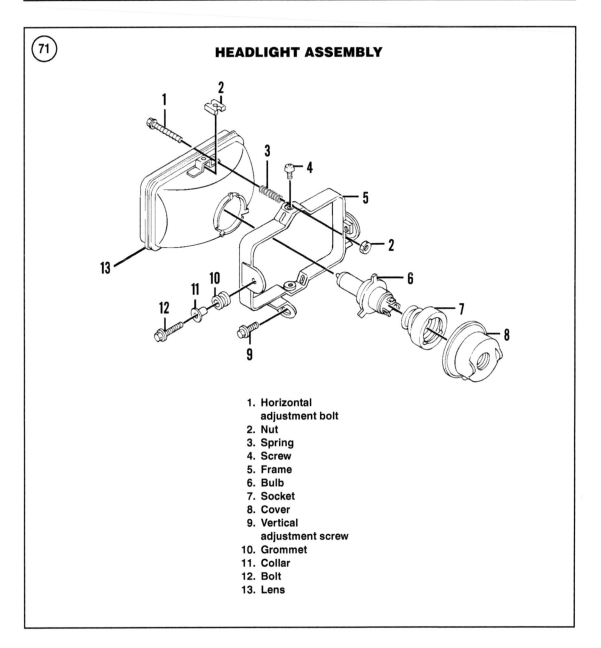

(71) **HEADLIGHT ASSEMBLY**

1. Horizontal
   adjustment bolt
2. Nut
3. Spring
4. Screw
5. Frame
6. Bulb
7. Socket
8. Cover
9. Vertical
   adjustment screw
10. Grommet
11. Collar
12. Bolt
13. Lens

vent heat dissipation from the bulb, which can cause shortened bulb life.

6. Seat the new bulb into the socket.

7. Seat the socket into the lens, then push in and turn the socket clockwise to lock it into the lens.

8. Orient the top mark (indicated on the cover), then install the cover onto the lens. Check that the cover is seated so the bulb terminals are fully exposed.

9. Plug the wire connector onto the bulb terminals.

10. Turn on the ignition switch and check that the bulb operates. Check both high and low beams.

11. Install the cowl.

Headlight Adjustment

1. Adjust vertical alignment by loosening the bolt at the rear of the light, inside the fairing (**Figure 73**).

2. Adjust horizontal alignment by turning the screw at the lower left corner of the light (**Figure 74**).

a. Turn the screw in to turn the light to the right.

b. Turn the screw out to turn the light to the left.

LEFT HANDLEBAR SWITCHES

Switch Tests

The left handlebar switch unit (**Figure 75**) houses the headlight high/low switch, turn signal switch and horn button. The switches can be checked for continuity using an ohmmeter, connected to the appropriate color-coded wires in the connector plug (D, **Figure 70**). The plug is located in front of the handlebar gauges, and behind the cover. Remove the cover to access the wiring. If necessary, remove the fairing.

Refer to the *Continuity Testing Guidelines* section in this chapter for the test procedure.

Table 1 ELECTRICAL SYSTEM GENERAL SPECIFICATIONS

Alternator	
Type	Three-phase AC
Output	14 volts/14 amps at 8000 rpm
Battery	12 volts/14 amp-hour
Ignition timing	
Idle	10° BTDC at 1300 rpm
Advanced	30° BTDC at 3300 rpm
(continued)	

Table 1 ELECTRICAL SYSTEM GENERAL SPECIFICATIONS (continued)

Light bulbs	
Headlight	12 volt, H4 55/60 watts
Taillight/brake light	
1987-2000	12 volt, 8/27 watts
2001-on	12 volt, 5/21 watts
Turn signals	12 volt, 23 watts
Spark plug	
Type	NGK DP8EA-9
	NGK DPR8EA-9
	ND X24EP-U9
	ND X24EPR-U9
Gap	0.8-0.9 mm (0.032-0.036 in.)

Table 2 ELECTRICAL SYSTEM TEST SPECIFICATIONS

Exciter coil resistance (white–to–red wires)	100-200 ohms
Fan relay coil resistance	65-85 ohms
Ignition coil resistance	
Primary coil	0.15-0.21 ohm
Secondary coil (spark plug lead cap removed)	3.8-5.8K ohms
Pickup coil resistance (black/yellow–to–black wires)	100-150 ohms
Radiator fan switch resistance	
Ambient temperature	Infinity
On position (above 94° C [201° F])	Infinity rising to 0.5 ohm
Off position (below 91° C [196° F])	Declining to infinity
Regulator/rectifier output voltage	12-15 volts
Stator coil resistance (yellow–to–yellow wires)	0.3-1.0 ohm
Starter	
Brush length	12-12.5 mm (0.47-0.49 in.)
Service limit	6 mm (0.24 in.)
Commutator outside diameter	28 mm (1.1 in.)
Service limit	27 mm (1.06 in.)
Starter circuit relay coil resistance	65-85 ohms
Water temperature gauge resistance	
Black/yellow–to–yellow/white wires	80-100 ohms
Brown–to–yellow/white wires	95-120 ohms
Water temperature sender resistance	
At 80° C (176° F)	47-57 ohms
At 100° C (212° F)	26-30 ohms

Table 3 ELECTRICAL SYSTEM TORQUE SPECIFICATIONS

	N•m	in.-lb.	ft.-lb.
Neutral switch	15	–	11
Radiator fan switch	7.4	65	–
Rotor bolt*			
To seat	120	–	88
Final	175	–	129
Spark plug	14	–	10
Starter clutch bolts	34	–	25
Starter motor mounting bolts	10	88	–
Upper engine mounting bolts	25	–	18
Water temperature sender	15	–	11
*Refer to text			

CHAPTER TEN

COOLING SYSTEM

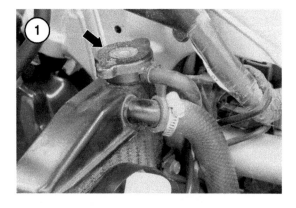

This chapter provides service procedures for the radiator, fan, fan switch, thermostat, water temperature sending unit and water pump. Read this chapter before repairing the cooling system. Become familiar with the procedures and photos to understand the skill and equipment required.

Refer to Chapter One for tool usage and techniques.

Refer to Chapter Nine for electrical test procedures.

SAFETY PRECAUTIONS

WARNING
Do not remove the radiator cap (Figure 1) immediately after or during en-

gine operation. When the engine has been operated, the liquid in the cooling system is scalding hot and under pressure. Removing the cap while the engine is hot can cause the coolant to spray violently from the radiator opening, possibly causing injury.

Wait for the engine to cool, then place a shop cloth over the cap. *Slowly* turn the cap to relieve any pressure. Turn the cap to the safety stop and check that all pressure is relieved. To remove the cap from the radiator, press down on the cap, then twist it free.

To prevent potential damage to the engine, change the coolant regularly, as described in Chapter Three. Always use an antifreeze solution. Antifreeze contains lubricants and rust inhibitors that protect the components of the cooling system. Always dispose of coolant in an environmentally safe manner.

RADIATOR AND FAN

The bike is equipped with one radiator and a thermostatically controlled fan. The fan is turned on and off by the fan switch and fan relay. During engine operation, hot coolant from the engine enters the radiator from the upper hose. The coolant loses heat

as it circulates to the bottom of the radiator. The coolant then returns to the engine by the lower hose. If the coolant returning to the engine is too hot, the radiator fan switch, located at the bottom of the radiator, turns on the fan. The fan draws air through the radiator to aid in lowering coolant temperature. The switch is set to turn on the fan when coolant temperature is 94-100° C (201-212° F). The fan shuts off when coolant temperature drops below 91° C (196° F).

Fan
Removal and Installation

1. Remove the fuel tank and radiator covers (Chapter Fifteen).
2. Disconnect the battery.

> *WARNING*
> *The fan is wired directly to the battery and does not require the ignition switch to be on in order to operate.*

3. Trace the wires leading from the fan, then disconnect the wires at the connector.
4. Remove the wires from the retainers (**Figure 2**).
5. Remove the three bolts securing the fan to the radiator (**Figure 2**).
6. If necessary, refer to Chapter Nine to test the fan.
7. Reverse this procedure to install the fan.

Radiator and Fan Switch
Removal and Installation

1. Drain the cooling system (Chapter Three).
2. Remove the fuel tank and radiator covers (Chapter Fifteen).
3. Remove the fan as described in this section.

4. Remove the wire from the fan switch (**Figure 3**). If necessary, remove the fan switch.
5. Remove the hoses from the radiator. If the hoses are seized to the fittings, cut and split the hoses so they can be peeled from the fittings. Avoid scoring the fittings. Replace the hoses.
6. Remove the two bolts at the front of the radiator (**Figure 4**) and the bolt at the back of the radiator.
7. Inspect the radiator as described in this chapter.
8. If necessary, refer to Chapter Nine to test the fan switch and fan relay.
9. Reverse this procedure to install the radiator. Note the following:

  a. Replace hoses that are hard, cracked or show signs of deterioration, both internally and externally. Hold each hose and flex it in several directions to check for damage. For a hose that is difficult to install on a fitting, dip the hose end in hot water until the rubber has softened, then install the hose.

  b. Install clamps in their original positions.

  c. If removed, install and tighten the fan switch to 7.4 N•m (65 in.-lb.).

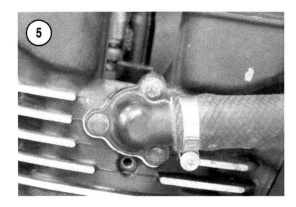

d. Fill and bleed the cooling system (Chapter Three).

e. Start the engine and allow it to warm up. Check for leaks.

Inspection

1. Clean the exterior of the radiator with a low-pressure water spray. Allow the radiator to dry.

2. Check for damaged cooling fins. Straighten bent fins with a screwdriver. If more than 20 percent of the cooling area is damaged, replace the radiator.

3. Check the seams and other soldered connections for corrosion (green residue). If corrosion is evident, there could be a leak in that spot. Perform a cooling system pressure check as described in *Cooling System Inspection* in Chapter Three. If the equipment is not available, take the radiator to a radiator repair shop to have it flushed and pressure checked.

4. Fill the radiator with water and check the flow rate out of the radiator. If the flow rate is slow, or if there is corrosion or other buildup, take the radiator to a radiator repair shop to have it flushed and pressure checked.

THERMOSTAT

The engine thermostat is located in a housing on the right side of the cylinder head (**Figure 5**). The thermostat is a temperature-sensitive valve that opens and closes, depending on the coolant temperature in the cylinder head. At startup, the thermostat is closed to retain coolant in the water jackets. When the cylinder head coolant temperature begins to exceed the ideal operating temperature, the thermostat opens and allows the coolant to pass to the radiator, where it is cooled. Since the temperature of the incoming coolant is low, the thermostat closes and reduces the flow to the radiator. As the temperature of the coolant rises, the cycle is repeated.

For the engine to run properly, the thermostat is necessary to maintain a specific amount of heat around the cylinder and cylinder head. Do not remove the thermostat, assuming that engine performance will be enhanced.

Removal, Inspection and Installation

1. Drain the cooling system (Chapter Three).

2. Remove the three bolts securing the thermostat housing to the cylinder head (**Figure 5**).

3. Remove the thermostat (**Figure 6**).

4. Visually inspect the valve in the thermostat. The valve should be closed when the thermostat is cold. If the valve is cold and open, replace the thermostat.

5. Wash the thermostat in cool water. If necessary, use a soft brush to scrub accumulation off the thermostat. If accumulation of rubber particles is evident, inspect the radiator hoses for internal deterioration.

6. Inspect the condition of the thermostat gasket.

7. If desired, open the drain bolt at the water pump and flush the cylinder head and cylinder.

8. Clean the bolts and threaded bores.

9. Test the thermostat as follows:

a. Suspend the thermostat and an accurate thermometer in a container of water (**Figure 7**). Do not allow the parts to touch the bottom or side of the container.

b. Slowly heat the water and observe the thermostat valve.

10

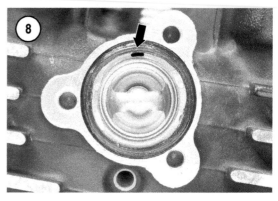

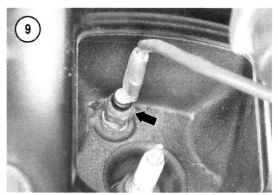

c. When the thermostat begins to open, observe the temperature on the thermometer. The thermostat should open between 69.5-72.5° C (157-162° F).

d. Continue to raise the temperature to 85° C (185° F). At this temperature, the thermostat valve should reach full lift, which is 3 mm (0.12 in.) or greater.

e. Replace the thermostat if it does meet the conditions of the tests.

5. Install the thermostat with the air bleed hole at the top of the opening (**Figure 8**). Seat the gasket in the bore.

6. Align and seat the housing, then bolt it into place.

7. Fill and bleed the cooling system (Chapter Three).

WATER TEMPERATURE SENDING UNIT

The water temperature sending unit is located on top of the cylinder head (**Figure 9**). The unit sends an electrical signal to the water temperature gauge, which reacts with a corresponding temperature reading. At a low coolant temperature, the sending unit has high resistance and the temperature gauge indicates a cold reading. As coolant temperature rises, the resistance of the sending unit lowers and the temperature gauge indicates a hotter reading.

Removal and Installation

1. Remove the wire from the sending unit connector.
2. Remove the sending unit from the cylinder head.
3. Clean the sending unit and the threads in the cylinder head.
4. Test the sending unit as described in Chapter Nine.
5. Apply waterproof sealant to the sending unit threads.
6. Tighten the sending unit to 15 N•m (11 ft.-lb.).

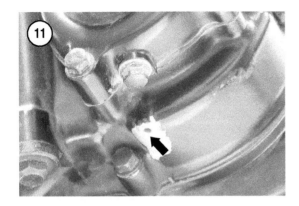

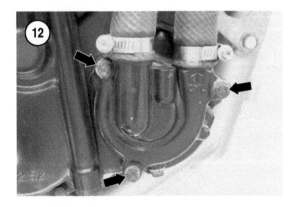

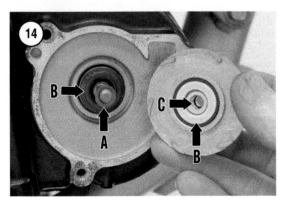

WATER PUMP

The water pump is located in the right crankcase cover. To inspect the condition of the impeller, mechanical seal or water pump cover, the cover can be removed without removing the right crankcase cover (**Figure 10**). The cooling system must be drained before removing the water pump cover.

An inspection hole for the water pump (**Figure 11**) is located at the bottom of the case. If there is water leakage at this hole, the pump mechanical seal is leaking. If there is oil leakage, the oil seal is leaking. Removal of the right crankcase cover is necessary to replace the seals.

Removal, Inspection and Installation

Remove and install the water pump cover and right crankcase cover as follows. Inspect and repair the water pump as described in this section.

1. Drain the cooling system (Chapter Three).

2. If necessary, remove the hoses from the water pump cover.

3. Remove the water pump cover as follows:

    a. Remove the three bolts securing the water pump cover (**Figure 12**).

    b. Lightly tap the cover to loosen it from the crankcase cover.

    c. Pull the cover straight out. Remove the gasket and two dowels beneath the cover.

4. Remove the nut and washer from the impeller (**Figure 13**).

5. Remove the impeller and the washer behind the impeller (A, **Figure 14**).

6. Inspect the parts.

    a. Inspect the impeller and cover for obvious damage.

    b. Inspect the face of the mechanical seal on the impeller and in the right crankcase cover (B, **Figure 14**). In order to seal properly, both faces must be smooth and free of scoring or damage. When installed, the impeller seal should fit firmly against the seal in the crankcase cover. Since the seal in the crankcase cover is spring loaded, it maintains pressure on the seals and compensates for wear. If necessary, replace the mechanical seal and oil seal as described in this section.

10

c. Install a new O-ring inside the impeller (C,
Figure 14). Lubricate the O-ring with waterproof grease.

7. If necessary, remove the right crankcase cover
(Chapter Six).

8. Reverse these steps to install the impeller assembly and water pump cover. Note the following:

   a. Apply waterproof grease to the impeller
   shaft.

   b. Tighten the impeller nut to 10 N•m (88
   in.-lb.).

   c. Install both dowels and a new gasket before
   installing the cover.

   d. Fill and bleed the cooling system (Chapter
   Three).

   e. Start the engine and allow it to warm up.
   Check for leaks.

Mechanical and Oil Seal Replacement

The water pump has a two-piece mechanical seal
(A, **Figure 15**) and an oil seal (B). The mechanical
seal prevents coolant in the pump chamber from
passing into the right crankcase cover, which contains oil. Likewise, the oil seal prevents oil in the
right crankcase cover from passing into the pump
chamber, which contains coolant. A drain hole (**Figure 11**) is located between the seals to allow any
water or oil leakage to drain to the outside of the engine. Whenever there is leakage at the drain hole,
replace the seals. The mechanical seal must be removed from the crankcase cover in order to remove
the oil seal.

1. Replace the mechanical seal in the impeller as
follows:

   a. Lift the impeller seal from the impeller (**Figure 16**). Clean the seal bore.

   b. Lightly lubricate the rubber edge of the new
   seal with waterproof grease.

   c. Seat the new seal into the impeller by hand.

2. In the crankcase cover, replace the remaining
half of the mechanical seal and the oil seal as follows:

   a. Place a narrow drift under the oil seal and
   onto the back of the mechanical seal (**Figure
   17**). Work around the seal and drive it from
   the bore. Avoid any contact with the surface
   of the bore. Do not pry the seal from the front
   side of the crankcase cover.

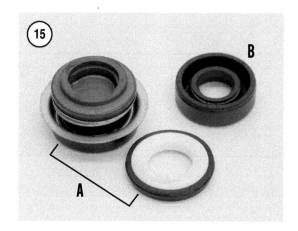

15

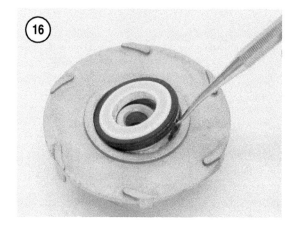

16

17

   b. Place a driver on the oil seal and drive it from
   the bore.

   c. Turn over the crankcase cover and clean the
   bore and drain hole.

   d. Pack molydisulfide grease into the lip of the
   oil seal.

e. Place the seal over the crankcase cover bore, with the closed side of the seal facing up.

f. Drive the seal into the cover using a driver that fits on the perimeter of the seal. Check that the seal is fully seated and the drain hole is visible.

g. Lightly lubricate the exterior of the mechanical seal.

h. Place the seal over the crankcase cover bore, then drive it into place. Use a driver that fits on the flange at the perimeter of the seal (**Figure 18**).

Table 1 COOLING SYSTEM SPECIFICATIONS

Antifreeze type	Ethylene glycol containing anti-corrosion inhibitors for aluminum engines
Coolant mixture	50/50 (antifreeze/distilled water)
Cooling system capacity	1.3 liters (1.4 U.S. qt.)
Radiator cap relief pressure	93-123 kPa (13.5-17.8 psi)
Thermostat valve opening temperature	69.5-72.5° C (157-162° F)
Thermostat minimum full open lift	3 mm at 85° C (0.12 in. at 185° F)

Table 2 COOLING SYSTEM TORQUE SPECIFICATIONS

	N•m	in.-lb.	ft.-lb.
Impeller nut	10	88	–
Radiator fan switch	7.4	65	–
Water temperature sending unit	15	–	11

10

CHAPTER ELEVEN

WHEELS, TIRES AND DRIVE CHAIN

This chapter provides service procedures for the wheels, speedometer drive unit, drive chain, sprockets and tires. Routine maintenance procedures for these components are found in Chapter Three. Refer to the tables at the end of this chapter for specifications.

FRONT WHEEL AND SPEEDOMETER DRIVE UNIT

Removal and Installation

1. Support the motorcycle so it is stable and the front wheel is off the ground.
2. Disconnect the speedometer cable from the right side of the hub (**Figure 1**).
3. Remove the cotter pin, nut and washer from the axle (**Figure 2**).
4. Pull up on the wheel to take the weight off the axle, then remove the axle from the wheel (**Figure 3**). Roll the wheel forward and out of the fork.

> *NOTE*
> *Do not operate the front brake lever while the wheel is removed. Insert a wooden block between the pads until the wheel is installed. This prevents the caliper piston from extending if operating the lever.*

5. Remove the speedometer drive unit from the right side of the hub (**Figure 4**). As the unit is removed, tilt it upward to prevent parts from falling from the housing.

6. Remove the axle spacer (**Figure 5**) from the left side of the hub.

7. Inspect and/or repair the wheel, axle assembly and speedometer drive unit as described in this chapter.

8. Reverse this procedure to install the wheel and speedometer drive unit. Note the following:

   a. If necessary, carefully spread the brake pads so the brake disc can enter the brake assembly.

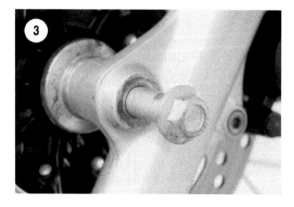

b. Apply waterproof grease to the axle, bearing seals, spacer, speedometer cable end, speedometer drive unit seal and speedometer drive unit tabbed washer.

c. When installing the speedometer unit onto the hub, check that the tabbed washer engages with the slots in the hub (**Figure 6**).

d. When installing the speedometer cable, the slot in the cable (**Figure 7**) must engage with the tab in the drive unit. If necessary, slowly rotate the front wheel until the cable housing fully seats (**Figure 8**) into the drive unit. Tighten the cable.

e. Tighten the axle nut to 78 N•m (58 ft.-lb.).

f. Check that the wheel spins freely and the brake operates properly.

g. Install a new cotter pin.

Inspection

The outer face of the bearings are permanently sealed at the time of manufacture. If either the bearing or seal is damaged, replace the complete bearing assembly.

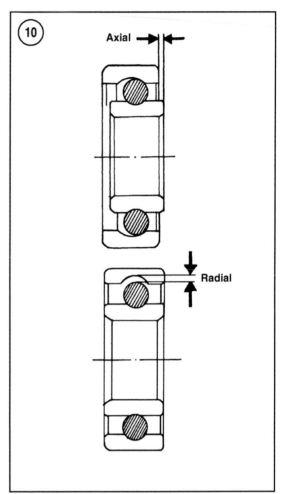

1. Inspect the seal (**Figure 9**) on both sides of the wheel. Check seals for:

   a. Nicked, damaged or missing rubber.

   b. Grease or water seepage from the seal.

2. Inspect the bearing on both sides of the wheel. Check bearings for:

   a. Roughness. Turn each bearing by hand and check for smooth, quiet operation.

   b. Radial and axial play (**Figure 10**). Try to push the bearing in and out to check for axial play. Try to push the bearing up and down to check for radial play. Any play should be difficult to feel. If play is easily felt, the bearing is worn out. Always replace bearings as a pair.

3. If seal or bearing damage is evident, refer to *Front and Rear Hubs* in this chapter for replacement procedures.

4. Clean the axle assembly and speedometer drive unit (**Figure 11**). Inspect for:

   a. Axle straightness.

   b. Damaged threads on the axle and nut.

   c. Damaged spacer or washer.

   d. Damaged speedometer drive unit seal.

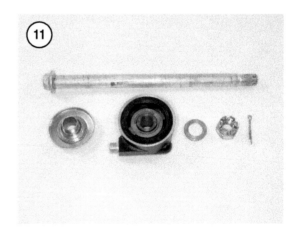

NOTE
If the speedometer drive unit contains water, dirt or dry grease, it should be disassembled and inspected as described in Step 5.

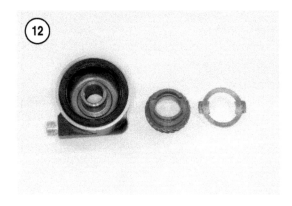

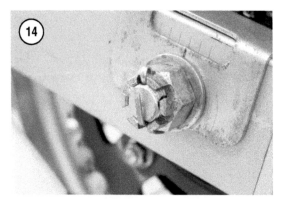

5. If necessary, service the speedometer drive unit as follows:

   a. Remove the tabbed washer and drive gear from the housing (**Figure 12**).

   b. If there are leaks around the seal, install a new seal after cleaning the housing.

   c. Clean all components in solvent, then inspect for obvious component damage.

NOTE
The tabbed washer can become damaged when installing the wheel and axle assembly. The tabs on the washer will bend if the washer is not seated properly before tightening the axle.

   d. Inspect the driven gear in the housing for smooth operation.

   e. Pack waterproof grease into the housing and driven gear, then insert and engage the drive gear.

   f. Pack waterproof grease around the drive gear, then install the tabbed washer.

6. Refer to *Rim and Spoke Service* in this chapter for inspecting and truing the rim.

REAR WHEEL

Removal and Installation

1. Support the motorcycle so it is stable and the rear wheel is off the ground.

2. Remove the two bolts securing the brake caliper to the caliper holder (**Figure 13**).

NOTE
Do not operate the rear brake pedal while the wheel is removed. Insert a wooden block between the pads until the wheel is installed. This prevents the caliper piston from extending if operating the pedal.

3. Remove the cotter pin, nut and washer from the axle (**Figure 14**).

4. Pull up on the rear wheel to take the weight off the axle, then remove the axle.

5. Lower the wheel and remove the chain from the sprocket. Roll the wheel out of the swing arm.

6. Remove the spacer from the right side (**Figure 15**) and left side of the hub (**Figure 16**).

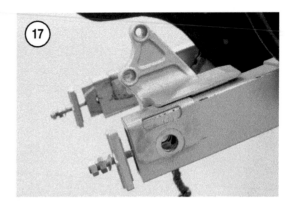

7. If necessary, remove the caliper holder and right axle adjuster assembly from the swing arm (**Figure 17**).

8. If necessary, remove the left axle adjuster assembly from the swing arm (**Figure 18**).

9. Inspect and/or repair the wheel and axle assembly as described in this chapter.

10. Reverse this procedure to install the wheel. Note the following:

a. Check that the adjustment plates are correctly oriented, so the adjustment marks on the swing arm are visible (A, **Figure 19**). Check that the bores in the plates are aligned with the adjuster bores.

b. Loosen the chain adjusters (B, **Figure 19**) so the chain and wheel can be mounted.

c. Apply waterproof grease to the axle, spacers, bearing seals and bores.

d. Loosely install the washer and axle nut.

e. If necessary, carefully spread the brake pads so the brake disc can enter the brake assembly. Tighten the caliper holder bolts to 25 N•m (18 ft.-lb.).

f. Adjust the chain (Chapter Three).

g. Check that the wheel spins freely and the brake operates properly.

Inspection

1. Pull the sprocket hub from the wheel. Account for the spacer in the back of the sprocket hub (**Figure 20**). Note that the small diameter of the spacer seats into the sprocket hub.

2. Inspect the seal on the right side of the wheel and on the outer face of the sprocket hub (**Figure 21**). Check seals for:

a. Nicked, damaged or missing rubber.

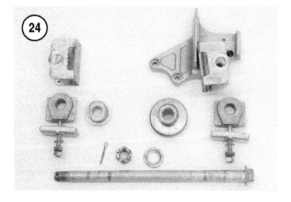

b. Grease or water seepage from the seal.

3. Inspect the bearings on both sides of the wheel, as well as the bearing in the sprocket hub. Check bearings for:

   a. Roughness. Turn each bearing by hand and check for smooth, quiet operation.

   b. Radial and axial play (**Figure 10**). Try to push the bearing in and out to check for axial play. Try to push the bearing up and down to check for radial play. Any play should be difficult to feel. If play is easily felt, the bearing is worn out. Always replace bearings as a set.

4. If seal or bearing damage is evident, refer to *Front and Rear Hubs* in this chapter for replacement procedures.

5. Inspect the grease in the left hub bearing (A, **Figure 22**) and in the sprocket hub bearing (**Figure 23**). If the grease is dirty or dry, clean the bearings and repack with waterproof grease.

> *CAUTION*
> *Do not remove bearings to check their condition or to lubricate. Bearing damage is likely to occur. If the bearings are removed, they should be replaced.*

6. Inspect the rubber damper in the wheel hub (B, **Figure 22**). The sprocket hub should fit firmly in the damper, with little or no play. A damaged damper creates excessive lash in the driveline, which can be felt during acceleration and deceleration.

7. Clean the axle assembly (**Figure 24**). Inspect the following:

   a. Axle straightness.

   b. Damaged threads on the axle, adjusters and locknuts.

   c. Damaged bores in the spacers and adjusters.

8. Refer to *Rim and Spoke Service* in this chapter for inspecting and truing the rim.

FRONT AND REAR HUBS

The wheel hubs contain bearings and a hub spacer. The bearings should be inspected anytime the wheel(s) are removed from the bike. This section describes the removal and installation of bearings.

Bearing Inspection

The bearings can be inspected with the wheels mounted on the bike. With the wheels mounted, a high amount of leverage can be applied to the bearings to detect wear. Also, the wheels can be spun to listen for roughness in the bearings. Use the following procedure to check the bearings while the wheels are mounted. If the wheels are dismounted, make the additional checks described in the wheel removal and inspection procedures in this chapter.

1. Support the bike with the wheel to be inspected off the ground. The axle nut must be tight. If inspecting the rear wheel, remove the chain from the sprocket.

2. Grasp the wheel, placing the hands 180° apart. Lever the wheel up and down, and side to side, to check for radial and axial play. Have an assistant apply the brake while repeating the test. Play will be detected in excessively worn bearings, even though the wheel is locked.

3. Spin the wheel and listen for bearing noise. A damaged bearing inconsistently sounds rough and smooth. An excessively worn bearing sounds consistently rough. In either case, replace the bearing.

NOTE
If the disc brake drags and the bearing cannot be heard, remove the wheel. Place the axle in the wheel, then support the axle so the wheel spins freely.

4. If damage is evident, replace the bearings as a set.

CAUTION
Do not remove bearings to check their condition or to lubricate. Bearing damage is likely to occur. If the bearings are removed, replace them.

Seal Inspection

Seals prevent the entry of moisture and dirt into the bearings. The seals on the wheel bearings and sprocket hub are permanent and cannot be replaced. If a seal is damaged, the bearing must be replaced.

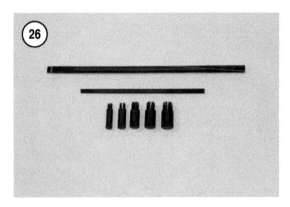

Bearing Replacement

Two methods for removing bearings from the wheel hub are provided in the following procedure. The first method (Step 3A) uses a wheel bearing removal set and the second method (Step 3B) uses common shop tools.

CAUTION
In the following procedure, do not allow the wheel to rest on the brake disc. Support the wheel to prevent pressure being applied to the disc.

1. Where used, remove any snap rings from the hub (**Figure 25**).

2. Examine the bearings. Note the following:

   a. Make note of any visible manufacturer's marks on the sides of the bearings. The new bearings must be installed with the marks in the same direction. Mark each bearing, indicating its original location in the hub. The replacement bearings can then be oriented correctly during installation.

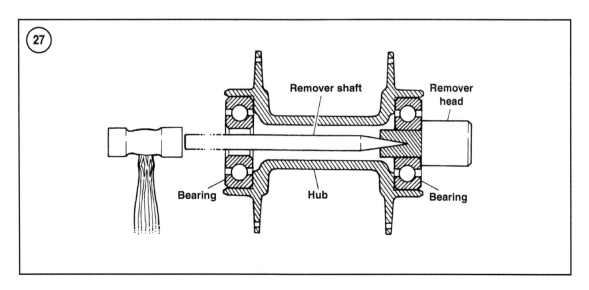

b. If the bearing is damaged, determine which bearing is damaged the least. Remove this bearing first.

3A. Remove the bearings using the wheel bearing removal set as follows:

> *NOTE*
> *The tools used in this procedure are part of the Kowa Seiki Wheel Bearing Remover Set (**Figure 26**). The set is distributed by K & L Supply Co., Santa Clara, CA. The set is designed so a proper-size remover head can be wedged against the inner bearing race. The bearing can then be driven from the hub (**Figure 27**).*

a. Select the appropriate size remover head. The small, split end of the remover must fit inside the bearing race.

b. Insert the split end of the remover head into the bearing (**Figure 28**). Seat the remover head against the bearing.

c. Insert the tapered end of the driver through the back side of the hub. Fit the tapered end into the slot of the remover head.

d. Position the hub so the remover head is against a solid surface, such as a concrete floor.

e. Strike the end of the driver so it wedges firmly in the remover head. The remover head should now be jammed tight against the inner bearing race.

f. Reposition the assembly so the remover head is free to move and the driver can be struck again.

g. Strike the driver, forcing the bearing (**Figure 29**) and hub spacer from the hub.

h. Remove the driver from the remover head.

11

i. Repeat the procedure to remove the remaining bearing(s).

3B. Remove the bearings using a hammer, drift and heat gun, or propane torch. The purpose for using heat is to slightly expand the hub bores so the bearings can removed with minimal resistance. Remove the bearings as follows:

WARNING
When using a heat gun or propane torch to heat the hub, care must be taken to prevent burning, finished or combustible surfaces. Work in a well-ventilated area and away from combustible materials. Wear protective clothing, including eye protection and insulated gloves.

a. Clean all lubricants from the wheel. For the rear wheel, remove the rubber damper.
b. Insert a long drift into the hub and tilt the hub spacer away from the bearing to be removed (**Figure 30**).
c. Heat the hub around the bearing to be removed. Keep the heat source moving at a steady rate and avoid heating the bearing. A large washer placed over the bearing helps insulate the bearing from the heat.
d. Turn the wheel over and use the drift to tap around the inner bearing race. Make several passes until the bearing is removed from the hub.
e. Remove the hub spacer.
f. Heat the hub around the remaining bearing. Drive out the remaining bearing using a large socket or bearing driver that fits on the outer bearing race.

4. Clean and dry the interior of the hub. Inspect the hub for:
a. Cracks, corrosion or other damage.
b. Fit of the new bearings. If a bearing fits loosely in the hub bore, replace the hub. The bearings must be a driven-fit.

5. Inspect the hub spacer for:
a. Cracks, corrosion or other damage.
b. Fit. Check the fit of the spacer against the back side of the bearings. It should fit flat against the bearings. Repair minor nicks and flaring with a file. Do not grind or shorten the spacer. The spacer must remain its full length, in order to prevent binding of the bearings when the axle is tightened.

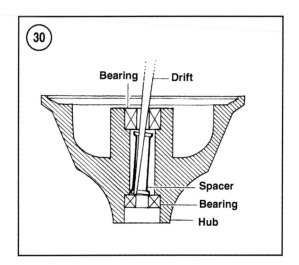

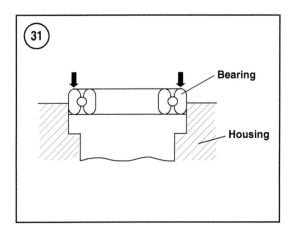

6. Before installing the new bearings, note the following:
a. Inspect the new bearings and determine which side faces out. This is usually the side with the manufacturer's marks and numbers. Bearings that are sealed on one side should be installed with the sealed side facing out.
b. Apply waterproof grease to bearings that are not lubricated by the manufacturer, or that are not sealed on both sides. Work the grease into the cavities between the balls and races.
c. Always support the bottom side of the hub, near the bore, when installing bearings.
d. To aid in driving the bearings, chill them in a freezer to temporarily reduce their diameter.

7. Heat the hub around the bearing bore.

8. Place a bearing *squarely* over the bearing bore.

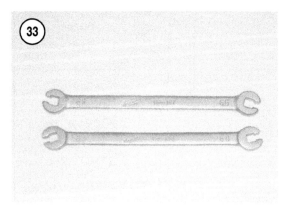

9. Place a suitable-size driver or socket over the bearing. The driver should seat against the outside diameter of the bearing (**Figure 31**).

10. Hold the driver, then squarely drive the bearing, seating it in the hub (**Figure 32**).

CAUTION
Do not press or strike the bearing directly. Bearing damage will occur.

11. Where used, install the snap ring.

12. Turn the hub over and install the hub spacer.

13. Drive in the remaining bearing, seating it in the hub.

RIM AND SPOKE SERVICE

The rim and hub must be concentric to ensure good handling and prevent damage to the parts. When the bike is new, all spokes are tensioned equally and the rim and hub are aligned and concentric. As the bike is used, the spoke tensions become unequal and the rim may become damaged. When this occurs the wheel develops radial (up and down) and lateral (side to side) runout. Wheel truing is the procedure that retensions the spokes, aligns the rim and hub, and makes the parts concentric. Regularly inspect and correct any problems with the wheel assembly.

Rim Inspection

Inspect the rims for flat spots, dents and warping. Also check the spoke holes for enlargement. Wheel dents are common to this type of motorcycle and cause the wheel to have excessive runout. Attempting to true a wheel with large dents can cause hub and rim damage, due to the overtightened spokes. If the dent is minor and runout is minimal, the rider may find it acceptable to continue to use the rim.

Spoke Inspection

Inspect the spokes for damage and proper tightness. For new wheels, or wheels that have been rebuilt, check the spokes frequently. After the tensions stabilize, check the spokes as recommended in the *Maintenance and Lubrication Schedule* in Chapter Three.

When tightening spokes, always use the correct size spoke wrench and do not exceed 2–4 N•m (17-35 in.-lb.) of torque. Use spoke wrenches (**Figure 33**) that grip the spoke on three sides. The spoke nipples can be rounded off or crushed if other types of tools are used. Do not true a wheel that has broken, bent or damaged spokes. Also, do not straighten bent spokes by excessive tightening. The spoke can crack the hub fitting and enlarge the rim hole.

In order to change spoke tension, spokes must be able to turn easily in the spoke nipples. If a spoke is seized in its nipple, apply penetrating lubricant to the threads. If the spoke does not free itself or turn smoothly, replace the spoke and nipple.

Wheel Truing

Wheels can be trued with the wheel on or off the bike. Before truing a wheel, check the condition of the wheel bearings. Accurate wheel truing is not possible with worn wheel bearings. Refer to **Table 1** for wheel runout specifications.

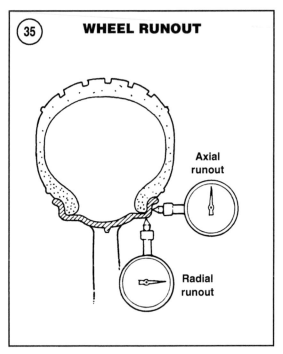

WHEEL RUNOUT

Axial
runout

Radial
runout

A wheel truing stand is used to measure runout (**Figure 34**), however, if runout appears minimal, the wheel can be left on the bike to make the check. Raise the wheel so it is level and free to spin. Solidly hold a pointer against the fork or swing arm. While the wheel is turned, move the pointer toward/away from the rim until maximum runout is determined. Measure the gap from the rim to the pointer. A more accurate check is to mount a dial indicator in the positions shown in **Figure 35**.

If the wheel needs major truing, mount the rim (tire and tube removed) on a truing stand and measure runout in both directions (**Figure 36**).

Correcting lateral runout

To move the rim to the left or right of the hub, loosen and tighten spokes as shown in **Figure 37**. The rim will move in the direction of the tightened spokes.

NOTE
Always loosen and tighten spokes equally. Loosen a minimum of three spokes, then tighten the opposite three spokes. If runout is over a large area, loosen and tighten a larger number of spokes.

Correcting radial runout

To make the rim concentric with the hub, loosen and tighten the spokes, as shown in **Figure 38**. The rim will move in the direction of the tightened spokes.

NOTE
Always loosen and tighten spokes equally. Loosen a minimum of three

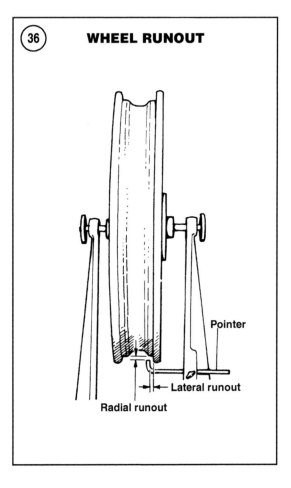

WHEEL RUNOUT

Pointer

Lateral runout

Radial runout

(37) LATERAL ADJUSTMENT

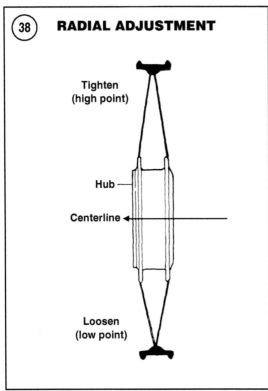

CENTERLINE

Rim

Tighten   Loosen

← To move rim

(38) RADIAL ADJUSTMENT

Tighten
(high point)

Hub

Centerline ←

Loosen
(low point)

(39)

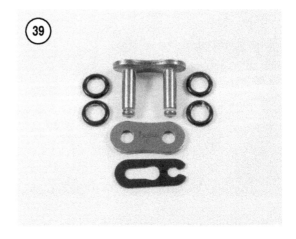

spokes, then tighten the opposite three spokes. If runout is over a large area, loosen and tighten a larger number of spokes.

DRIVE CHAIN

Refer to Chapter Three for drive chain cleaning, lubrication, adjustment and measurement. Refer to **Table 2** in this chapter for chain specifications.

When checking the condition of the chain, also check the condition of the sprockets, as described in Chapter Three. If either the chain or sprockets are worn, replace all drive components. Using new sprockets with a worn chain, or a new chain on worn sprockets will shorten the life of the new part.

When new, the motorcycle comes with an endless O-ring type chain. This type of chain is recommended for this motorcycle since it is internally lubricated and requires minimal maintenance. To remove the chain, the swing arm must be partially disassembled so the chain can pass out of the swing arm pivot.

The following procedure describes the removal and installation of the chain out of the swing arm. If the chain has been replaced with one that uses a master link and spring clip (**Figure 39**), refer to that procedure in this section.

Chain With No Master Link
Removal and Installation

1. Remove the rear wheel as described in this chapter.

2. Remove the sprocket guard so the chain can be removed from the drive sprocket (A, **Figure 40**).

3. Pull the chain forward and remove it from the sprocket. Clearance will be minimal when working the chain downward and past the sprocket shaft.

4. Disconnect the lever arms at either end (**Figure 41**).

5. Position the chain so it can be removed from the chain guard and guide (**Figure 42**) when the swing arm is lowered.

6. Loosen the swing arm bolt (B, **Figure 40**). Support the swing arm while removing the bolt.

7. Lower the swing arm and remove the chain.

8. Reverse this procedure to install the chain. Note the following:

   a. Clean and inspect the swing arm and lever arm bores before assembly. Apply waterproof grease to the parts before installing.

   b. Seat the lever arm bolt head in the counterbore of the left arm.

   c. Tighten the lever arm bolt and swing arm bolt to 98 N•m (72 ft.-lb.).

   d. Adjust the chain (Chapter Three).

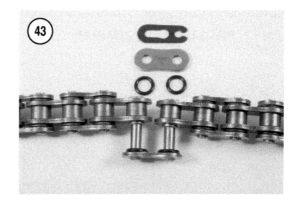

Chain With Clip Master Link Removal and Installation

1. Support the bike so it is stable and the rear wheel is off the ground.

2. Shift the transmission into neutral.

3. Find the master link on the chain. Remove the spring clip with pliers, then press the link out of the chain. Account for the O-rings under the side plates (**Figure 43**).

4. Remove the chain from the bike.

5. Clean and inspect the chain (Chapter Three).

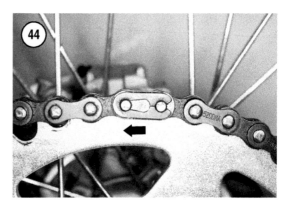

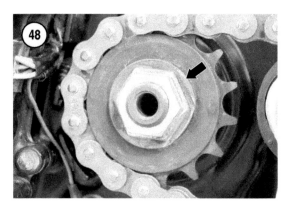

45

**DRIVE SPROCKET
1987-1996 (TO SERIAL
NO. KL650AE032209)**

48

46

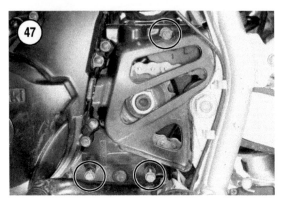

47

6. Reverse this procedure to install the chain. Note the following:

a. Install the O-rings under both side plates.

b. Install a new spring clip on the master link. The spring clip must be installed so the closed end of the clip points toward the direction of travel (**Figure 44**).

c. Adjust the chain (Chapter Three).

SPROCKETS

Check the condition of both sprockets and the drive chain, as described in Chapter Three. If either the chain or sprockets are worn, replace all drive components. Using new sprockets with a worn chain, or a new chain on worn sprockets will shorten the life of the new part.

Drive Sprocket
Removal and Installation

Two methods of locking the drive sprocket to the output shaft have been used. On 1987-1996 models (to engine serial number KL650AE032209), the sprocket is secured with a locking plate and two bolts (**Figure 45**). Later models use a lockwasher and nut (**Figure 46**).

1. Support the motorcycle so it is stable and secure. Keep the rear wheel on the ground.

2. Put the transmission in gear.

3. Remove the sprocket guard (**Figure 47**).

4. Loosen the bolts/nut securing the sprocket to the shaft. Note the following:

    a. On later models, flatten the lockwasher (**Figure 48**) before removing the nut.

    b. If the rear wheel turns, have an assistant lock the brakes while the bolts/nut are loosened. The wheel is more likely to rotate on later models, since the nut is highly torqued.

5. After loosening the bolts/nut, raise the rear wheel and loosen the rear axle and chain adjusters, Remove the chain from the rear sprocket.

6. Remove the bolts/nut and locking plate/lockwasher from the shaft.

7. Pull the chain forward, then lift it from the sprocket while removing the sprocket.

11

8. Clean and inspect the output shaft spacer and seal (**Figure 49**). If there is damage or leakage, remove the spacer and internal O-ring (**Figure 50**) as follows:

   a. With the transmission in gear, twist the spacer from the shaft. If necessary, warm the seal with a heat gun, then use locking pliers to grip the spacer near its outer edge, away from the seal surface.

   b. Remove the O-ring from the groove, located past the seal lip.

9. Inspect the sprocket (Chapter Three).

10. Inspect the sprocket guard assembly (**Figure 51**) for damage.

11. Clean and inspect the output shaft and seal area.

12. Reverse this procedure to install the drive sprocket. Note the following:

   a. If removed, install a new, lubricated O-ring onto the output shaft. The O-ring must seat in the groove, located past the seal lip (A, **Figure 52**).

   b. If removed, install the output shaft spacer with the notched edge facing in (B, **Figure 52**).

   c. Install the sprocket with the cupped side facing out.

   d. Fit the sprocket into the chain before sliding the sprocket into position.

   e. On later models, install a new lockwasher with the cupped side facing out (**Figure 53**). Install the locknut with the small diameter facing out.

   f. Torque the bolts/nut to the value listed in **Table 3**.

   g. On later models, flatten two sides of the lockwasher against the nut (**Figure 54**).

   h. Adjust the chain (Chapter Three).

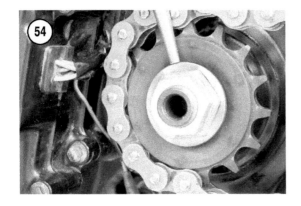

Driven Sprocket
Removal and Installation

1. Remove the rear wheel as described in this chapter.

2. Support the wheel so it does not rest directly on the brake disc, or remove the sprocket and hub from the wheel (**Figure 55**). It may be easier to leave the sprocket hub in the wheel until the sprocket nuts are loosened.

3. Remove the nuts securing the sprocket to the hub.

4. Inspect the sprocket (Chapter Three).

5. Inspect the hub as described in *Rear Wheel Inspection*.

6. Reverse this procedure to install the driven sprocket and rear wheel. Note the following:

    a. Mount the sprocket so the tooth identification number (**Figure 56**) faces out.

    b. Tighten the bolts in several passes, working in a crossing pattern. Tighten the bolts to 32 N•m (24 ft.-lb.).

    c. Adjust the chain (Chapter Three).

TIRE CHANGING

Removal

When changing a tire, work over a pad to prevent damage to the wheel assembly. Do not allow the wheel to rest on the brake disc.

1. Remove the core from the valve stem and deflate the tire.

2. Press the entire bead on both sides of the tire into the rim.

3. Lubricate the beads with soapy water.

4. Insert a tire iron under the bead, next to the valve stem (**Figure 57**). Pry the bead over the rim, while forcing the bead on the opposite side of the tire into the rim.

5. Insert a second tire iron next to the first (**Figure 58**). While holding the tire with one iron, work around the perimeter of the rim with the second iron, prying the tire over the rim.

NOTE
If the inner tube must be reused, be careful to not pinch the tube as the tire is being removed from the rim. It is common practice to replace the tube when a new tire is being mounted.

6. Remove the nut from the valve stem and remove the inner tube from the tire.

7. Pry the second tire bead over the rim (**Figure 59**).

Inspection

1. If the tire must be reused, inspect the inside and outside of the tire for damage and objects that could cause a puncture.

2. Inspect the rim for damage.

3. Check that the spokes do not protrude through the spoke nipples.

4. Inspect the rim band for deterioration. If a new rim band is installed, place the roughest side of the band against the rim. If water is entering the rim, an alternative to the rim band is to wrap the rim with two revolutions of duct tape. Punch a hole for the valve stem.

Installation

> *NOTE*
> *Installation is easier if the tire is warm and pliable. This can be achieved by placing the tire in the sun or an enclosed vehicle.*

1. Sprinkle talcum powder around the interior of the tire casing. Distribute the powder so it is on all surfaces that will touch the inner tube. The powder minimizes chafing and helps the tube distribute itself when inflated.

> *NOTE*
> *Depending on the make and type of tire installed, check the sidewall and determine if it must be installed in a specific direction. A direction arrow is*

often embossed in the sidewall. Also check for a dot or mark that indicates the light side of the tire. This mark should align with the valve stem.

2. Lubricate one of the tire beads, then push it onto the rim (**Figure 60**). Use a tire iron to lever the final section of bead onto the rim.

3. Install the core into the valve stem, then insert the tube into the tire. Check that the tube is not twisted while tucking it into the tire. Install the valve stem nut loosely.

4. Inflate the tube until it is rounded and no longer wrinkled. Too much air makes tire installation diffi-

ommended seating pressure. If none is indicated, inflate the tire to 172–207 kPa (25–30 psi).

> *WARNING*
> *If the tire does not seat at the recommended pressure, do not continue to overinflate the tire. Deflate the tire and reinflate to the recommended seating pressure. Relubricate the beads, if necessary.*

8. Finger-tighten the valve stem nut.

9. Bleed the tire pressure to the recommended setting in **Table 1**.

10. Install the valve stem cap.

11. Balance the wheel as described in this section.

Wheel Balancing

Whenever a new tire is installed it should be inspected for proper balance. An unbalanced wheel shortens tire life and puts avoidable wear on the wheel bearings. Ride quality is also diminished. For maximum tire life, check wheel balance whenever the wheel is removed from the bike. Balance changes over the life of the tire. If the proper equipment is not available, have the wheel balanced by a dealership.

> *NOTE*
> *Do not balance a wheel that has damaged bearings. Balance will not be accurate. Replace the bearings before balancing.*

1. Place the wheel and axle assembly in a truing stand, or a similar fixture that allows the axle to be level and the wheel to freely spin (**Figure 63**). Install the necessary spacers to keep the wheel from moving laterally.

2. Spin the wheel and allow it to come to a complete stop. Mark the sidewall at the top of the tire. This is the light side of the wheel. Repeat the step several times to ensure an accurate reading. If the wheel consistently stops at a different location, balance is acceptable and additional weight is not required.

3. When the light side of the wheel is verified, lightly clamp a weight (**Figure 64**) to the spoke nearest the light mark. Start with the smallest increment of weight. Weights are available from a dealership.

cult and too little air increases the chance of pinching the tube.

5. Lubricate the second tire bead, then start installation opposite the valve stem. Hand-fit as much of the tire as possible (**Figure 61**). Before final installation, check that the valve stem is straight and the inner tube is not pinched. If necessary, relubricate the bead. Use tire irons to hold and pry the remaining section of bead onto the rim (**Figure 62**).

6. Check the bead for uniform fit, on both sides of the tire.

7. Lubricate both beads and inflate the tire to seat the beads onto the rim. Check the sidewall for a rec-

4. Continue to spin and check the tire for balance. Continue to add/subtract weight to the spokes until the wheel consistently stops at a different location. When balance is achieved, tightly clamp the weight(s) to the spoke(s).

NOTE
Adhesive-backed weights are not recommended for this type of motorcycle, since the weights are more vulnerable to falling off in off-road conditions.

Table 1 FRONT AND REAR WHEEL SPECIFICATIONS

Tire size	
Front	90/90-21 54S
Rear	130/80-17 65S
Tire pressure (standard)	
Front	150 kPa (22 psi)
Rear	
Loads up to 97.5 kg (215 lb.)	150 kPa (22 psi)
Loads 97.5-182 kg (215-400 lb.)	200 kPa (29 psi)
Wheel rim runout (radial and lateral)	2.0 mm (0.08 in.)
Wheel rim size	
Front	21 × 1.60
Rear	17 × 2.50

Table 2 DRIVE CHAIN AND SPROCKET SPECIFICATIONS

Drive chain	
Size	520
Type	O-ring
Number of links	106
Drive chain slack	50-60 mm (2.0-2.4 in.)
Drive chain length wear limit (20 links/21 pins)	323 mm (12.7 in.)
Sprocket sizes (standard, front/rear teeth)	15/43

Table 3 WHEEL AND DRIVE TORQUE SPECIFICATIONS

	N•m	in.-lb.	ft.-lb.
Axle nut			
Front	78	–	58
Rear	93	–	69
Drive sprocket			
bolts (1987-1995)	10	88	–
nut (1996-on)	98	–	72
Driven sprocket bolts	32	–	24
Rear brake caliper			
holder bolts	25	–	18
Swing arm pivot bolt	98	–	72
Swing arm lever bolts	98	–	72
Spokes	2-4	17-35	–

FRONT SUSPENSION AND STEERING

This chapter provides service procedures for the front suspension and steering components. This includes the handlebar, steering head and fork. Refer to the tables at the end of the chapter for specifications, capacities and torque requirements.

HANDLEBAR

Removal and Installation

Use the following procedure to remove and install the handlebar into the holders. If the handlebar only needs to be repositioned, adjust by loosening the lower holder cap bolts. Tilt the handlebar to the desired position, then retorque the bolts. If complete disassembly of the handlebar is necessary, disassemble the handlebar before removing it from the holders.

1. Support the motorcycle so it is stable and secure.
2. Lift the plastic caps (**Figure 1**) from the holder bolts.
3. Remove the bolts from the holders (**Figure 2**).
4. Secure the handlebar either forward or back from the holders. Keep the brake fluid reservoir upright.
5. Clean the handlebar and holders.
6. Reverse this procedure to install the handlebar. Note the following:
   a. Position the holder caps so the arrows on the caps point up (**Figure 3**).
   b. Lightly tighten the holder bolts and check the riding position.
   c. Tighten the upper bolt on each cap, then tighten the lower bolts to 24 N•m (18 ft.-lb.).

Disassembly and Assembly

1. Support the motorcycle so it is stable and secure.
2. At the right handlebar:

12

a. Remove the two screws securing the throttle assembly (A, **Figure 4**). Slide the throttle assembly off the handlebar. If the throttle cables prevent the assembly from sliding off the handlebar, remove the throttle after the handlebar is removed from the bike. Do not allow the throttle cables to kink during removal.

b. Disconnect the brake switch (B, **Figure 4**).

c. Remove the screw securing the stop switch (C, **Figure 4**).

d. Remove the brake hose from the handlebar clamps.

e. Remove the bolts securing the master cylinder assembly (D, **Figure 4**).

CAUTION
Do not allow the master cylinder to hang by its hose. Keep the master cylinder in an upright position so fluid cannot leak out of the cap. This also prevents air from getting into the system. Wrap the master cylinder with a clean shop cloth, then secure it to the bike until after handlebar reassembly.

3. At the left handlebar:

a. Remove the two screws securing the switch assembly (A, **Figure 5**).

b. Remove the two screws securing the clutch lever assembly (B, **Figure 5**).

c. Remove the ties securing the wiring to the handlebar.

4. If the handlebar will be replaced, remove the left grip. If necessary, insert a compressed air nozzle between the grip and handlebar. As air expands the grip, twist and pull the grip from the handlebar.

5. Remove the handlebar as described in this section.

6. Reverse this procedure to install handlebar. Note the following:

a. If the throttle was removed after the handlebar was removed, install the throttle before mounting the handlebar.

b. Install the master cylinder clamp so *UP* and the arrow are facing up. Tighten the upper bolt first, then the bottom bolt to 9 N•m (80 in.-lb.).

c. If removed, install the left grip before installing the clutch and switch assemblies. Clean the handlebar and grip with solvent. Apply adhesive to the grip, then twist it into position on the handlebar.

Inspection

1. Inspect the handlebar for cracks, bending or other damage. If the handlebar is made of aluminum, check closely where the handlebar is clamped to the fork, and at the clutch lever. If cracks, scores or other damage is found, replace the handlebar.

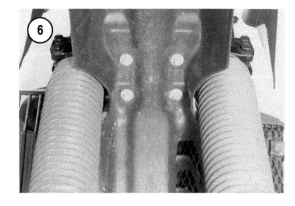

Damage in these areas may cause the handlebar to break.

> **WARNING**
> *Never straighten, weld or heat a damaged handlebar. The metal can weaken and possibly break when subjected to the shocks and stresses that occur when riding the motorcycle.*

2. Inspect the threads on the mounting bolts and in the holders. Clean all residue from the threads. Replace damaged bolts.

3. Clean the handlebar, holders and caps with solvent or electrical contact cleaner. Use a stiff brush to clean the residue from the knurled areas on the handlebar. Use a soft brush on aluminum handlebars.

STEERING STEM AND HEAD

The steering stem pivots in the steering head on tapered roller bearings. The bearings are at the top and bottom of the steering stem. Do not remove the inner races (mounted in the frame) and the lower bearing (mounted on the steering stem) unless they require replacement. The bearings should be lubricated as specified in Chapter Three. Before disassembling the steering head, perform the *Steering Play Check and Adjustment* procedures in this section. The checks will help determine if the bearings and races are worn, or if they only require adjustment.

Disassembly

1. Remove the front wheel (Chapter Eleven).

2. Remove the bolts securing the front fender (**Figure 6**).

3. Remove the fuel tank, radiator covers and fairing (Chapter Fifteen).

4. Remove the handlebar as described in this chapter.

5. Loosen the stem nut (A, **Figure 7**).

6. Remove the fork legs as described in this chapter.

7. Remove the gauges and ignition switch from the upper fork bridge.

8. Remove the clamps from the left side of the upper and lower fork bridges.

9. Remove the stem nut and washer.

10. Remove the upper fork bridge (B, **Figure 7**).

11. Hold the steering stem so it cannot fall, then remove the claw washer and adjust nut (**Figure 8**).

> **NOTE**
> *If tight, loosen the nut with a stem nut wrench (part No. 57001-1100) or similar tool. Order the tool from a Kawasaki dealership.*

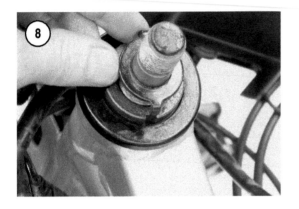

12. Lower the steering stem assembly out of the steering head, then remove the seal and bearing at the top of the steering head.

> *CAUTION*
> *The bottom bearing is pressed onto the steering stem. Do not remove the bearing unless it needs replacement. Damage will occur during removal.*

Inspection

1. Clean the bearings and races with solvent.
2. Inspect the steering head (on the frame) for cracks or other damage. If damage is evident, have a qualified welding shop make the repairs.
3. Inspect the steering stem assembly (**Figure 9**). Inspect the clamping and pivot areas for cracks or damage.
4. Inspect the steering stem threads, stem nut, washers and adjust the nut for damage (**Figure 10**).
5. Inspect the seals for damage.
6. Inspect the bearings for pitting, scratches, corrosion or discoloration (**Figure 11**).
7. Inspect the bearing races in the frame for pitting, galling or corrosion (**Figure 12**). If a race is worn or damaged, replace both races and bearings as described in this section.
8. If the bearings are to be reused, pack the bearings with waterproof grease.

Outer Bearing Race Replacement

Only remove the bearing races from the frame when installing new races. When installing new races, always install new bearings. The bearing races are recessed in the bores, and installation requires the Kawasaki drivers (part No. 57001-1106

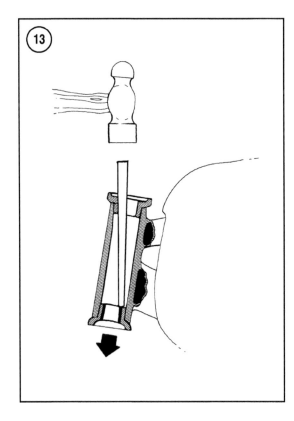

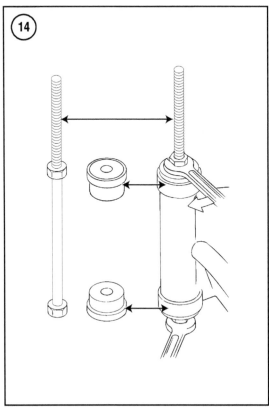

and part No. 57001-1076) and driver press shaft (part No. 57001-1075). The tool is a threaded rod with sized disc drivers fitted to each end. A similar tool can be fabricated with common hardware. If this is done, the parts used as the drivers must fit at the outside edge of the races, and still be capable of entering the steering head bore. The drivers cannot contact the surface of the races.

1. Insert an aluminum drift into the steering head and position it on the edge of the lower race (**Figure 13**). Carefully drive out the race. To prevent binding, make several passes around the perimeter of the race. Repeat the procedure to remove the upper race.

2. Clean the race bores and inspect them for damage.

3. To install the upper race, do the following:
   a. Place a new, lubricated race squarely into the bore opening, with its wide side facing out.
   b. Assemble the tool, seating a driver at the outer edge of the race. The lower driver can rest on the perimeter of the steering head, or in the stepped bore (**Figure 14**).
   c. Hold the lower nut with a wrench and tighten the upper nut to seat the race.
   d. Remove the tool from the frame and check that the race is fully seated.
   e. Repeat the procedure to install the second race. During installation, do not allow the driver or shaft to contact the face of the first race. Damage will occur.

4. Lubricate the races with waterproof bearing grease.

Steering Stem Bearing Replacement

The steering stem bearing is press-fitted into place. Perform the following steps to replace the bearing.

1. Thread the stem nut (**Figure 15**) onto the steering stem to protect the threads.

2. Stabilize the steering stem, then remove the bearing and seal, using a hammer and chisel (**Figure 15**). To prevent binding, make several passes around the perimeter of the bearing.

> *WARNING*
> *Wear safety glasses when using the hammer and chisel.*

3. Clean and inspect the steering stem.

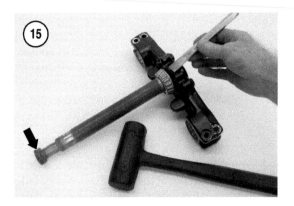

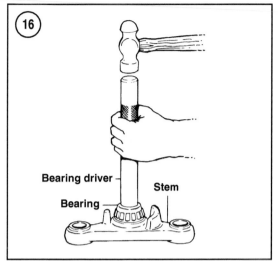

4. Pack the new bearing and seal with waterproof grease.

5. Slide the new seal and bearing onto the steering stem.

6. Drive the bearing into place as follows:

   a. Support the steering stem directly below the bearing.

   b. Slide a bearing driver or long pipe over the steering stem, seating the tool on the *inner* bearing race (**Figure 16**).

   c. Drive the bearing onto the steering stem until it is seated.

NOTE
If available, a press can also be used to drive the bearing and seal.

Assembly and Adjustment

1. Check that the upper and lower bearing races are seated in the frame.

2. Check that bearings, races and seals are lubricated with waterproof grease.

3. Guide the steering stem through the bottom of the frame (**Figure 17**).

4. Install the upper bearing into its race.

5. Install the bearing seal and adjust nut. Finger-tighten the adjust nut.

6. Seat the bearings as follows:

   a. Tighten the adjust nut with a spanner (**Figure 18**) to seat the bearings, then loosen the nut. Do not use excessive force when seating the bearings.

   b. Retighten the adjust nut as necessary, while checking for horizontal and vertical play. Stop adjusting the nut when play is eliminated in both directions. If properly adjusted, the steering stem will pivot to the lock positions

under its own weight, after an initial assist. Final play is rechecked in a later step when more leverage can be applied to the steering stem assembly.

7. Install the claw washer, upper fork bridge, washer and stem nut. Finger-tighten the nut until the fork legs are installed.

8. Install the gauges and ignition switch onto the upper fork bridge.

9. Loosely install the handlebar so wiring, hoses and cables can be routed and handled.

10. Install the fork legs as described in this chapter. Do not tighten the lower bridge pinch bolts until the steering stem nut has been torqued. The lower bridge will move slightly as the nut is tightened.

11. Remove the handlebar and tighten the stem nut to 39 N•m (29 ft.-lb.).

12. Tighten the lower bridge pinch bolts just enough to grip the fork legs.

13. Check bearing play again, as follows:

a. Turn the steering stem from lock to lock. The steering stem should turn smoothly and freely. If binding is felt, the steering stem is too tight.

b. Grasp the fork legs near the axle. Lever the fork legs in all directions and feel for play. If play is felt or heard, the steering stem is too loose. If necessary, perform substeps c-h to eliminate or add play in the steering stem.

c. Loosen the lower bridge pinch bolts.

d. Loosen the steering stem nut.

e. Loosen or tighten the adjust nut (**Figure 19**).

f. Retighten the stem nut to 39 N•m (29 ft.-lb.).

g. Tighten the lower bridge nuts.

h. Recheck bearing play. Adjust as necessary.

14. Tighten the lower bridge pinch bolts to 25 N•m (18 ft.-lb.).

15. Install the handlebar as described in this chapter.

16. Install the front fender.

17. Install the front wheel (Chapter Eleven).

18. Install the fairing, fuel tank and radiator covers (Chapter Fifteen).

Steering Play Check and Adjustment

Steering adjustment takes up any play in the steering stem and bearings and allows the steering stem to operate with free rotation. Excessive play or roughness in the steering stem makes steering imprecise and causes bearing damage. These conditions are usually caused by improper bearing lubrication and steering adjustment. Improperly routed control cables can also affect steering operation.

1. Support the motorcycle so the front wheel is off the ground.

2. Turn the handlebar from lock to lock and check for roughness or binding. Movement should be smooth with no resistance.

3. Position the handlebar so the front wheel points straight ahead. Lightly push the end of the handlebar. The front end should fully turn to the side from the center position, under its own weight. Check in both directions. Note the following:

a. If the steering stem moves roughly or stops before reaching the frame stop, check that all cables are routed properly.

b. If cable routing is correct and the steering binds, the steering adjustment may be too tight. This condition can also occur if the bearings and races require lubrication or replacement. Perform the remaining checks before adjusting the steering in Step 5.

c. If the steering stem moves from side to side correctly, perform Step 4 to check for excessive looseness.

4. Position the fork so it points straight ahead. Have an assistant hold the bike, then grasp the fork legs near the axle. Lever the fork legs in all directions and feel for play.

a. If movement can be felt at the steering stem, tighten the steering as described in Step 5.

b. If no excessive movement can be felt and the steering turns from side to side correctly, the steering is adjusted properly and in good condition.

5. Adjust the steering as follows:

a. Remove the handlebar as described in this chapter.

b. Loosen the lower bridge pinch bolts (**Figure 20**).

c. Loosen the steering stem nut (A, **Figure 7**).

d. Loosen or tighten the adjust nut (**Figure 19**) with a pin spanner wrench.

12

e. Retighten the stem nut to 39 N•m (29 ft.-lb.).

f. Tighten the lower bridge nuts.

g. Recheck bearing play. Adjust as necessary.

6. Tighten the lower bridge pinch bolts to 25 N•m (18 ft.-lb.).

7. Install the handlebar as described in this chapter.

FORK REMOVAL AND INSTALLATION

Removal

Although removing the front fairing, fender, fuel tank and radiator covers is not required, removing some of these parts will make it easier to access and handle the fork.

1. Depress the air valve (**Figure 21**) and bleed any air pressure from both fork legs.

2. If the fork legs will be disassembled, the oil can be drained at this time, or at time of disassembly. If drained now, there will be less oil on the components as the legs are disassembled. Remove the drain screw and O-ring (A, **Figure 22**) from both fork legs and allow the oil to drain. Pump the fork to assist in draining. If possible, allow the legs to drain overnight.

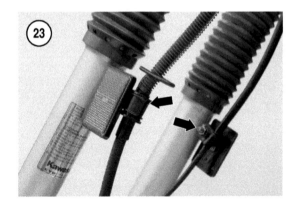

3. Support the motorcycle so it is stable and the front wheel is off the ground.

4. Remove the front wheel (Chapter Eleven).

5. Remove the brake caliper bolts (B, **Figure 22**) (Chapter Fourteen).

NOTE
When the caliper is removed, insert a spacer block between the brake pads. This prevents the pad from being forced out of the caliper if the brake lever is accidentally operated. Hang

the caliper with a piece of wire. Do not allow it to hang by the hose.

6. Remove the brake hose and speedometer cable guides from the fork (**Figure 23**).

7. Remove the brake hose clamp and tie (**Figure 24**). Retie the caliper away from the fork.

8. If the forks are to be disassembled, loosen (do not remove) the fork caps (**Figure 25**).

9. If the steering stem assembly will be removed:

    a. Remove the handlebar and loosen the steering stem nut.

    b. Install the handlebar and lightly bolt it into place.

10. Loosen the upper bridge pinch bolts on each fork leg (**Figure 26**).

11. Loosen the lower bridge pinch bolts (**Figure 27**), then twist and remove the fork tube from the upper and lower fork bridges. Repeat for the remaining fork tube.

12. Clean the fork tubes, fork bridges and clamping surfaces.

13. Remove and clean the pinch bolts. Replace damaged bolts.

Installation

1. Install the left fork leg (with brake caliper mount) into the lower and upper fork bridges.

2. Adjust the fork leg clearance at the upper fork bridge. The top surface of the upper bridge should be 1 mm (0.040 in.) below the top edge of the fork tube (**Figure 28**). Do not include the fork cap in the measurement.

3. Lightly tighten the upper and lower bridge bolts (**Figure 26** and **Figure 27**) to hold the fork leg in place. Repeat Steps 1-3 to install the right fork leg.

4. Tighten the bridge pinch bolts to 25 N•m (18 ft.-lb.).

5. Install the brake hose clamp and a new tie (**Figure 24**).

6. Install the brake hose and speedometer cable guides (**Figure 23**).

7. Install the brake caliper (B, **Figure 22**) (Chapter Fourteen).

8. Install the front wheel (Chapter Eleven).

9. If the fork legs were rebuilt:

    a. Align the fork boots and clamp into place.

    b. Tighten the fork caps to 29 N•m (21 ft.-lb.).

12

FORK SERVICE

Fork Tools

The following tools are required to disassemble and reassemble the fork legs:

1. Fork seal driver (38 mm). Kawasaki part No. 57001-1219, or equivalent.
2. Handle and adapter (**Figure 29**). Kawasaki part No. 57001-183 and part. No. 57001-1057, or equivalent. Refer to the text for an optional tool that can be fabricated.
3. 10 mm Allen socket, to loosen and torque the dampers.

Disassembly

Preferably, rebuild one fork leg before rebuilding the remaining leg. Parts should not be intermixed. Refer to **Figure 30**.

1. Loosen the fork boot clamp, then remove the boot from the outer tube (**Figure 31**).
2. Clean the fork tubes and fork cap area.

> *NOTE*
> *In the following step, the removed parts will be oily. Lay the parts on absorbent paper or cloth that can be disposed.*

3. Hold the fork leg upright, then remove the cap, spring spacer, spring seat and spring (**Figure 32**). If the spring is progressively wound, make note of which end of the spring is up.
4. Drain the oil from the fork tubes. Pump the fork tubes to aid in draining the oil.
5. Loosen the fork damper as follows:

> *NOTE*
> *The Kawasaki adapter (A, **Figure 33**) is tapered and square. The tapered end seats into the damper, which is smooth and round. The edges of the adapter must wedge into the damper in order to hold it steady. If working alone, the tool can easily dislodge itself while handling the fork leg. An alternative is to use the 12-point hex (B) at the top of the damper. A 24 mm nut (C) locks into the fitting and makes fork handling much easier. The nut can be welded to a T-handle (approximately 460 mm [18 in.] long), or, the*

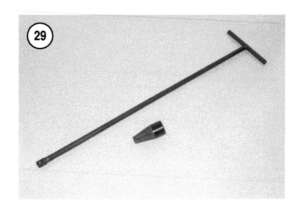

*Kawasaki adapter can be driven into the nut (**Figure 34**), then mounted on the handle.*

a. Push the fork tubes together, then seat the damper tool into the damper (**Figure 35**). If the Kawasaki adapter is used (no nut), tap the handle to seat the tool.

b. Lock the axle holder in a vise fitted with soft jaws, or invert the fork and lock the tool handle into a floor joint or other stable device. The handle must not be able to turn.

> *CAUTION*
> *If a vise is used, do not overtighten the vise. Damage or cracking of the axle holder could occur. Removal will also be easier if an assistant can stabilize the fork leg and damper tool.*

6. Remove the damper bolt with a 10 mm Allen socket (**Figure 36**).

7. Remove the fork damper and rebound spring (**Figure 37**).

8. Separate the fork tubes as follows:

a. Remove the stop ring from the top of the outer tube (**Figure 38**).

b. Before disassembling the fork tube and slider, slide the tubes together and check for roughness and binding. If the action is not smooth, inspect the tubes for damage when they are disassembled.

c. There is an interference-fit between the outer fork tube and the guide bushing and seals. To separate the tubes, as well as the parts, quickly and firmly pull the fork tubes apart (**Figure 39**). Repeat as necessary until the parts are separated.

FORK LEG ASSEMBLY

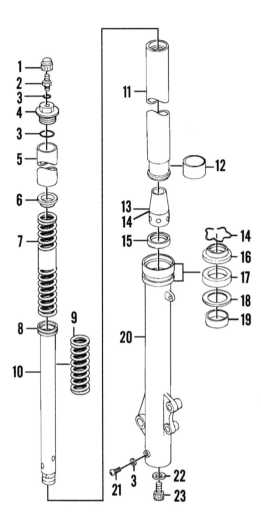

1. Cap
2. Air valve
3. O-ring
4. Fork cap
5. Spring spacer
6. Spring seat
7. Fork spring
8. Bushing
9. Rebound spring*
10. Fork damper*
11. Inner tube
12. Slide bushing
13. Valve*
14. Stop ring*
15. Collar*
16. Dust seal
17. Oil seal
18. Backup ring
19. Guide bushing
20. Outer tube
21. Drain screw
22. Seal washer
23. Damper bolt

*All noted parts only available together, as a single part number.

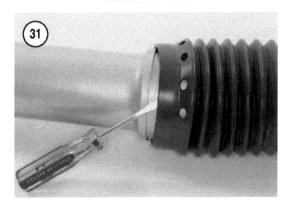

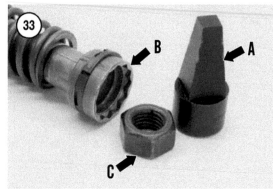

9. Remove the valve and collar (**Figure 40**). Note the orientation of the collar. During tube separation, the parts will likely have fallen from the inner tube.

10. *Carefully* spread open the slide bushing, then remove it from the fork tube (**Figure 41**). Do not spread the bushing more than necessary. Remove the remaining parts from the fork leg (**Figure 42**).

> *NOTE*
> *Do not disassemble the remaining parts in the inner fork tube and fork damper. Attempted removal of the parts may cause damage. The parts are integral to the tube and damper and are not available separately.*

11. Clean and inspect the parts as described in this section.

Inspection

Inspect each set of fork parts as follows. Refer to **Figure 30**.

1. Initially clean all parts in solvent, checking that the solvent is compatible with the painted tube, rub-

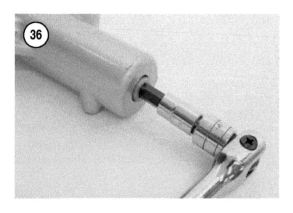

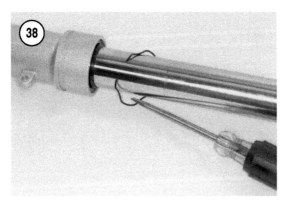

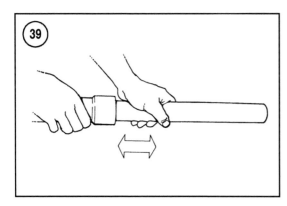

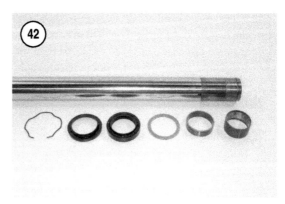

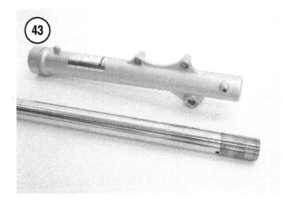

ber parts and the coating on the bushings. Immediately wash the parts a second time with soap and water. Rinse the parts with clean water, then dry with compressed air.

CAUTION
Carefully handle and clean the parts. Harsh cleaning habits and some solvents can remove or damage the coating that is on the bushings.

2. Inspect the fork tubes (**Figure 43**) for:

a. Nicks, rust, chrome flaking or creasing along the length of the inner tube. These conditions damage the dust and oil seals. Repair minor roughness with 600 grit sandpaper and solvent.

b. Pitting or abrasion in the bore of the outer tube (**Figure 44**).

c. Wear and stress cracks at the axle holder.

d. Damage at the seal and bushing areas.

e. Damaged threads.

3. Inspect the fork cap and threaded fasteners (**Figure 45**) for:

a. Damaged threads.

b. Damaged air valve and cap.

c. Install new O-rings on the drain screw and fork cap. Install a new seal washer on the damper bolt.

4. Inspect the spring, spring seat and spring spacer (**Figure 46**) for damage. Kawasaki does not specify service limits for the spring.

5. Inspect the rebound spring and top of the fork damper.

a. Inspect the small holes in the damper (**Figure 47**) for cleanliness.

b. Inspect the damper bushing for damage. Do not remove the bushing. If damaged, replace the entire fork damper assembly. The part is not available separately.

c. Inspect the spring for damage.

6. Inspect the valve, collar and lower end of the fork damper.

a. Inspect the holes in the damper and valve for cleanliness (**Figure 48**).

b. Check the valve for dents. Fit the valve on the damper and check that it smoothly and fully seats.

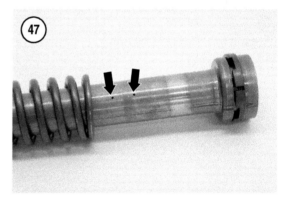

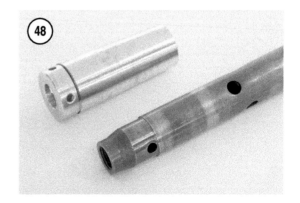

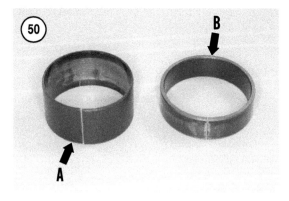

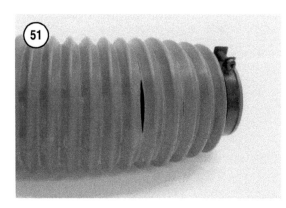

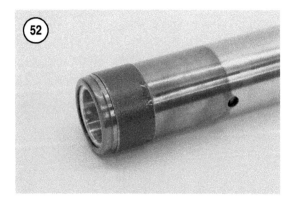

c. Check that the valve stop ring is tight and fully seated. Do not attempt to remove the stop ring. If damaged, replace the entire fork damper assembly. The part is not available separately.

d. Inspect the collar (**Figure 49**) for damage.

7. Inspect the slide bushing (A, **Figure 50**) and guide bushing (B) for scoring, scratches and excessive wear. Inspect the exterior of the slide bushing and interior of the guide bushing. Minor metallic buildup can be removed with fork oil and a nylon brush. Replace the bushings if discolored, or if the coating is excessively worn and the metal base is visible below the coating. Replace both bushings as a set.

8. Inspect the backup ring and stop ring for damage.

9. Inspect the fork boot (**Figure 51**). If damaged, replace the boot.

Assembly and Fork Oil Filling

The work area and all parts must be clean before starting assembly. Refer to **Figure 30**.

1. Lubricate the parts with fork oil as they are assembled. Use the same fork oil that is used for refilling the fork. Refer to **Table 2** for the recommended fork oil, or equivalent.

2. Install the drain screw and O-ring into the outer tube.

3. Install the slide bushing onto the end of the inner tube (**Figure 52**). Spread the bushing only far enough to slip it onto the tube. Avoid scratching the coating on the bushing. Seat the bushing in the groove.

4. Install the rebound spring and fork damper into the inner tube (**Figure 37**).

NOTE
The manufacturer does not specify the orientation of the collar. The fork shown had the deeper bore in the collar seated against the stop ring.

5. Extend the fork damper out the bottom of the inner tube, then install the valve and collar on the damper (**Figure 53**). The side of the collar with the deeper bore must fully seat against the stop ring on the valve.

6. With the damper assembly fully extended (**Figure 54**), hold the inner tube horizontally and guide it into the outer tube. Fully seat and hold the inner tube, keeping the tubes horizontal. The collar and

valve must not fall from the damper as the tubes are seated together. Keep the tubes horizontal and place them on the workbench so the damper bolt hole is accessible. Leave the tubes in this position until the damper bolt is installed. Proper assembly can be verified by looking into the damper bolt hole. The valve should be tight against the outer tube. If space between the parts is visible, the collar is probably jammed. Disassemble the parts and repeat the assembly process.

7. Apply threadlocking compound to the damper bolt threads, then install the seal washer and bolt into the bottom of the fork leg assembly. Finger-tighten the bolt into the damper. When seated, the fork leg can be handled. When tilted end to end, only the rebound spring should be heard.

8. Insert the tool to hold the fork damper, then torque the damper bolt to 39 N•m (29 ft.-lb.).

9. Slide the guide bushing and backup ring onto the inner tube (**Figure 55**).

10. Install the guide bushing and backup ring as follows:

   a. Hold the fork leg vertical, then align the bushing so the gap (in the bushing) faces to the left or right of the outer tube, and not to the front or back.

   b. Fit a 38 mm fork seal driver around the inner tube and against the backup ring.

   c. Use the fork seal driver like a slide hammer and drive the guide bushing and backup ring into the fork tube (**Figure 56**). Check that the parts are completely seated.

11. Install the oil seal and dust seal as follows:

   a. Cover the end of the inner tube with plastic, then coat it with fork oil. The plastic prevents the edge of the tube from tearing the dust and oil seals during installation.

   NOTE
   Before installing the seals, thoroughly lubricate the seals with fork oil. If preferred, lubricate the parts with one of the anti-stiction lubricants that is specifically for fork parts.

   b. Slide the oil seal and dust seal down the inner tube (**Figure 57**). The lip on both seals should face up.

   c. Position the seals squarely against the bore.

   d. Use the fork seal driver like a slide hammer and drive both seals together into the fork

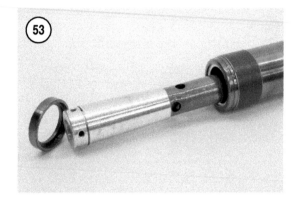

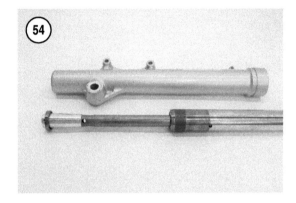

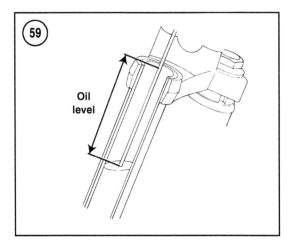

Oil level

15. Accurately measure and pour the oil into the fork leg (**Figure 58**).

16. Pump the leg for several minutes to distribute the oil and purge air pockets.

17. Measure the oil level as follows:

  a. Compress the fork leg to the bottom of its travel.

  b. Measure the distance from the top edge of the fork leg to the surface of the oil (**Figure 59**). Refer to **Table 2** for the required distance.

  c. If necessary, add or remove oil to bring the levels to the required distance. Both legs must have identical levels.

NOTE
The oil level establishes the air pocket that is above the oil. Since the air pocket contributes to fork damping, slight variation of the oil level is permissible to meet the riding conditions. Excess oil in the fork leg (small air pocket), makes damping harder. This condition can also lead to fork seal leakage. Too little oil in the fork leg (large air pocket), makes damping softer. This condition can cause the fork to bottom and possibly damage fork components.

18. Extend the fork leg, then install the spring and spring seat (**Figure 60**). If the spring is progressively wound, install the correct end up.

19. Install the spring spacer and fork cap. Tighten the cap enough to prevent the fork leg from leaking. The cap will be tighten to the torque specification after the leg is installed on the motorcycle.

20. Install the fork boot. Seat the boot into the outer tube, but do not tighten the clamp until the fork leg has been installed on the motorcycle.

tube. Drive the seals until the stop ring groove is visible above the seal.

12. Install and seat the stop ring in the groove.

13. Slide the fork tubes together several times and check for smooth operation. If there is roughness or binding, disassemble the parts and determine the cause.

14. Refer to **Table 2** for the recommended fork oil and capacity.

12

Table 1 STEERING AND FRONT SUSPENSION SPECIFICATIONS

Fork make/type	Kayaba/38 mm, air-adjustable
Fork air pressure (standard)	0 (atmospheric pressure)
Steering	
Caster (rake angle)	28°
Trail	112 mm (4.4 in.)
Turning radius (minimum)	2.4 m (94.5 in.)
Fork travel	230 mm (9.1 in.)

Table 2 FORK OIL LEVEL AND CAPACITY

Fork oil grade	Kayaba G-10, or equivalent 10 weight fork oil
Fork oil capacity (each leg)	
Oil change	355 cc (12.0 U.S. oz.)
Fork rebuild (all parts dry)	416-424 cc (14.0-14.3 U.S. oz.)
Fork tube oil level (from top edge of inner tube)	188-192 mm (7.40-7.56 in.)

Table 3 FRONT SUSPENSION TORQUE SPECIFICATIONS

	N•m	in.-lb.	ft.-lb.
Air valves	12	106	–
Bridge pinch bolts	25	–	18
Damper bolts (socket head)	39	–	29
Fork caps	29	–	21
Front axle nut	78	–	58
Handlebar holder bolts	24	–	18
Master cylinder mounting bolts	9	80	–
Steering stem nut	39	–	29

CHAPTER THIRTEEN

REAR SUSPENSION

This chapter provides service procedures for the rear shock absorber, swing arm and linkage assembly. Refer to the tables at the end of the chapter for specifications, recommended shock absorber settings and torque values.

LINK-TYPE SUSPENSION SYSTEM

The link suspension includes the swing arm, single shock absorber and a three-piece linkage system. The lower end of the shock absorber is connected to the swing arm by the linkage system. The design of the linkage system combines with the spring and damper to provide good suspension performance over the operating range of the swing arm.

The function of the linkage is to vary the speed of shock absorber compression, depending on the position of the swing arm. This change in compression speed varies the *damping curve* of the shock absorber. Small bumps cause the swing arm to compress slightly, and the shock absorber provides a compliant ride. The damping curve is relatively flat and soft under this condition. As riding conditions become more aggressive (large bumps) and swing arm travel increases, the same damping curve is no

longer effective. The damping curve must rise, in order to prevent *bottoming* of the suspension. To raise the curve, shock absorber speed must be *increased*, to raise hydraulic resistance (damping) in the shock absorber.

This increase in speed is achieved by the linkage system. During the transition from low to high swing arm movement, the linkage system pivots, increasing its leverage on the shock absorber. This increases shock absorber speed, which causes a progressively firmer damping action.

SHOCK ABSORBER

The single shock absorber (**Figure 1**) is a spring-loaded, hydraulically damped unit with an integral oil/nitrogen reservoir. The shock absorber damper is not rebuildable. To adjust the shock absorber, refer to *Rear Suspension Adjustment* in this chapter.

Removal and Installation

Read all procedures before removing the shock absorber. If the motorcycle has been modified,

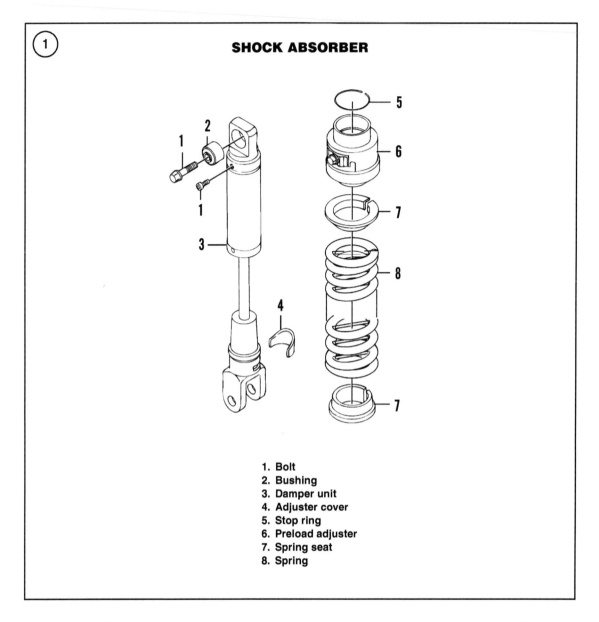

SHOCK ABSORBER

1. Bolt
2. Bushing
3. Damper unit
4. Adjuster cover
5. Stop ring
6. Preload adjuster
7. Spring seat
8. Spring

some disassembly procedures may not be necessary.

1. Remove the rear wheel (Chapter Eleven).

2. Remove the brake hose guide from the swing arm (**Figure 2**).

3. Remove the lever arms from the swing arm (**Figure 3**). When the bolt is removed, the swing arm pivots freely. Use wire to secure the swing arm in the up position.

4. Remove the nut from the lower shock absorber mounting bolt (**Figure 4**). Do not remove the bolt at this time.

5. Remove the left side cover (Chapter Fifteen).

6. Remove the breather hose (**Figure 5**) near the top of the shock absorber. For California models, first remove the liquid/vapor separator.

7. Remove the upper mounting bolt (**Figure 6**).

8. Remove the lower mounting bolt, then remove the shock absorber from the bike. Inspect the unit as described in this chapter.

9. Refer to **Table 2** for recommended shock absorber settings.

10. Reverse this procedure to install the shock absorber. Note the following:

   a. Lubricate the bores and mounting bolts with waterproof grease.

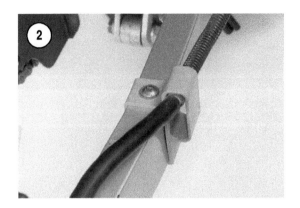

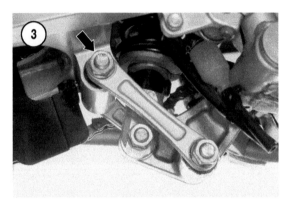

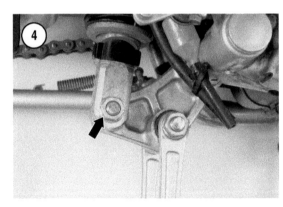

b. Install the shock absorber so the preload adjuster (**Figure 7**) is on the left side of the motorcycle.

c. Tighten the bolts to the specifications in **Table 3**.

Inspection

Individual parts are available for repairing the shock absorber. If damper leakage is evident, replace the complete shock absorber, as the damper is not available separately and is not rebuildable. If disassembly of the shock absorber is required, it is recommended that it be disassembled by a dealership. Because of the design of the shock absorber, a common spring compressor will not work in removing the spring.

1. Inspect the upper bushing (A, **Figure 8**) and mounting bolt for wear or damage. The bolt must be a firm fit in the bore. The bushing must be tight in the rubber mounting. If the bushing is worn, press or drive out the bushing. Install a new bushing and mounting bolt.

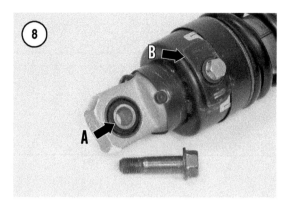

2. Inspect the spring preload adjuster and housing (B, **Figure 8**) for damage. Turning the adjuster bolt clockwise should increase preload and a higher number should be visible on the housing.

3. Inspect the damper rod for leakage at the seal (**Figure 9**). The damper rod should be smooth and shiny.

4. Inspect the spring for cracks.

5. Remove the cover from the rebound damper adjuster (**Figure 10**). Clean any dirt from the adjuster cavity.

6. Turn the rebound damper adjuster (**Figure 11**) and check for proper operation. Only turn the adjuster to the right, as indicated by the arrow on the adjuster. Each setting should have a perceptible click. When installing the adjuster cover, the arrow on the cover should point upward.

7. Inspect the lower clevis (**Figure 12**) and mounting bolt for wear or damage. The bolt must be a firm fit in the bore.

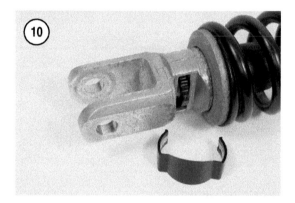

SHOCK ABSORBER LINKAGE

The shock linkage consists of the lever, lever arms, pivot bolts, seals, pivot spacers and needle bearings (**Figure 13** and **Figure 14**). The lever is joined to the swing arm by the lever arms. The linkage should be disassembled and lubricated at the intervals indicated in Chapter Three. If the linkage is often subjected to harsh riding conditions, service the linkage more frequently.

NOTE
In the following procedures, whenever grease is referenced, a molydisulfide or waterproof grease should be used. Kawasaki recommends molydisulfide grease for all

⑬ **SHOCK ABSORBER LINKAGE**

1. Cap
2. Nut
3. Lever arm
4. Seal
5. Bearing
6. Pivot spacer
7. Pivot bolt
8. Lever

⑭

linkage bearings. This grease has excellent antiwear characteristics when subjected to extreme pressure. Waterproof grease, which is very durable, has a high tack and is very resistant to washout when subjected to wet conditions. Which grease to use is a preference of the rider and the conditions in which the bike is operated. If the linkage is regularly maintained, either grease will perform well.

Lever and Lever Arms Removal and Installation

The lever and lever arms can be removed for service without removing the swing arm or rear wheel.

This procedure details the removal and separation of the lever and lever arm assemblies.

Note the direction of all bolts being removed. During assembly, install bolts in their original direction.

When disassembling the components, do not remove or allow the pivot spacers to slide out of the bearings. The original bearings have rollers that are locked into the bearing housing. However, if aftermarket replacement bearings have been installed, the needle bearings may be held in place only by the grease on the bearings. Keep any removed rollers with their respective bearing housing.

Refer to **Figure 13**.

1. Support the motorcycle so it is stable and secure. The rear wheel must be slightly off the ground, so the suspension is fully extended.

2. Remove the nuts and pivot bolts connecting the lever arms to the swing arm and lever (A, **Figure 15**). If desired, the pivot bolt connecting the lever arms to the lever can be removed at the workbench. However, the bolt is highly torqued and must be held stable in order to remove the nut.

3. Remove the nut and pivot bolt from the shock absorber and lever (B, **Figure 15**).

4. Remove the cap from both sides of the frame, then remove the nut and pivot bolt (**Figure 16**) from the frame.

5. Pull the shock absorber back and tap the lever out of the frame.

6. Inspect and service the lever and lever arms as described in this section.

7. Reverse these steps to install the parts. Note the following:

   a. Lubricate all bearings, seals and pivot bolts with grease.

   b. Clean the frame mounting bores (**Figure 17**) and drain holes, then lubricate with grease.

   c. After the lever has been tapped into the frame, install the shock absorber mounting bolt (A, **Figure 18**). This holds the assembly in place while installing the pivot bolt (B, **Figure 18**) through the frame.

   d. The lever arm pivot bolt heads must seat into the counterbores of the left lever arm.

   e. Check that all pivot spacers are in place before joining the components.

   f. Install and finger-tighten all bolts before torquing.

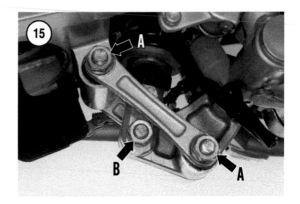

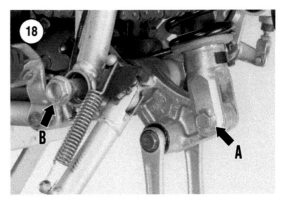

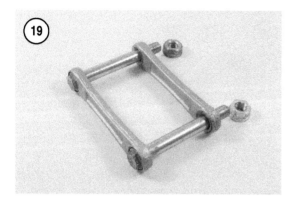

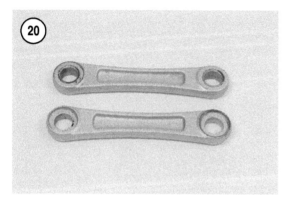

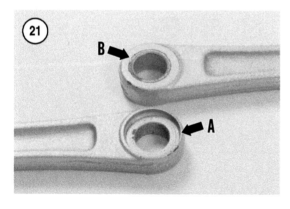

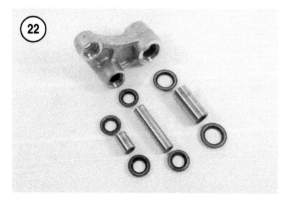

g. Tighten all linkage bolts to 98 N•m (72 ft.-lb.).

Lever Arms
Inspection

The lever arm assembly does not contain any bearings and can be visually inspected. Parts that show wear or damage should be replaced, in order for the suspension to operate properly and safely. Refer to **Figure 13**.

1. Inspect for worn pivot bolts. Check the fit of the bolts in the lever arms (**Figure 19**). The bolts should be a snug fit. Replace nuts and bolts that have rounded flats. Proper torquing may not be achieved if the nuts and bolts cannot be gripped by a socket.

2. Inspect the lever arms for bends or twisting. The arms should lay flat (**Figure 20**).

3. Inspect the lever arm bores for damage and elongation. If worn or damaged, also replace the pivot bolts. Note that the left arm is counterbored to accept the pivot bolt heads (A, **Figure 21**).

4. Inspect the contact point where the lever arm and pivot spacer mate (B, **Figure 21**). If the area is worn or ground away, the pivot bolts have been inadequately torqued.

5. Install the lever arms as described in this section.

Lever
Inspection and Repair

1. At each bearing, remove the pivot spacer and seals (**Figure 22**). Pry the seals at their outer edge.

NOTE
Inspect the bearing rollers to determine if they are removable. Original equipment bearings are not removable. Some aftermarket bearings may be removable. If the rollers can be removed, put the rollers in a marked container so they can be reinstalled in their original housing.

2. Clean the lever assembly in clean solvent, then carefully dry all parts.

3. Inspect the following:

   a. Inspect the lever for cracks, particularly around the bearing bores.

13

b. Check the frame pivot bolt and nut for scoring, wear and other damage. Replace the nut and bolt if it has rounded flats. Proper torquing may not be achieved if the nut and bolt cannot be gripped by a socket.

c. Check the seals for cracks, wear or other damage.

d. Check the pivot spacers for scoring, wear or other damage.

e. Check the needle bearings (**Figure 23**) for wear, flat spots, rust or discoloration. If the rollers are blue, overheating has occurred.

f. Lightly lubricate the bearings and pivot spacers, then insert each spacer into its respective bearing(s) (**Figure 24**). The parts should turn freely and smoothly with no play. If play or roughness exists, replace the bearing(s) as described in Step 4. For bores containing two bearings, always replace both bearings and the pivot spacer. If the bearing(s) are in good condition, go to Step 5.

4. Replace the needle bearing(s) in the lever as follows:

a. Apply penetrating oil to the bearing(s) and bore.

b. It is recommended to remove the bearing(s) with a press. If the bearing(s) and bore are not corroded, a drawbolt can be used, as demonstrated in *Swing Arm Bearing Replacement* in this chapter. Whichever method is used, the lever is supported against the open side of a large socket (A, **Figure 25**), and a driver (or appropriate-size socket) (B) is used to drive the bearing(s) into the lower socket.

NOTE
The driver must be capable of passing through the bore and be longer than the bore depth. The lower socket must fit on the perimeter of the bore, but also be large enough to accept the removed bearing.

c. Clean and inspect the mounting bore.

d. Lubricate the new bearing(s) with grease.

e. Fit the bearing squarely over the bore (A, **Figure 26**), with the manufacturer's marks facing out. Fit or assemble a driver tool squarely against the end of the bearing (B, **Figure 26**). The lower socket only acts to support the lever around the bore.

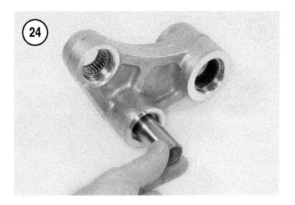

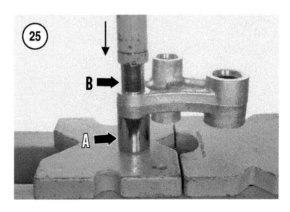

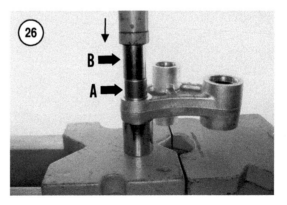

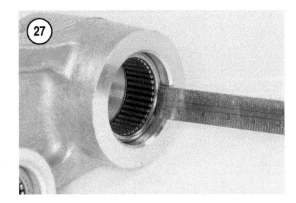

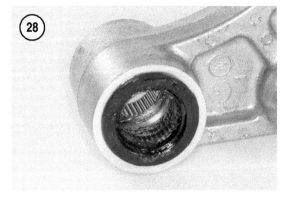

f. Begin driving the bearing. After the bearing has entered the bore, frequently check the bearing depth (**Figure 27**). All bearings in the lever should be driven to 5 mm below the outer edge of the bore. The depth is required so the seals can be seated in the bore.

g. Repeat for any remaining bearings.

5. Pack the bearings, bearing bores and seals with grease. Also apply grease to the pivot spacers and bolts.

6. Press the seals into position by hand (**Figure 28**). If the seals do not seat, inspect the bearing depth. If necessary, adjust the depth of the bearing(s).

7. Install the pivot spacers into the lever. Check that all seals remain seated as the pivot spacers pass through the seals.

8. Install the lever as described in this section.

SWING ARM

Bearing Inspection

The swing arm bearings can be inspected with the swing arm mounted on the bike. Periodically check the bearings for play, roughness or damage.

1. Remove the rear wheel (Chapter Eleven).

2. Loosen the swing arm pivot nut (**Figure 29**), then tighten it to 98 N•m (72 ft.-lb.).

3. Remove the pivot bolt (**Figure 30**) from the swing arm and lever arms.

4. Separate the linkage so the swing arm action is only influenced by the swing arm pivot bolt.

5. Check the bearings as follows:

a. Have an assistant steady the motorcycle.

b. Grasp the ends of the swing arm and leverage it from side to side. There should be no detectable play in the bearings.

c. Pivot the swing arm up and down, through its full travel. The bearings must pivot smoothly.

d. If there is play or roughness in the bearings, remove the swing arm and inspect the bearing and pivot assembly for wear.

6. Reinstall the lever arms and tighten the pivot bolt to 98 N•m (72 ft.-lb.).

Removal and Installation

If the components of the shock absorber linkage will be removed and inspected, remove or loosen all pivot bolts before removing the swing arm. The

13

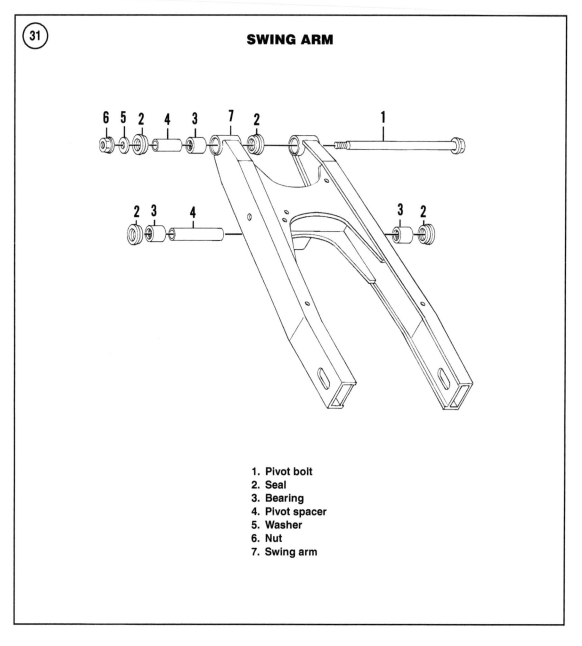

SWING ARM

1. Pivot bolt
2. Seal
3. Bearing
4. Pivot spacer
5. Washer
6. Nut
7. Swing arm

bolts are easier to loosen while the linkage is mounted on the bike. Make note of the direction of all pivot bolts that are removed. Refer to **Figure 31** as needed.

1. Remove the rear wheel (Chapter Eleven).

2. Remove the brake hose guide from the swing arm (**Figure 32**).

3. Remove the pivot bolt (**Figure 30**) from the swing arm and lever arms.

4. Remove the hose guide (**Figure 33**).

5. Remove the sprocket guard (**Figure 34**).

6. Remove the nut and washer from the swing arm pivot bolt (**Figure 35**).

7. Have an assistant hold the swing arm while the pivot bolt is pulled from the swing arm. If using a drift to drive out the bolt, avoid damaging the bearing assemblies.

8. Route the chain from behind the guards (**Figure 36**), then pull the swing arm out of the frame.

9. Inspect and service the swing arm as described in this chapter. If necessary, refer to *Shock Absorber*

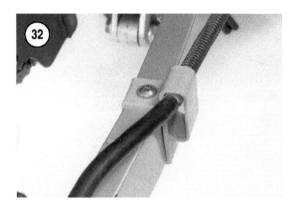

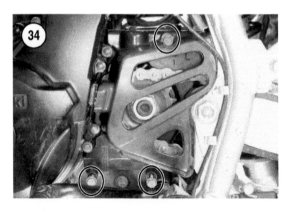

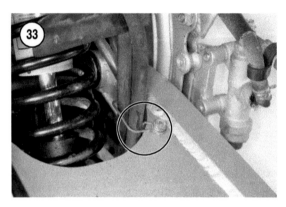

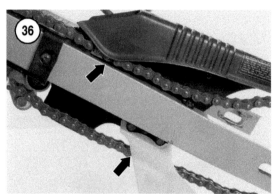

Linkage in this chapter for inspection and repair of the linkage.

10. Reverse these steps to install the swing arm. Note the following:

    a. Lubricate all bearings, seals and pivot bolts with grease.

    b. Check that the chain passes over and under the swing arm pivot bolt.

    c. Install the pivot bolts in their correct direction.

    d. Tighten the pivot bolt to 98 N•m (72 ft.-lb.).

    e. If the complete shock absorber linkage needs to be installed, raise the swing arm and wire it into position. This creates additional work space for installing the linkage. Optionally, install the linkage before installing the swing arm. Refer to *Shock Absorber Linkage* in this chapter for installation of the linkage.

SWING ARM SERVICE

NOTE
In the following procedures, whenever grease is referenced, a molydisulfide or waterproof grease should be used. Kawasaki recommends molydisulfide grease for all swing arm bearings. This grease has excellent antiwear characteristics when subjected to extreme pressure. Waterproof grease, which is very durable, has a high tack and is very resistant to washout when subjected to wet conditions. Which grease to use is a preference of the rider and the conditions in which the bike is operated. If the swing arm is regularly maintained, either grease will perform well.

13

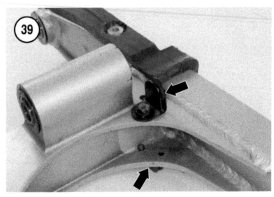

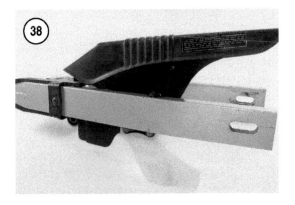

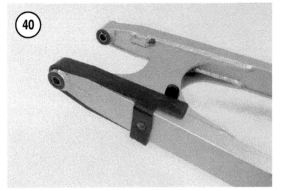

Inspection and Repair

When inspecting and cleaning the components, do not remove the bearing pivot spacers until the bearings will be inspected. The original bearings have rollers that are locked into the bearing housing. However, if aftermarket replacement bearings have been installed, the needle bearings may be held in place only by the grease on the bearings. Check that the spacers are firmly in place before inspecting. If necessary, tape the ends of the bearing bores.

Refer to **Figure 31**.

1. Clean the swing arm, particularly around all bearings.

2. Inspect the hose guide (**Figure 37**). Replace the guide if it is corroded or damaged.

3. Inspect the chain guard and chain guide (**Figure 38**).

    a. Check that all fasteners are tight.

    b. Check that the chain guard is locked into the upper and lower brackets (**Figure 39**) at the front of the swing arm.

    c. Check the chain guide for excessive wear.

4. Inspect the chain slider and mounting bolts (**Figure 40**). Replace the slider if it is worn to more than half its thickness (**Figure 41**). Damage to the swing arm can occur if the slider wears through. Although not part of the swing arm, a small slider is located below the swing arm bolt, as the chain returns to the rear wheel. Check the condition of this slider.

NOTE
Before servicing the bearings in the following steps, handling the swing

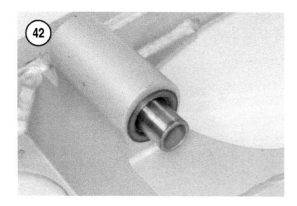

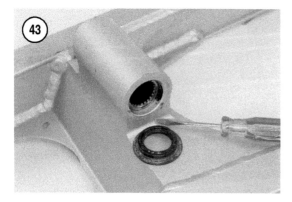

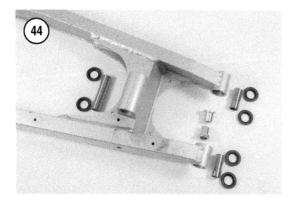

arm will be easier if the chain guard and chain guide are removed.

5. When working with each bearing(s), do the following:

a. Remove the pivot spacer (**Figure 42**).

b. Remove the seals (**Figure 43**). Pry the seals at their outer edge.

c. Inspect the bearing rollers to determine if they are removable. Original equipment bearings are not removable. Some aftermarket bearings may be removable. If the rollers can be removed, put the rollers in a marked container so they can be reinstalled in their original housing.

6. Clean the bearings, spacers, seals and bores (**Figure 44**) in clean solvent, then carefully dry all parts.

7. Inspect the following:

a. Inspect the swing arm for cracks, particularly around the bearing bores.

b. Check the swing arm pivot bolt and nut for scoring, wear and other damage. Replace the nut and bolt if it has rounded flats. Proper torquing may not be achieved if the nut and bolt cannot be gripped by a socket.

c. Check the fit of the pivot bolt in the bushings, located at the back of the engine (**Figure 45**). If worn, use a drift to drive out both bushings.

d. Check the seals for cracks, wear or other damage.

e. Check the pivot spacers for scoring, wear or other damage.

f. Check the needle bearings for wear, flat spots, rust or discoloration. If the rollers are blue, overheating has occurred.

g. Lightly lubricate the bearings and pivot spacers, then insert each spacer into its respective bearing(s) (**Figure 46**). The parts should turn freely and smoothly with no play. If play or roughness exists, replace the bearing set as described in *Swing Arm Bearing Replacement* in this section. When replacing bearing sets in the swing arm pivots, always replace the bushing (**Figure 47**) that fits into each side of the engine case.

8. Pack grease into the bearings and bores (**Figure 48**). Also apply grease to the seals, pivot spacers swing arm bushings (at back of engine) and bolts.

13

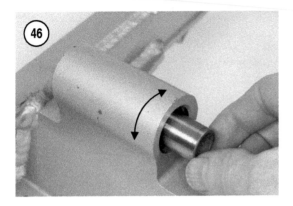

9. Press the seals into position by hand. If the seals do not seat, inspect the bearing depth. If necessary, adjust the depth of the bearing(s).

10. Install the pivot spacers. Grip the seals (**Figure 49**) while twisting the pivot spacers into place.

11. If removed, install the chain guard and chain guide.

12. Install the swing arm (**Figure 50**) as described in this chapter.

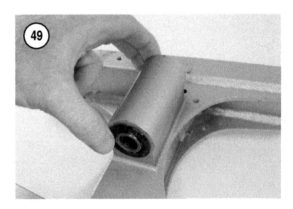

Swing Arm Bearing Replacement

For the lever arm bore, always replace both bearings. For the swing arm pivot bores, always replace the bearing in each bore. Mixing new and worn bearings on the same pivot bolt will shorten the life of the new part.

It is recommended to remove the bearings with a press. If the bearings and bores are not corroded, hand tools and a drawbolt can be used. The following procedures describe removing and installing of the bearings using both methods. Do not perform the following procedures until all seals, spacers and guards have been removed from the swing arm. Read both procedures to determine which method is most practical. If in doubt, take the swing arm to a dealership or machine shop to have the bearings replaced.

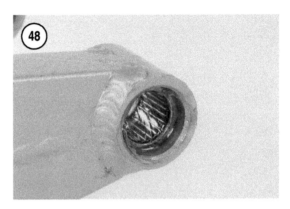

WARNING
If heat will be used to ease the removal of the bearings, take care to prevent burning finished or combustible surfaces. Remove any parts that could be damaged by heat. The purpose for using a heat gun or propane torch heat is to slightly expand the bores in order to remove the bearings

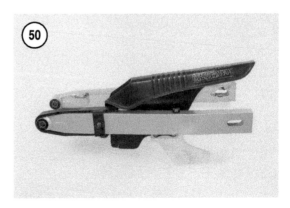

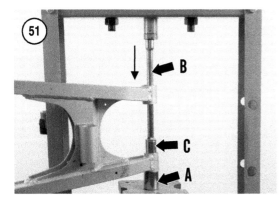

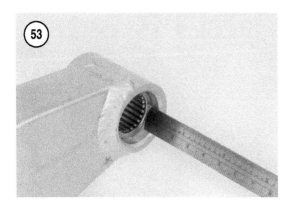

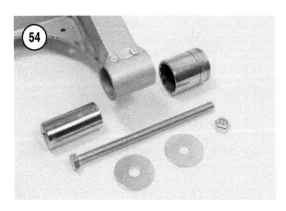

with minimal resistance. This is particularly helpful when removing heavily corroded bearings. Always keep the heat source moving at a steady rate and avoid heating the bearing. Work in a well-ventilated area and away from combustible materials. Wear protective clothing, including eye protection and insulated gloves.

Press method

1. Apply penetrating oil to the bearing(s) and bore.
2. If necessary, heat the immediate area around the bearing(s) to be removed.
3. Support the swing arm in a press. Place the bearing bore over a large socket or similar tool (A, **Figure 51**) so the bearing can be driven out of the bore. The lower socket must fit on the perimeter of the bore, but also be large enough to accept the removed bearing.
4. Pass a driver through the upper swing arm bore and to the lower bore (B, **Figure 51**).
5. Place a socket or driver (C, **Figure 51**) squarely against the bearing. The driver must be capable of passing through the bore, and be longer than the bore depth.
6. Press the bearing out of the arm. Turn the swing arm over and repeat for the other arm.
7. Clean and inspect the mounting bores.
8. Lubricate the new bearings with grease.
9. Support the swing arm bore on a flat stable surface. The lower socket is not required.
10. Fit the new bearing squarely over the bore with the manufacturer's marks facing out.
11. Place a socket or driver squarely against the bearing and drive the bearing into the swing arm. As the bearing driver begins to enter the bore (**Figure 52**), frequently check the bearing depth (**Figure 53**). All bearings in the swing arm should be driven to 5 mm below the outer edge of the bore. The depth is required so the seals can be seated in the bore.
12. Drive the remaining bearings.
13. Refer to *Inspection and Repair* in this section to complete the assembly.

Hand tool method

In the following procedure, a bearing removal/installation tool (drawbolt) (**Figure 54**) is made from

13

a bolt, nut, washers and sockets. The driver is a socket that is capable of passing through the bore, and is longer than the bore depth. The larger socket fits on the perimeter of the bore, but is also large enough to accept the removed bearing(s). A similar tool is made by Motion Pro (part No. 08-213) (**Figure 55**).

1. Apply penetrating oil to the bearing(s) and bore.

2. If necessary, heat the immediate area around the bearing(s) to be removed.

3. Assemble the tool as shown in **Figure 56**.

4. Hand-tighten the nut until the assembly is squarely positioned against the bearing and swing arm contact points.

5. Turn the nut and drive the bearing(s) into the large socket.

6. Clean and inspect the bearing bore.

7. Lubricate the new bearings with grease.

8. Align the bearing squarely on the outside face of the bore. The manufacturer's marks on the bearing must face out. For bores that contain two bearings, install each bearing from its nearest end of the bore.

9. Reverse the direction of the tool and hand-tighten the nut until the tool and bearing are squarely positioned with the bore. Note that a large-diameter, thick washer can now be substituted for the large socket. Handling the assembly is now easier.

10. Drive the bearing into the swing arm. As the bearing driver begins to enter the bore, frequently check the bearing depth (**Figure 53**). All bearings in the swing arm should be driven to 5 mm below the outer edge of the bore. The depth is required so the seals can be seated in the bore.

11. Drive the remaining bearings.

12. Refer to *Inspection and Repair* in this section to complete assembly.

SHOCK ABSORBER
ADJUSTMENT

Shock Spring Preload Adjustment

Shock spring preload affects handling and ride quality, and should be adjusted to accommodate the load on the motorcycle. Spring preload can be adjusted with the shock absorber mounted on the motorcycle.

1. Refer to **Table 2** for the recommended settings. Higher settings increase preload.

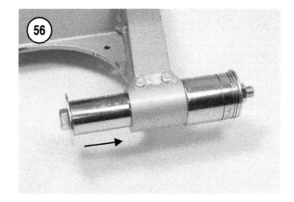

2. To access the preload adjuster, near the top of the shock absorber, do the following:

   a. Remove the left side cover (Chapter Fifteen).

   b. For California models, remove the liquid/vapor separator.

3. Turn the adjuster (**Figure 57**) to the desired numerical setting.

4. Install the removed parts.

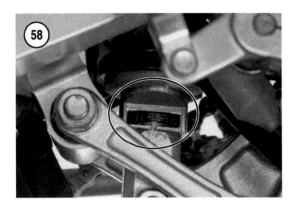

Rebound Damping Adjustment

Rebound damping controls the rate of extension of the shock absorber after it has been compressed. This setting has no affect on the compression rate of the shock. If rebound damping is set too high, the rear suspension will not extend quickly enough to prevent *bottoming* on subsequent bumps. Rebound damping that is set too low can cause the rear wheel to *kick up* and the handling to be unstable.

1. Refer to **Table 2** for the recommended settings. Higher settings increase damping force.

2. At the bottom right side of the shock, remove the snap-on cover from the adjuster (**Figure 58**).

3. Clean any dirt from the adjuster cavity.

4. Turn the rebound damper adjuster to the desired numerical setting. Only turn the adjuster to the right, as indicated by the arrow on the adjuster. Each setting should have a perceptible click.

5. Install the cover so the arrow on the cover points upward.

Table 1 REAR SUSPENSION SPECIFICATIONS

Suspension	Uni-Trak, link-type
Suspension travel	230 mm (9.1 in.)
Shock absorber	Nitrogen-charged, adjustable for preload and rebound damping
Pivot spacer outside diameter (swing arm and linkage assembly)	
Small spacers	19.979-20.0 mm (0.7866-0.7874)
Service limit	19.95 mm (0.7854 in.)
Large spacer	27.979-28.0 mm (1.1015-1.1024)
Service limit	27.95 mm (1.1004 in.)

Table 2 SHOCK ABSORBER SETTINGS

	Rider only	Rider and passenger	Rider, passenger and load
Spring preload	*1 or 2	2 or 3	3, 4 or 5
Rebound damper	*I or II	II or III	III or IIII

*Settings 1 and I are the softest settings and are for riders up to 68 kg (150 lb.).

Table 3 REAR SUSPENSION TORQUE SPECIFICATIONS

	N•m	ft.-lb.
Shock absorber mounting bolts		
Lower	98	72
Upper	59	44
Swing arm pivot bolt	98	72
Lever–to–frame bolt	98	72
Lever arm bolts	98	72

CHAPTER FOURTEEN

BRAKES

This chapter provides service procedures for the front and rear brake systems. This includes brake pads, master cylinders, calipers and discs.

Refer to Chapter Three for brake fluid level inspection, brake pad/disc inspection, brake light adjustment, and brake pedal and lever adjustment.

DISC BRAKE FUNDAMENTALS

The front and rear brakes are hydraulically actuated and therefore do not require cables and mechanical linkages to operate. When pressure is applied to the brake pedal or lever, the brake fluid is compressed in the brake line and pushes one brake pad toward the brake disc. Since the caliper and second brake pad are not locked in a stationary position, the assembly slides toward the disc when the first pad contacts the disc. This motion allows both pads to contact the disc, as well as to center the caliper around the disc. When pressure is relieved, the first pad and the caliper assembly slightly retract from the disc, allowing the wheel to spin freely. As the pads wear, the piston in the caliper extends, automatically keeping the pads adjusted and centered around the disc. It is important to not only ensure that the piston can extend and retract, but that the caliper is free to move.

A hydraulic brake system is not maintenance-free, nor indestructible. Observe the following practices when maintaining or working on a hydraulic brake system:

1. Keep brake fluid off painted surfaces, plastic and decals. The fluid will damage these surfaces. If fluid does contact these surfaces, flush the surface thoroughly with clean water.

2. Keep the fluid reservoirs closed except when changing the fluid.

3. Replace brake fluid often. The fluid absorbs moisture from the air and will cause internal corrosion of the brake system. Fresh fluid is clear to slightly yellow. If the fluid is obviously colored, it is contaminated.

4. Do not reuse brake fluid, or use new fluid that has been in a partially-used container for any length of time.

5. When rebuilding brake system components, lubricate new parts with fresh fluid before assembly. Do not use petroleum-based solvents. These can swell and damage rubber components.

6. Bleed the brake system whenever a banjo bolt or other connector in the brake line has been loosened. Air will be in the system and brake action will be spongy.

FRONT BRAKE CALIPER

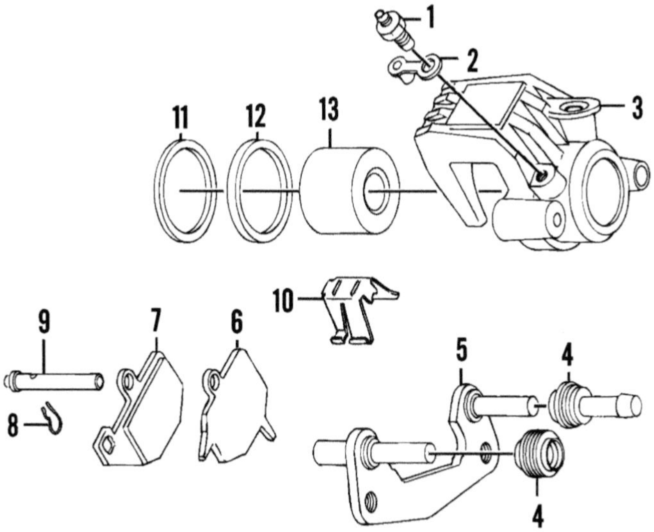

1. Bleeder valve
2. Cap
3. Caliper
4. Boot
5. Caliper holder
6. Inner brake pad
7. Outer brake pad
8. Lock pin
9. Pad pin
10. Pad spring
11. Piston boot
12. Piston seal
13. Piston

14

FRONT BRAKE PADS

Brake pad life depends on the riding habits of the rider and the material of the brake pads. Replace the pads when they are worn to within 1 mm (0.040 in.) of the backing plate, or have been contaminated with oil or other chemicals.

Removal and Installation

The brake pads can only be replaced by removing the caliper from the fork leg. The brake hose does not have to be removed from the caliper. The caliper can be hand held during the following procedure, however, a small work surface placed next to the wheel makes handling easier. Keep the caliper supported and do not allow it to hang by the brake hose.

If the caliper will be rebuilt, or if other damage is detected during this procedure, the pads can be removed when the caliper is at the workbench. Refer to *Front Brake Caliper* in this chapter for complete removal, repair and installation. Refer to **Figure 1**.

1. Remove the caliper mounting bolts (**Figure 2**).

2. Remove the lock pin (**Figure 3**) from the pad pin.

3. Press up on the pads to relieve the pressure on the pad pin, then remove the pin (**Figure 4**).

4. Pivot and remove the outer pad.

5. Press up on the the inner pad to release and remove it from the caliper (**Figure 5**).

6. Remove the pad spring (**Figure 6**).

> *CAUTION*
> *In the following step, monitor the level of fluid in the master cylinder reservoir. Brake fluid will back flow to the reservoir when the caliper piston is pressed into the bore. Do not allow brake fluid to spill from the reservoir, or damage can occur to painted and plastic surfaces. Immediately wash any spills with water.*

7. Grasp the caliper and press the caliper piston down into the bore, creating room for the new pads.

> *NOTE*
> *Do not operate the brake lever with the pads removed. The caliper piston can come out of the bore.*

8. Clean the interior of the caliper and inspect for the following:

   a. Leakage or damage around the piston, bleeder valve and hose connection.

   b. Leakage at the fork seals. Fork oil contaminates brake pads.

   c. Damaged or missing boots.

   d. Excessive drag of the caliper holder when it is moved in and out of the caliper. If there is corrosion or water around the slide pin boots, clean and lubricate the parts with silicone brake grease.

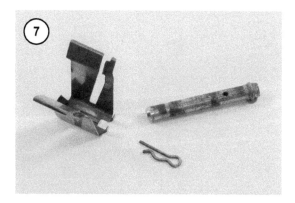

9. Inspect the pad pin, lock pin and pad spring (**Figure 7**). The pin and spring must be in good condition to allow the inner pad to slightly move when installed. Check that all small tabs on the spring are not corroded or missing.

10. Inspect the pads for contamination, scoring and wear.

   a. Replace the pads when they are worn to within 1 mm (0.040 in.) of the backing plate, or have been contaminated with oil or other chemicals.

   b. If the pads are worn unevenly, the caliper is probably not sliding correctly on the slide pins. The caliper must be free to *float* on the pins. Buildup or corrosion on the pins can hold the caliper in one position, causing brake drag and excessive pad wear.

11. Install the pad spring with the narrow arms facing out. Seat the spring into the caliper.

12. Install the inner pad, seating it into the pad spring and caliper, then pivot the outer pad into the caliper (**Figure 8**). Seat the pad between the small tabs at the top and bottom of the pad spring.

13. Press up on the pads so the pad pin can be aligned and installed.

14. Insert the lock pin into the pad pin (**Figure 9**).

15. Position the caliper over the brake disc and guide the disc between the pads.

16. Install and torque the caliper mounting bolts to 25 N•m (18 ft.-lb.).

17. Operate the brake lever several times to seat the pads.

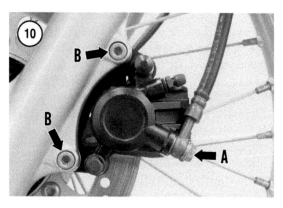

18. Check the brake fluid reservoir and replenish or remove fluid, as necessary.

19. With the front wheel raised, check that the wheel spins freely and the brake operates properly.

FRONT BRAKE CALIPER

Removal and Installation

Use the following procedure to remove the caliper from the motorcycle. Refer to **Figure 1**.

1A. If the caliper will be disassembled, drain the system as described in this chapter. After draining, loosen the brake hose banjo bolt (A, **Figure 10**) while the caliper is stable on the fork. Leave the bolt finger-tight. It will be removed in a later step.

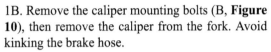

1B. Remove the caliper mounting bolts (B, **Figure 10**), then remove the caliper from the fork. Avoid kinking the brake hose.

2A. If the caliper is left attached to the brake hose, but not disassembled and serviced:

    a. Attach a wire to the caliper and hang the caliper on the motorcycle. Do not let the caliper hang by the brake hose.

    b. Insert a small wooden block between the brake pads. This prevents the caliper piston from extending out of the caliper if the brake lever is inadvertently applied.

2B. If the caliper will be disassembled, do the following:

    a. Remove the banjo bolt and seal washers from the brake hose. Have a shop cloth ready to absorb excess brake fluid that drips from the hose.

    b. Wrap the hose end to prevent brake fluid from damaging other surfaces.

    c. Drain excess brake fluid from the caliper.

3. Repair the caliper as described in this section.

4. Reverse this procedure to install the caliper. Note the following:

    a. Position the caliper over the brake disc and guide the disc between the pads.

    b. Install new seal washers on the banjo bolt (**Figure 11**).

    c. Install and tighten the caliper mounting bolts and banjo bolt to 25 N•m (18 ft.-lb.).

NOTE
If the caliper was rebuilt, or the brake hose disconnected from the caliper, fill and bleed the brake system as described in this chapter.

5. Operate the brake lever several times to seat the pads.

6. Check the brake fluid reservoir and replenish or remove fluid, as necessary.

7. With the front wheel raised, check that the wheel spins freely and the brake operates properly.

Repair

Use the following procedure to disassemble, inspect and assemble the brake caliper, using new seals. Refer to **Figure 1**.

1. Remove the caliper as described in this section.

2. Remove the lock pin and pad pin (**Figure 12**). Press down on the pads to relieve the pressure on the pad pin when removing it.

3. Pivot and remove the outer pad, then remove the inner pad and pad spring (**Figure 13**).

4. Remove the caliper holder and boots (**Figure 14**).

5. Remove the piston from the caliper bore using compressed air. To perform this technique, an air nozzle is tightly held in the brake hose fitting and air pressure ejects the piston. Do not pry the piston out

of the caliper. Read the following procedure entirely before removal.

   a. Place the caliper on a padded work surface.

   b. Close the bleeder valve on the caliper so air cannot escape.

   c. Place a strip of wood, or similar pad, in the caliper. The pad cushions the piston when it comes out of the caliper.

WARNING
Wear eye protection when using compressed air to remove the piston. Keep fingers away from the piston discharge area. Injury can occur if an attempt is made to stop the piston with the fingers.

   d. Lay the caliper so the piston discharges downward.

   e. Insert an air nozzle into the brake hose fitting. If the nozzle does not have a rubber tip, wrap the nozzle with tape. This allows the nozzle to seal tightly and prevent thread damage.

   f. Place a shop cloth over the entire caliper to catch any spray that may discharge from the caliper.

   g. Apply pressure and listen for the piston to *pop* from the caliper (**Figure 15**).

6. Remove the piston boot and seal from the cylinder bore (**Figure 16**).

7. Remove the bleeder valve and cap from the caliper.

8. Inspect the caliper assembly.

   a. Clean all parts that will be reused with fresh brake fluid or isopropyl (rubbing) alcohol. Use a wood or plastic-tipped tool to clean the caliper seal and boot grooves. Use clean brake fluid to clean the piston, bore and seal grooves.

   b. Inspect the cylinder bore and piston (**Figure 17**) for wear, pitting or corrosion. Service limits are not specified by Kawasaki.

   c. Inspect the caliper holder and slide pins (**Figure 18**) for wear, pitting or corrosion.

   d. Inspect the boots for deterioration.

   e. Inspect the pad pin, lock pin and pad spring (**Figure 19**). The pin and spring must be in good condition to allow the inner pad to slightly move when installed. Replace the parts if worn, corroded or damaged.

14

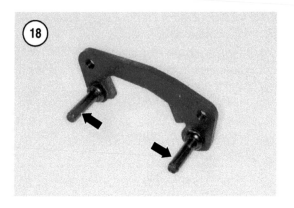

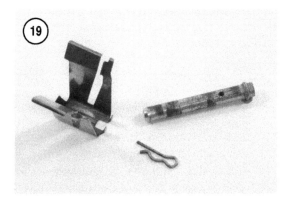

f. Inspect all threaded parts (**Figure 20**). Check the threads and seats for corrosion and damage.

g. Inspect the pads for contamination, scoring and wear. Replace the pads when they are worn to within 1 mm (0.040 in.) of the backing plate, or are contaminated with oil or other chemicals. If the pads are worn unevenly, the caliper is probably not sliding correctly on the slide pins. The caliper must be free to *float* on the pins. Buildup or corrosion on the pins can hold the caliper in one position, causing brake drag and excessive pad wear.

NOTE
Use new brake fluid (rated DOT 4) to lubricate the parts in the following steps.

9. Install the new piston seal (A, **Figure 21**) and boot (B) as follows:

a. Soak the seal and boot in brake fluid for 15 minutes.

b. Coat the caliper bore and piston with brake fluid.

c. Seat the seal, then the boot (**Figure 22**) in the caliper grooves. The piston seal goes in the back groove.

d. Install the piston with the insulator facing out (**Figure 23**). Twist the piston past the seal and boot, then press the piston to the bottom of the bore.

10. Apply silicone brake grease to the interior of the boots, then install the boots onto the caliper. If necessary, apply a light coat of grease on the exterior of the large boot to help pass it through the caliper.

11. Lubricate the slide pins on the caliper holder, then insert them into the boots. Press the holder against the boots and carefully lift the boots onto the holder seats (**Figure 24**).

12. Install the pad spring with the narrow arms facing out (**Figure 25**). Seat the spring into the caliper.

13. Install the inner pad, seating it into the pad spring and caliper (**Figure 26**).

14. Install and pivot the outer pad into the caliper. Seat the pad between the small tabs at the top and bottom of the pad spring.

15. Press down on the pads so the pad pin can be aligned and installed (**Figure 27**).

16. Insert the lock pin into the pad pin (**Figure 28**), then lock it into position.

17. Install the bleeder valve and cap.

18. Install the caliper as described in this section.

14

FRONT MASTER CYLINDER

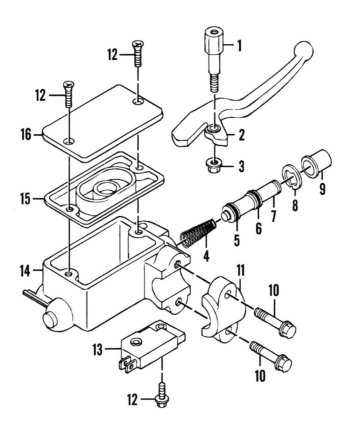

1. Pivot bolt
2. Lever
3. Nut
4. Spring
5. Primary seal
6. Secondary seal
7. Piston
8. Snap ring
9. Boot
10. Bolt
11. Holder
12. Screw
13. Brake switch
14. Reservoir/cylinder
15. Diaphragm
16. Cap

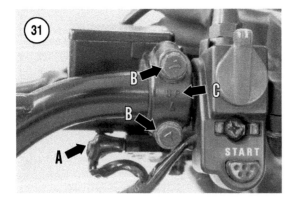

FRONT MASTER CYLINDER

Removal and Installation

Use the following procedure to remove the front master cylinder/brake fluid reservoir from the motorcycle. Refer to *Repair* in this section to make internal repairs to the master cylinder. Refer to **Figure 29**.

1. Cover and protect the fuel tank and area surrounding the master cylinder.

CAUTION
Do not allow brake fluid to splash from the reservoir or hose. Brake fluid can damage painted and plastic surfaces. Immediately clean up any spills, flooding the area with water.

2. Drain the brake system as described in this chapter.
3. Remove the brake hose banjo bolt from the master cylinder as follows:
   a. Pull back the rubber boot.
   b. Remove the banjo bolt and seal washers from the brake hose (**Figure 30**). Have a shop cloth ready to absorb excess brake fluid that drips from the hose.
   c. Wrap the hose end to prevent brake fluid from damaging other surfaces.
4. Remove the brake switch wires (A, **Figure 31**).
5. Remove the bolts (B, **Figure 31**) securing the master cylinder to the handlebar, then remove the master cylinder, holder and hand guard.
6. Repair the master cylinder as described in this chapter.
7. Reverse this procedure to install the master cylinder. Note the following:
   a. The mounting bracket must be installed so *UP* and the arrow (C, **Figure 31**) are facing up. Tighten the upper bolt first, then the bottom bolt. Tighten the bolts to 9 N•m (80 in.-lb.).
   b. Apply dielectric grease to the brake switch connectors.
   c. Position the brake hose fitting so it is above the support (**Figure 32**).
   d. Install new seal washers on the banjo bolt. Tighten the bolt to 25 N•m (18 ft.-lb.).
8. Fill the brake fluid reservoir and bleed the brake system as described in this chapter.

NOTE
Brake lever adjustment is not required. If properly bled, the brake lever automatically adjusts.

Repair

Use the following procedure to disassemble, inspect and assemble the master cylinder using new parts. The piston, seals and spring are only available as a complete assembly. Refer to **Figure 29**.

14

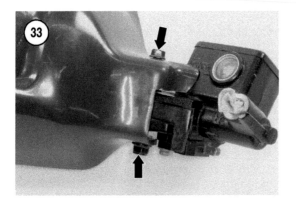

1. Remove the master cylinder as described in this section.

2. Remove the nut and bolt securing the hand guard (**Figure 33**). Carefully flex the guard to remove it from the master cylinder.

3. Remove the brake switch (**Figure 34**).

4. Remove the pivot bolt and lever (**Figure 35**).

5. Remove the cap and diaphragm from the reservoir. Drain and wipe excess fluid from the reservoir.

6. Remove the boot from the piston (**Figure 36**). The boot is a friction fit. To avoid damaging the boot on removal, apply penetrating lubricant around the perimeter of the boot. Carefully pull the bottom edge back so the lubricant can loosen the boot.

7. Remove the snap ring from the master cylinder (**Figure 37**) as follows:

    a. Lock the cylinder in a vise with soft jaws. A shop cloth placed between the jaws will help absorb any drips. Do not overtighten the vise or cylinder damage could occur.

    b. Press down on the piston to relieve pressure on the snap ring (**Figure 38**), then remove the snap ring.

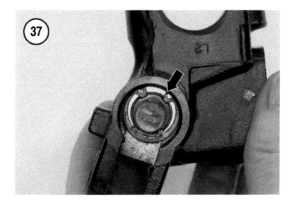

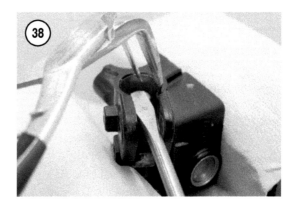

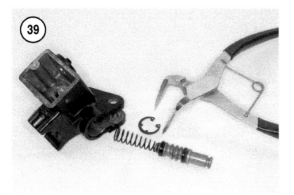

c. Slowly relieve the pressure on the piston.

8. Remove the piston assembly from the bore (**Figure 39**).

9. Inspect the master cylinder assembly.

    a. Clean all parts that will be reused with fresh brake fluid or isopropyl (rubbing) alcohol.

    b. Inspect the cylinder bore for wear, pitting or corrosion (**Figure 40**).

    c. Inspect and clean the threads and orifices in the reservoir (**Figure 41**). Clean with compressed air.

    d. Inspect the brake lever bore and pivot bolt for wear (**Figure 42**).

    e. Inspect the diaphragm and reservoir cap for damage (**Figure 43**).

    f. Inspect the boot, snap ring and mounting hardware (**Figure 44**) for corrosion and damage.

    g. Inspect the brake switch. Clean the switch with electrical contact cleaner by spraying into the holes in the case (**Figure 45**). Operate the switch while flushing the contacts. If the switch condition is not known, attach an ohmmeter to the terminals on the switch. The me-

ter should indicate continuity with the switch out, and no continuity when the switch is pressed.

10. Assemble the piston, seals and spring as follows. **Figure 46** shows the piston seals, before and after assembly.

    a. Soak the primary seal (A, **Figure 46**) and secondary seal (B) in fresh brake fluid (rated DOT 4) for 15 minutes. This softens and lubricates the seals.

    b. Apply brake fluid to the piston so the seals can slide over the ends.

    c. Identify the wide (open) side of the primary seal. When installed, the wide side of the seal *must* face in the direction of the arrow (**Figure 46**). Install the primary seal onto the piston.

    d. Identify the wide (open) side of the secondary seal. When installed, the wide side of the seal *must* face in the direction of the arrow (**Figure 46**). Install the secondary seal on the piston.

    e. Install and seat the *small end* of the spring onto the piston.

11. Install the piston and snap ring into the master cylinder (**Figure 47**) as follows:

    a. Lock the cylinder in a vise with soft jaws. Do not overtighten the vise or cylinder damage could occur.

    b. Lubricate the cylinder bore and piston assembly with brake fluid.

    c. Rest the piston assembly in the cylinder.

    d. Place the snap ring over the end of the piston, resting it on the edge of the bore. The flat side of the snap ring must face out.

    e. Place a screwdriver over the end of the piston and compress the snap ring with snap ring pliers.

NOTE
In the following step, after the piston seals have entered the cylinder, the piston should be held in place until the snap ring is installed. Anytime the seals come out of the cylinder there is a chance of damaging the seal lips during the reinsertion process. This should be avoided.

    f. Press the piston into the cylinder while guiding the snap ring into position. If the snap ring does not easily seat, release the snap ring and

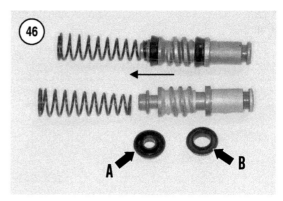

use the tip of the pliers to press it into the groove. Keep the screwdriver in position until the snap ring seats.

12. Remove the cylinder from the vise.

13. Apply silicone brake grease to the end of the piston and inside the boot. Seat the boot into the cylinder.

14. Screw the brake switch to the master cylinder. Seat the contact points on the switch with the points on the cylinder (**Figure 48**).

15. Install the lever and pivot bolt (**Figure 35**). Lubricate the pivot point with waterproof grease.

16. Loosely screw the diaphragm and cap onto the reservoir.

17. Install the hand guard, carefully flexing it onto the pivot bolt. Install the nut and bolt (**Figure 33**) to secure the guard.

18. Install the master cylinder as described in this section.

REAR BRAKE PADS

Brake pad life depends on the riding habits of the rider and the material of the brake pads. Replace the

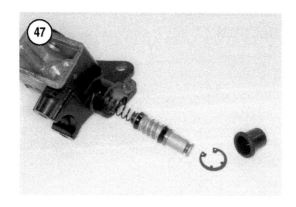

pads when they are worn to within 1 mm (0.040 in.) of the backing plate, or have been contaminated with oil or other chemicals.

Replacement

The brake pads can only be replaced by removing the caliper from the swing arm. The brake hose does not have to be removed from the caliper. The caliper can be hand held during the following procedure, however, a small work surface placed next to the wheel makes handling easier. Keep the caliper supported and do not allow it to hang by the brake hose.

If the caliper will be rebuilt, or if other damage is detected during this procedure, the pads can be removed when the caliper is at the workbench. Refer to *Rear Brake Caliper* in this chapter for complete removal, repair and installation. Refer to **Figure 49** as needed.

1. Remove the caliper mounting bolts and hose guide (**Figure 50**). Avoid kinking the brake hose.

CAUTION
In the following step, monitor the level of fluid in the master cylinder reservoir. Brake fluid will back flow to the reservoir when the caliper piston is pressed into the bore. Do not allow brake fluid to spill from the reservoir, or damage can occur to painted and plastic surfaces. Immediately wash any spills with water.

2. Grasp the caliper and press the inner pad and caliper holder (**Figure 51**) against the caliper. This presses the piston into the bore, and compresses the holder so the pads can be removed. Keep the caliper holder compressed until the outer pad is removed.

3. Tilt the outer pad toward the caliper, then disengage the pad from the slide pins and pad spring (**Figure 52**).

4. Press down on the inner pad and disengage it from the caliper holder and pad spring.

NOTE
Do not operate the brake lever with the pads removed. The caliper piston can come out of the bore.

5. Clean the interior of the caliper and inspect for the following:
   a. Leakage or damage around the piston, bleeder valve and hose connection.
   b. Damaged or missing boots.
   c. Excessive drag of the caliper holder when it is moved in and out of the caliper. If there is corrosion or water around the slide pin boots, clean and lubricate the parts with silicone brake grease.
   d. Corroded or damaged pad spring. Check that all small tabs on the spring are not corroded or missing. The spring must be slightly arched and flexible.

6. Inspect the pads for contamination, scoring and wear.
   a. Replace the pads when they are worn to within 1 mm (0.040 in.) of the backing plate, or have been contaminated with oil or other chemicals.
   b. If the pads are worn unevenly, the caliper is probably not sliding correctly on the slide pins. The caliper must be free to *float* on the pins. Buildup or corrosion on the pins can

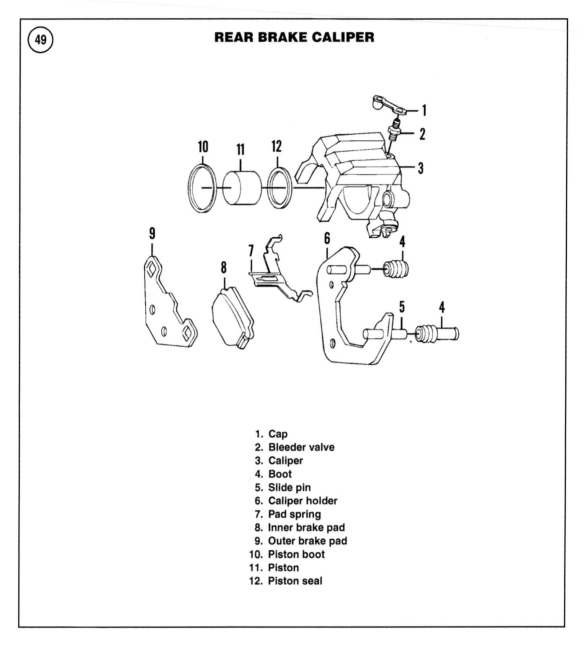

REAR BRAKE CALIPER

1. Cap
2. Bleeder valve
3. Caliper
4. Boot
5. Slide pin
6. Caliper holder
7. Pad spring
8. Inner brake pad
9. Outer brake pad
10. Piston boot
11. Piston
12. Piston seal

hold the caliper in one position, causing brake drag and excessive pad wear.

7. Compress and hold the caliper holder against the caliper (**Figure 53**) so the pads can be installed.

8. Install the inner pad, seating it into the pad spring and the notches in the caliper holder (**Figure 54**). Press the pad against the piston.

9. Tilt the outer pad and engage it with the slide pins and pad spring (**Figure 55**).

10. Position the caliper over the brake disc and guide the disc between the pads.

11. Install and tighten the caliper mounting bolts to 25 N•m (18 ft.-lb.).

12. Install the hose guide.

13. Operate the brake pedal several times to seat the pads.

14. Check the brake fluid reservoir and replenish or remove fluid, as necessary.

15. With the rear wheel raised, check that the wheel spins freely and the brake operates properly.

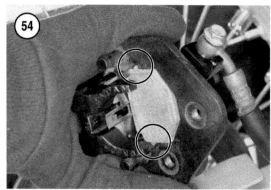

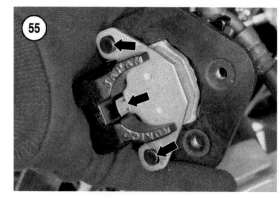

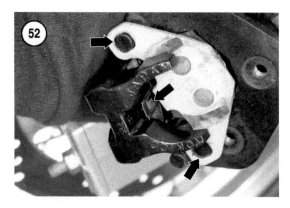

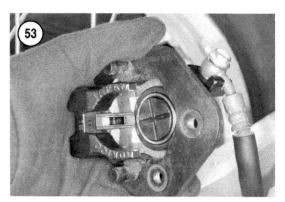

REAR BRAKE CALIPER

Removal and Installation

Use the following procedure to remove the caliper from the motorcycle. Refer to **Figure 49**.

1A. If the caliper will be disassembled, drain the system as described in this chapter. After draining, loosen the brake hose banjo bolt (A, **Figure 56**) while the caliper is stable on the swing arm. Leave the bolt finger-tight. It will be removed in a later step.

14

1B. Remove the caliper mounting bolts (B, **Figure 56**), then remove the caliper from the swing arm. Avoid kinking the brake hose.

2A. If the caliper will be left attached to the brake hose, but not disassembled and serviced:

    a. Attach a wire to the caliper and hang the caliper on the motorcycle. Do not let the caliper hang by the brake hose.

    b. Insert a small wooden block between the brake pads. This prevents the caliper piston from extending out of the caliper, if inadvertently operating the brake pedal.

2B. If the caliper will be disassembled, do the following:

    a. Remove the banjo bolt and seal washers from the brake hose. Have a shop cloth ready to absorb excess brake fluid that drips from the hose.

    b. Wrap the hose end to prevent brake fluid from damaging other surfaces.

    c. Drain excess brake fluid from the caliper.

3. Repair the caliper as described in this section.

4. Reverse this procedure to install the caliper. Note the following:

    a. Position the caliper over the brake disc and guide the disc between the pads.

    b. Install new seal washers on the banjo bolt (A, **Figure 57**).

    c. Position the brake hose fitting so it is behind the support (B, **Figure 57**).

    d. Install and tighten the caliper mounting bolts and banjo bolt to 25 N•m (18 ft.-lb.).

> *NOTE*
> *If the caliper was rebuilt, or the brake hose disconnected from the caliper, fill and bleed the brake system as described in this chapter.*

5. Operate the brake pedal several times to seat the pads.

6. Check the brake fluid reservoir and replenish or remove fluid, as necessary.

7. With the rear wheel raised, check that the wheel spins freely and the brake operates properly.

Repair

Use the following procedure to disassemble, inspect and assemble the brake caliper, using new seals. Refer to **Figure 49**.

1. Remove the caliper as described in this section.

2. Grasp the caliper and press the inner pad and caliper holder (**Figure 58**) against the caliper. This presses the piston into the bore, and compresses the holder so the pads can be removed. Keep the caliper holder compressed until the outer pad is removed.

3. Tilt the outer pad toward the caliper, then disengage the pad from the slide pins and pad spring (**Figure 59**).

4. Press down on the inner pad and disengage it from the caliper holder and pad spring (**Figure 60**).

5. Remove the caliper holder, pad spring and boots (**Figure 61**).

6. Remove the piston from the caliper bore using compressed air. To perform this technique, an air nozzle is tightly held in the brake hose fitting and air pressure ejects the piston. Do not pry the piston out of the caliper. Read the following procedure entirely before removal.

    a. Place the caliper on a padded work surface.

    b. Close the bleeder valve on the caliper so air cannot escape.

    c. Place a strip of wood, or similar pad, in the caliper. The pad will cushion the piston when it comes out of the caliper.

> *WARNING*
> *Wear eye protection when using compressed air to remove the piston. Keep fingers away from the piston discharge area. Injury can occur if an attempt is made to stop the piston with the fingers.*

    d. Lay the caliper so the piston discharges downward.

    e. Insert an air nozzle into the brake hose fitting. If the nozzle does not have a rubber tip, wrap the nozzle with tape. This allows the nozzle to seal tightly and prevent thread damage.

    f. Place a shop cloth over the entire caliper to catch any spray that may discharge from the caliper.

    g. Apply pressure and listen for the piston to *pop* from the caliper (**Figure 62**).

7. Remove the piston boot and seal from the cylinder bore (**Figure 63**).

8. Remove the bleeder valve and cap from the caliper.

9. Inspect the caliper assembly.

    a. Clean all parts that will be reused with fresh brake fluid or isopropyl (rubbing) alcohol. Use a wood or plastic-tipped tool to clean the caliper seal and boot grooves. Use clean brake fluid to clean the piston, bore and seal grooves.

    b. Inspect the cylinder bore and piston (**Figure 64**) for wear, pitting or corrosion.

    c. Inspect the caliper holder and boots. Check the slide pins (**Figure 65**) for wear, pitting or corrosion. Inspect the boots for deterioration.

14

d. Inspect the pad spring (**Figure 66**). Check that all small tabs on the spring are not corroded or missing. The spring must be slightly arched and flexible.

e. Inspect all threaded parts (**Figure 67**). Check the threads and seats for corrosion and damage.

f. Inspect the pads for contamination, scoring and wear. Replace the pads when they are worn to within 1 mm (0.040 in.) of the backing plate, or are contaminated with oil or other chemicals. If the pads are worn unevenly, the caliper is probably not sliding correctly on the slide pins. The caliper must be free to *float* on the pins. Buildup or corrosion on the pins can hold the caliper in one position, causing brake drag and excessive pad wear.

NOTE
Use new brake fluid (rated DOT 4) to lubricate the parts in the following steps.

10. Install the new piston seal, boot and piston as follows:

a. Soak the seal and boot in brake fluid for 15 minutes.

b. Coat the caliper bore and piston with brake fluid.

c. Seat the seal (**Figure 68**) in the caliper groove. The piston seal goes in the back groove.

d. Install the boot onto the back end of the piston. The small diameter of the boot (**Figure 69**) will fit into the caliper groove.

e. Carefully seat the boot into its groove, beginning at the bottom of the caliper (**Figure 70**).

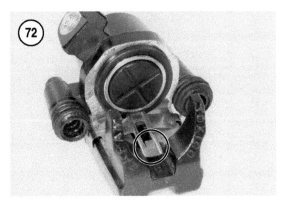

Support the piston and use a small screwdriver to work the boot into the groove. When the boot is fully seated, twist the piston past the seal and press the piston to the bottom of the bore. Check that the boot seats into the piston (**Figure 71**).

11. Apply silicone brake grease to the interior of the boots, then install the boots onto the caliper. If necessary, apply a light coat of grease on the exterior of the large boot, to help pass it through the caliper.

12. Install the pad spring with the narrow end facing out (**Figure 72**).

13. Lubricate the slide pins on the caliper holder, then insert them into the boots. Press the holder against the boots and carefully lift the boots onto the holder seats (**Figure 73**).

14. Compress and hold the caliper holder against the caliper so the pads can be installed.

15. Install the inner pad, seating it into the pad spring and the notches in the caliper holder. Press the pad against the piston.

16. Tilt the outer pad and engage it with the slide pins and pad spring (**Figure 74**).

REAR MASTER CYLINDER

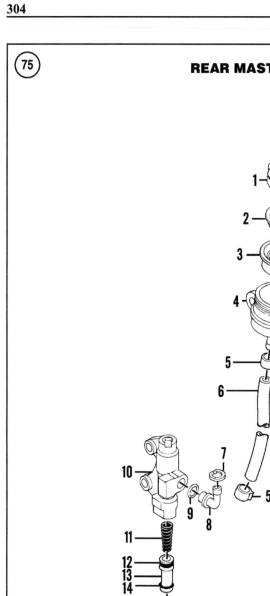

1. Cap
2. Diaphragm plate
3. Diaphragm
4. Reservoir
5. Clamp
6. Hose
7. Snap ring
8. Hose fitting
9. O-ring
10. Master cylinder
11. Spring
12. Primary seal
13. Piston
14. Secondary seal
15. Boot
16. Pushrod
17. Clevis
18. Cotter pin
19. Clevis pin

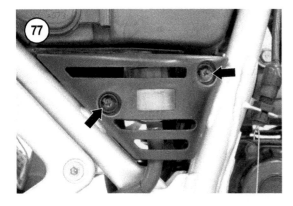

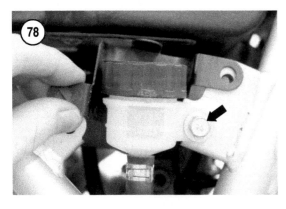

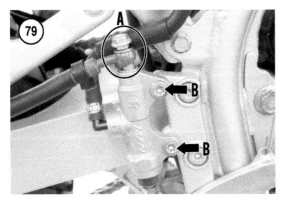

17. Install the bleeder valve and cap.

18. Install the caliper as described in this section.

REAR MASTER CYLINDER

Removal and Installation

Use the following procedure to remove the rear master cylinder and brake fluid reservoir from the motorcycle. Refer to *Repair* in this section to make internal repairs to the master cylinder. Refer to **Figure 75**.

1. Drain the brake system as described in this chapter.

2. Remove the cotter pin and clevis pin (**Figure 76**) that secure the master cylinder clevis to the brake pedal.

3. Remove the reservoir cover (**Figure 77**).

4. Remove the back plate and reservoir bolt (**Figure 78**).

5. Remove the banjo bolt and seal washers from the brake hose (A, **Figure 79**). Have a shop cloth ready to absorb excess brake fluid that drips from the hose. Wrap the hose end to prevent brake fluid from damaging other surfaces.

6. Remove the master cylinder mounting bolts (B, **Figure 79**).

7. Repair the master cylinder as described in this section.

8. Reverse this procedure to install the master cylinder and reservoir. Note the following:

   a. Tighten the master cylinder mounting bolts to 23 N•m (17 ft.-lb.).

   b. Install new seal washers on the banjo bolt. Tighten the bolt to 25 N•m (18 ft.-lb.).

   c. Install a new cotter pin on the clevis pin.

9. Fill the brake fluid reservoir and bleed the brake system as described in this chapter.

Repair

Use the following procedure to disassemble, inspect and assemble the master cylinder, using new parts. The piston, seals and spring are only available as a complete assembly. Refer to **Figure 75**.

1. Remove the master cylinder and reservoir as described in this section.

2. Remove the snap ring that retains the hose and fitting against the master cylinder (**Figure 80**). Remove the hose assembly and internal O-ring.

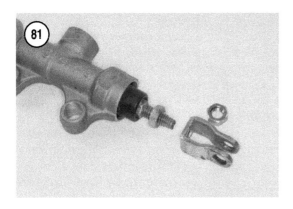

3. Remove the clevis and locknuts (**Figure 81**).

4. Remove the boot from the pushrod (**Figure 82**). The boot is a friction fit. To avoid damaging the boot on removal, apply penetrating lubricant around the perimeter of the boot. Carefully pull the bottom edge back so the lubricant can loosen the boot.

5. Remove the snap ring from the master cylinder as follows:

    a. Lock the cylinder in a vise with soft jaws.

    b. Thread a locknut onto the end of the pushrod (**Figure 83**). The locknut creates a larger and more comfortable area when depressing the pushrod.

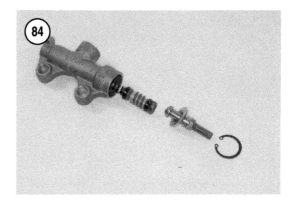

    c. Press down on the pushrod to relieve pressure on the snap ring, then remove the snap ring with snap ring pliers.

    d. Slowly relieve the pressure on the piston.

6. Remove the piston and pushrod assembly from the bore (**Figure 84**).

7. Inspect the master cylinder assembly.

    a. Clean all parts that will be reused with fresh brake fluid or isopropyl (rubbing) alcohol.

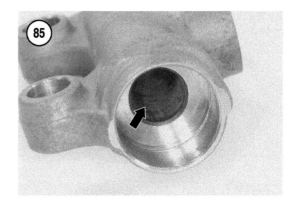

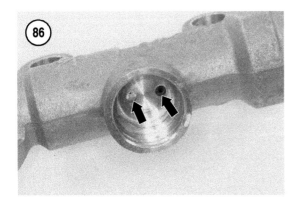

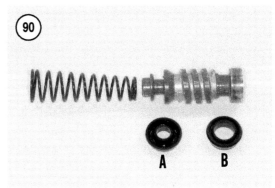

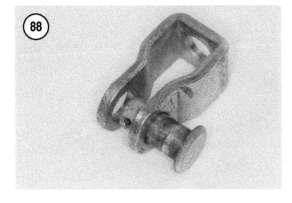

b. Inspect the cylinder bore for wear, pitting or corrosion (**Figure 85**).

c. Inspect and clean the threads and orifices in the master cylinder (**Figure 86**). Clean with compressed air.

d. Inspect the boot, snap ring and pushrod (**Figure 87**). Check the parts for corrosion and damage.

e. Inspect the clevis and pin (**Figure 88**) for wear at their contact points.

f. Inspect the diaphragm, diaphragm plate and reservoir cap for damage (**Figure 89**).

g. Inspect the reservoir hose and fittings for damage.

8. Assemble the piston, seals and spring as follows.

a. Soak the primary seal (A, **Figure 90**) and secondary cup (B) in fresh brake fluid (rated DOT 4) for 15 minutes. This softens and lubricates the seals.

b. Apply brake fluid to the piston so the cups can slide over the ends.

c. Identify the wide (open) side of the primary seal. When installed, the wide side of the seal *must* face in the direction of the arrow (**Figure 91**). Install the primary seal onto the piston.

d. Identify the wide (open) side of the secondary seal. When installed, the wide side of the seal *must* face in the direction of the arrow (**Figure 91**). Install the secondary seal on the piston.

e. Install and seat the *small end* of the spring onto the piston.

9. Install the piston, pushrod and snap ring into the master cylinder as follows:

a. Lubricate the cylinder bore and piston assembly with brake fluid.

14

b. Apply a small amount of silicone brake grease to the contact area of the pushrod.

c. Insert and seat the piston into the cylinder (**Figure 92**).

d. Lock the cylinder in a vise with soft jaws. Do not overtighten the vise or cylinder damage could occur.

e. Thread a nut onto the end of the pushrod, then rest the pushrod in the cylinder (**Figure 83**). The nut will provide a more comfortable surface when the pushrod is depressed.

f. Place the snap ring over the end of the pushrod, resting it on the edge of the bore. The flat side of the snap ring must face out.

g. Compress the snap ring with snap ring pliers.

h. Press the pushrod into the cylinder while guiding the snap ring into position. If the snap ring does not easily seat, release the snap ring and use the tip of the pliers to press it into the groove. Keep the pushrod compressed until the snap ring seats.

10. Remove the cylinder from the vise, then remove the nut from the pushrod.

11. Apply silicone brake grease to the inside the boot. Seat the boot into the cylinder.

12. Install the clevis and locknuts. For proper brake pedal height, adjust and lock the clevis so the pushrod threads protrude 3-3.5 mm (0.12-0.14 in.) below the nut (**Figure 93**).

13. Install a new, lubricated O-ring into the master cylinder, then lock the hose and fitting into place (**Figure 94**). Install the snap ring so the flat side faces out.

14. Loosely install the diaphragm, diaphragm plate and cap onto the reservoir.

15. Install the master cylinder and reservoir as described in this section.

REAR BRAKE PEDAL

Removal and Installation

Use the following procedure to remove and install the rear brake pedal. Refer to **Figure 95**.

1. Remove the brake light switch spring from the pedal.

2. Remove the cotter pin and clevis pin (A, **Figure 96**) that secure the master cylinder clevis to the brake pedal.

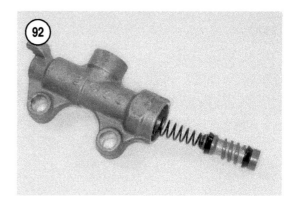

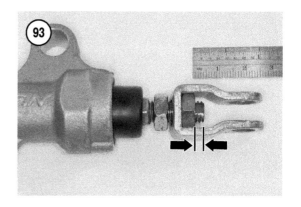

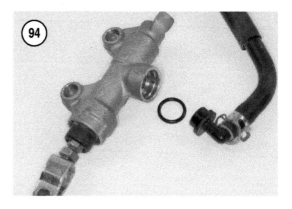

⑨⑤ **REAR BRAKE PEDAL**

2

1

3

3

7   6   5   4

1. Brake switch spring
2. Pedal
3. Bolt
4. Bushing
5. Brake stay
6. Pedal return spring
7. Pedal shaft

14

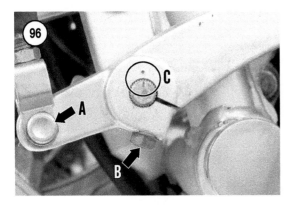

3. Completely remove the bolt securing the pedal to the shaft (B, **Figure 96**).

4. Before removing the pedal, note how the spring contacts the post on the brake stay and pedal.

5. Hold the pedal shaft in place, then pull the pedal from the shaft.

6. Remove the pedal shaft and spring.

7. Clean and inspect the parts.

   a. Inspect the pedal shaft and bore bushing for scoring, damage or the entry of water and dirt. If the bushing is worn, replace it.

b. Inspect the clevis pin and pedal bore. The pin must be a firm fit in the clevis and pedal.

c. Inspect the spring and shaft bolt for corrosion and damage.

d. Check that the brake stay is tightly fastened.

8. Install the pedal as follows:

a. Apply waterproof grease to the bushing and pedal shaft.

b. Install the spring, seating it against the brake stay. The straight end of the spring rests against the back of the brake stay post.

c. Install the pedal shaft. Engage the hooked end of the spring with the narrow arm on the pedal shaft.

d. Hold the shaft in place, then turn it forward to preload the spring.

e. Install the pedal, guiding it into the clevis, and aligning the index marks on the pedal and shaft (C, **Figure 96**).

f. Bolt the pedal to the shaft.

g. For proper brake pedal height, check that the pushrod threads protrude 3-3.5 mm (0.12-0.14 in.) below the nut in the clevis (**Figure 93**). Adjust the clevis position if necessary.

h. Install the clevis pin and a new cotter pin.

i. Attach the brake light switch spring.

j. Check pedal height and operation. If the pedal is not level with the footpeg, inspect for bent or damaged parts.

k. Check brake light operation. If necessary, adjust the switch (Chapter Three).

BRAKE SYSTEM DRAINING

To drain the brake fluid from the system, have a 10 mm wrench, tip-resistant container and a length of clear tubing that fits snugly on the bleeder valve. Use the following procedure to drain either the front or rear brakes.

CAUTION
Brake fluid can damage painted and finished surfaces. Use water to immediately wash any surface that becomes contaminated with brake fluid.

1. Attach one end of the tubing to the bleeder valve and place the other end into the container (**Figure 97**).

2. Open the bleeder valve so fluid can pass into the tubing.

3. Pump the brake lever/pedal to force the fluid from the system.

4. When the system no longer drips fluid, close the bleeder valve.

5. Dispose the brake fluid in an environmentally safe manner.

BRAKE SYSTEM BLEEDING

Whenever the brake fluid is replaced, or if the brake lever or pedal feels spongy, bleed the brakes to purge all air from the system. Before bleeding the brakes, determine where the air is entering the system. Check all brake components for leakage, and fittings and hoses for deterioration, damage or looseness. The brake system can be bled manually or by using a vacuum pump. Both methods are described in this section.

CAUTION
Before bleeding brakes, always secure the bike so it is stable and locked in place, particularly the fork. This minimizes the chance of spilled fluid from an open reservoir.

Brake Fluid Reservoirs

Regardless of the bleeding method used, the reservoir cap of the master cylinder being bled must be removed so the reservoir can be filled with brake fluid. The reservoirs must not be over or under filled. Note the following when working with each reservoir.

1. Front brake reservoir.

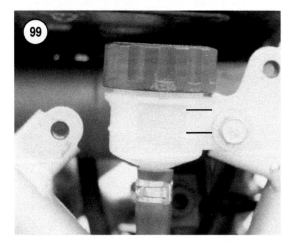

a. After removing the cap, remove the diaphragm from the reservoir before filling with fluid.

b. Keep the reservoir filled between the top of the sight glass and the lower mark on the reservoir (**Figure 98**) during the bleeding procedure.

c. After bleeding, replenish the reservoir to the upper mark, then install the diaphragm and cap.

2. Rear brake reservoir.

a. Remove the reservoir guard and backing plate. Remove the cap, diaphragm plate and diaphragm.

b. Keep the reservoir filled between the upper and lower marks on the reservoir during the bleeding procedure (**Figure 99**).

c. After bleeding, replenish the reservoir to the upper mark, then install the diaphragm, diaphragm plate and cap.

d. Install the backing plate and reservoir guard.

Manual Bleeding

To manually bleed the brake system, have a 10 mm wrench, tip-resistant container and a length of clear tubing that fits snugly on the brake bleeder. Bleeding the system is much easier if two people are available to perform the procedure. One person can open and close the bleeder valve while the other person operates the brake lever or pedal. Use the following procedure to bleed either the front or rear brake.

CAUTION
Brake fluid can damage painted and finished surfaces. Use water and immediately wash any surface that becomes contaminated with brake fluid.

1. Attach one end of the tubing to the bleeder valve and place the other end into the container (**Figure 97**).

2. Fill the reservoir to the upper level with DOT 4 brake fluid.

CAUTION
Do not use brake fluid from an unsealed container. It will have absorbed moisture from the air and already be contaminated. Use DOT 4 brake fluid from a sealed container.

3. Apply pressure (do not pump) to the brake lever or pedal, then open the bleeder valve. As the fluid is forced from the system, the lever/pedal travels its full length of operation. When the lever/pedal can move no farther, hold the lever/pedal in the *down* position and close the bleeder valve. Do not allow the lever or pedal to return to its *up* position before the bleeder valve is closed. Air will be drawn into the system.

NOTE
In the following step, release the lever/pedal slowly. This minimizes the chance of fluid splashing out of the reservoir, as excess fluid in the brake line is returned to the reservoir.

4. When the bleeder valve is closed, release the lever/pedal so it returns to its *up* position. Check the

fluid level in the reservoir and replenish, if necessary.

NOTE
During the bleeding process, the reservoir must contain fluid during the entire procedure. If the reservoir is allowed to become empty, air will be in the system and the bleeding process will have to be repeated.

5. Repeat Step 3 and Step 4 until clear fluid (minimal air bubbles) is seen passing out of the bleeder valve. Unless the bleeder valve threads are wrapped with Teflon tape, or coated with silicone brake grease, a small amount of air will enter the system when the bleeder valve is opened.

NOTE
If small bubbles (foam) remain in the system after several bleeding attempts, close the reservoir and allow the system to stand undisturbed for a few hours. The system will stabilize and the air can be purged as large bubbles.

6. The bleeding procedure is completed when the feel of the lever/pedal is firm.
7. Check the brake fluid reservoir and fill the reservoir to the upper level, if necessary.
8. Tighten the bleeder valve to 8 N•m (70 in.-lb.). Do not overtighten.
9. Dispose the waste brake fluid in an environmentally safe manner.

Vacuum Bleeding

To vacuum-bleed the brake system, have a 10 mm wrench and a vacuum pump, such as the Mityvac pump shown in **Figure 100**. Use the following procedure to bleed either the front or rear brake.

CAUTION
Brake fluid can damage painted and finished surfaces. Use water and immediately wash any surface that becomes contaminated with brake fluid.

1. Check that the banjo bolts are tight at the master cylinder and caliper.
2. Attach the brake bleeder to the bleeder valve (**Figure 100**). Suspend the tool with wire. This al-

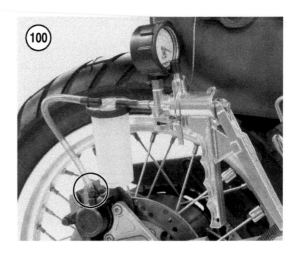

lows the tool to be released when the fluid reservoir needs to be refilled.

3. Fill the reservoir to the upper level with DOT 4 brake fluid.

CAUTION
Do not use brake fluid from an unsealed container. It will have absorbed moisture from the air and already be contaminated. Use DOT 4 brake fluid from a sealed container.

NOTE
During the bleeding process, the reservoir must contain fluid during the entire procedure. If the reservoir is allowed to become empty, air will be in the system and the bleeding process will have to be repeated.

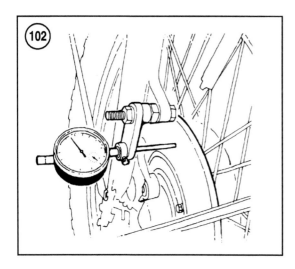

4. Pump the handle on the brake bleeder to create a vacuum.

5. Open the bleeder valve and draw the air and fluid from the system. Close the valve *before* the fluid stops moving. If the vacuum pump is equipped with a gauge, close the bleeder before the gauge reads 0 in. Hg. Replenish the fluid level in the reservoir.

6. Repeat Step 4 and Step 5 until clear fluid (minimal air bubbles) is passing out of the bleeder. Unless the bleeder valve threads are wrapped with Teflon tape, or coated with silicone brake grease, a small amount of air will enter the system when the bleeder valve is opened. The bleeding procedure is completed when the feel of the lever/pedal is firm.

7. Check the brake fluid reservoir and fill the reservoir to the upper level, if necessary.

8. Tighten the bleeder valve to 8 N•m (70 in.-lb.). Do not overtighten.

9. Dispose the waste brake fluid in an environmentally safe manner.

BRAKE DISC

The condition of the brake discs and pads are often a reflection of one another. If disc scoring is evident, inspect the pads and disc as soon as possible. Visually inspect the discs and pads with the wheels mounted on the motorcycle (Chapter Three). If damage is detected, perform the inspections described in this section.

NOTE
Do not true a deeply scored or warped disc. Removing disc material causes

the disc to overheat rapidly and warp. Maintain the discs by keeping them clean and corrosion-free. Use a solvent, that is not oil-based, to wipe grit that accumulates on the discs and at the edge of the pads.

Thickness and Runout Inspection

1. Measure the thickness of each disc at several locations around its perimeter (**Figure 101**). Refer to **Table 1** for the service limit. Replace the disc if it is out of specification.

2. Measure disc runout as follows:
   a. Mount a dial indicator on a stable surface and in contact with the disc (**Figure 102**).
   b. Zero the gauge.
   c. Turn the wheel and watch the amount of runout measured on the gauge.
   d. Refer to **Table 1** for the service limit. Replace the disc if it is out of specification.

NOTE
If the disc runout is out of specification, check the condition of the wheel bearings before replacing the disc. If the bearings are not in good condition, the bearings should be replaced before disc runout is determined.

Removal and Installation

The discs are mounted to the hubs with bolts. Remove and install either disc as follows:

1. Remove the wheel from motorcycle (Chapter Eleven).

2. Remove the bolts (**Figure 103**) that secure the disc to the hub.

14

3. Clean the bolts and mounting holes.

4. Reverse this procedure to install the discs. Note the following:

a. Install the disc with the thickness marking and direction arrow (**Figure 104**) facing out.

b. Apply nonpermanent threadlocking compound to the bolt threads.

c. Tighten the bolts in several passes and in a crossing pattern.

d. Tighten the bolts to 23 N•m (17 ft.-lb.).

e. Check the disc for runout as described in this section.

Table 1 BRAKE SPECIFICATIONS

	New mm (in.)	Service limit mm (in.)
Disc thickness		
Front	3.8-4.1 (0.15-0.16)	3.5 (0.14)
Rear	4.8-5.1 (0.19-0.20)	4.5 (0.18)
Disc runout	0-0.2 (0-0.008)	0.3 (0.012)
Pad lining thickness	4.5 (0.18)	1.0 (0.040)
Brake fluid type	DOT 4	

Table 2 BRAKE SYSTEM TORQUE SPECIFICATIONS

	N•m	in.-lb.	ft.-lb.
Brake bleeder valve	8	70	–
Brake disc bolts	23	–	17
Brake hose banjo bolts	25	–	18
Brake lever pivot nut	6	52	–
Caliper mounting bolts	25	–	18
Master cylinder mounting bolts			
Front	9	80	–
Rear	23	–	17

CHAPTER FIFTEEN

BODY

This chapter provides removal and installation procedures for the side covers, seat, fuel tank, radiator covers, skid plate, front fender, fairing and subframe. When removing or installing the bodywork, do not use excessive force. Although the bodywork is flexible, it will discolor when it is excessively flexed or pinched.

SIDE COVERS

Removal and Installation

1. Remove the screws at the front and rear of the cover (**Figure 1**).

2. Pull the bottom of the cover out to remove the friction barb from the grommet.

3. Remove the cover.

4. To install the cover:

    a. Insert the rear edge of the cover first. The cover must fit under the rear fender.

    b. Align the friction barb with the grommet, then press the cover into place.

    c. Install and tighten the screws.

SEAT

Removal and Installation

1. Remove the side covers as described in this chapter.

2. Remove the bolt (**Figure 2**) from both sides of the seat.

3. Grasp the front and rear of the seat, then pull it backwards to disengage it from the fuel tank (**Figure 3**).

4. Reverse this procedure to install the seat. Check that the seat is secure at the front.

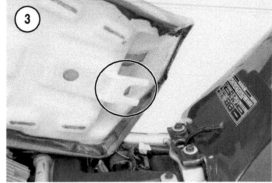

FUEL TANK

Removal and Installation

1. Remove the seat as described in this chapter.
2. Turn the fuel valve off.
3. Remove the fuel hose (A, **Figure 4**) and vacuum hose (B) from the fuel valve assembly.
4. Remove the bolts and hose(s) at the rear of the fuel tank (**Figure 5**).
5. Remove the screw at the top of the radiator cover (**Figure 6**), then pull the friction barb from the grommet in the fuel tank. Do not excessively flex the cover.
6. Remove the fuel tank, moving it toward the rear of the motorcycle.
7. Reverse this procedure to install the fuel tank. Align the fuel tank with all bolt and screw holes before tightening the fasteners.

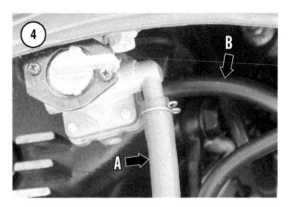

RADIATOR COVERS

At the left side of the motorcycle, the cover shrouds the radiator. At the right side, the cover shrouds the coolant reserve tank. If desired, the covers can be removed with the fuel tank. Only the fasteners at the front of the covers need to be removed. If the covers remain attached to the fuel tank, support the fuel tank after removal, so it does not rest on the flexible covers.

Removal and Installation

1. At the front of the cover, remove the bolt/screw at the top and bottom of the cover (**Figure 7**).
2. Remove the screw at the top of the radiator cover, then pull the friction barb from the grommet

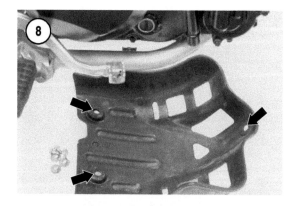

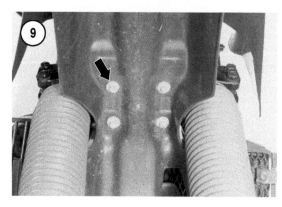

in the fuel tank (**Figure 6**). Do not excessively flex the cover.

3. Reverse this procedure to install the radiator covers.

SKID PLATE

The skid plate is attached to the frame by three bolts and collars (**Figure 8**). If using the motorcycle in rough terrain, inspect the skid plate often. Consider a metal aftermarket skid plate when operating the motorcycle in rough conditions.

FRONT FENDER

The front fender is attached to the fork by four bolts and washers (**Figure 9**).

FAIRING

The fairing is frame-mounted and houses the headlight and turn signals. Aftermarket windshields are available that provide more coverage than the stock windshield. A larger windshield is more susceptible to damage if the motorcycle is taken off-road.

Removal and Installation

1. Cover the front fender or remove the fender.

2. Remove the upper mounting bolt from the frame (**Figure 10**). If necessary, loosen the lower bolt so the fairing can be tilted forward.

3. Tilt the fairing forward and remove the wire ties securing the wiring to the framework.

4. Disconnect the turn signal wires and unplug the headlight. Keep the turn signal wires identified. Also, remove the wires from the frame tab.

5. Remove the lower mounting bolt from the frame. Support the fairing while removing the bolt.

6. Inspect the fairing and frame for damage.

7. Reverse this procedure to install the fairing. Clean all electrical connections, then lightly apply dielectric grease to the connectors.

15

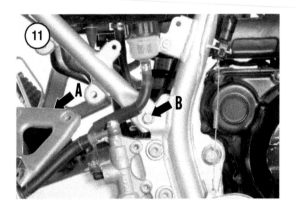

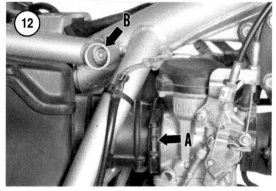

SUBFRAME

Removal and Installation

1. Remove the side covers and seat as described in this chapter.

2. Remove the muffler from the exhaust pipe and frame (Chapter Four).

3. Remove the right passenger footpeg (A, **Figure 11**).

4. Loosen the carburetor clamp (A, **Figure 12**).

5. Remove the battery (Chapter Three).

6. Disconnect all hoses and wires passing from the front to the rear of the motorcycle (**Figure 13**). Route the hoses and wires out of the subframe.

7. Remove the lower subframe bolts (B, **Figure 11**).

CAUTION
Have an assistant aid in the final step, by holding the subframe and preventing it from falling when the upper subframe bolts are removed.

8. Remove the upper subframe bolts (B, **Figure 12**).

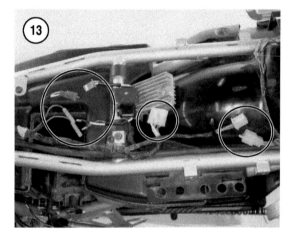

9. To install the subframe, reverse this procedure. Note the following:

    a. Inspect and replace subframe mounting bolts that are corroded or damaged.

    b. Have an assistant aid in installing the subframe.

    c. Finger-tighten all subframe mounting bolts, then tighten the bolts to 25 N•m (18 ft.-lb.).

INDEX

16

16

16

WIRING DIAGRAMS

U.S. AND CANADA MODELS

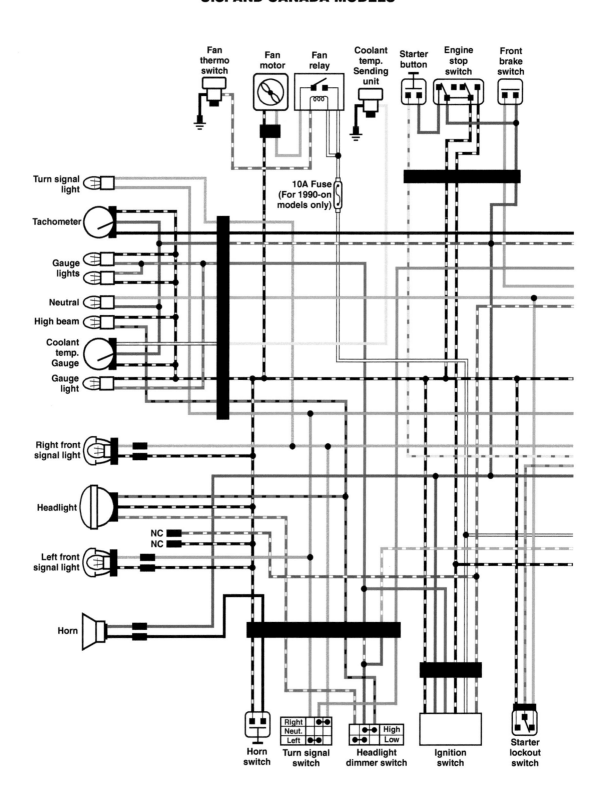

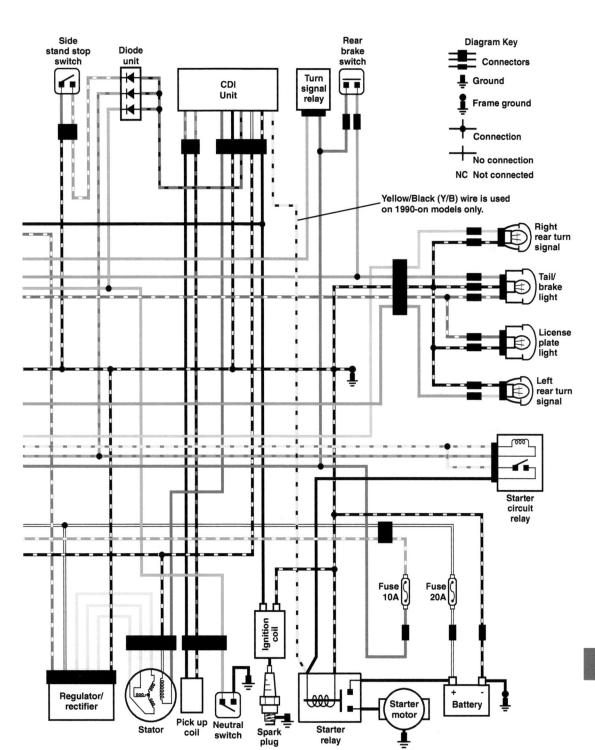

Side stand stop switch

Diode unit

CDI Unit

Rear brake switch

Turn signal relay

Diagram Key

Connectors

Ground

Frame ground

Connection

No connection

NC Not connected

Yellow/Black (Y/B) wire is used on 1990-on models only.

Right rear turn signal

Tail/ brake light

License plate light

Left rear turn signal

Starter circuit relay

Fuse 10A

Fuse 20A

Regulator/ rectifier

Stator

Pick up coil

Neutral switch

Ignition coil

Spark plug

Starter relay

Starter motor

Battery

17

ALL MODELS EXCEPT U.S. AND CANADA

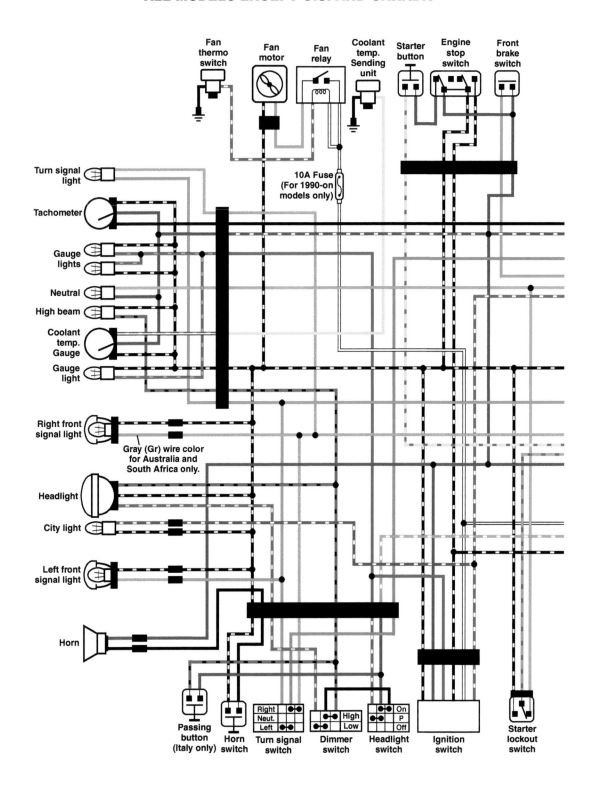

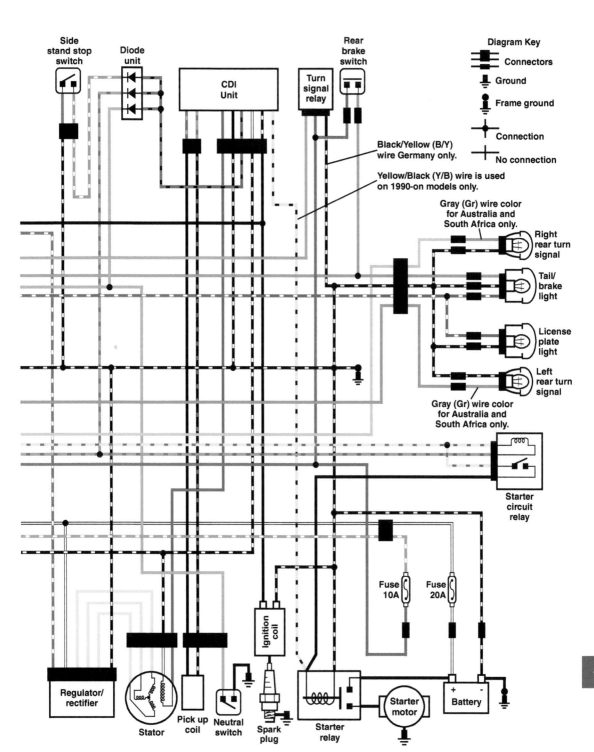

NOTES

NOTES

NOTES